Lecture Notes of the Institute for Computer Sciences, Social Informatics and Telecommunications Engineering 257

More information about this series at http://www.springer.com/series/8197

Trung Q. Duong · Nguyen-Son Vo (Eds.)

Industrial Networks and Intelligent Systems

14th EAI International Conference, INISCOM 2018
Da Nang, Vietnam, August 27–28, 2018
Proceedings

 Springer

Editors
Trung Q. Duong
Queen's University Belfast
Belfast, UK

Nguyen-Son Vo
Duy Tan University
Da Nang, Vietnam

ISSN 1867-8211   ISSN 1867-822X (electronic)
Lecture Notes of the Institute for Computer Sciences, Social Informatics
and Telecommunications Engineering
ISBN 978-3-030-05872-2   ISBN 978-3-030-05873-9 (eBook)
https://doi.org/10.1007/978-3-030-05873-9

Library of Congress Control Number: 2018964093

This Springer imprint is published by the registered company Springer Nature Switzerland AG
The registered company address is: Gewerbestrasse 11, 6330 Cham, Switzerland

Preface

We are delighted to introduce the proceedings of the 2018 European Alliance for Innovation (EAI) International Conference on Industrial Networks and Intelligent Systems (INISCOM). This conference brought together researchers, developers, and practitioners from around the world who are leveraging and developing industrial networks and intelligent systems. The theme of INISCOM 2018 was "Internet of Things Can Improve Citizens' Lives."

The technical program of INISCOM 2018 consisted of 27 full papers, including one invited paper in oral presentation sessions at the main conference tracks. The conference tracks were: Track 1, Telecommunications Systems and Networks; Track 2, Industrial Networks and Applications; Track 3, Hardware and Software Design and Development; Track 4, Information Processing and Data Analysis; Tack 5, Signal Processing; and Track 6, Security and Privacy. Aside from the high-quality technical paper presentations, the technical program also featured two keynote speeches. The two keynote speeches were by Prof. Marco di Renzo, Paris-Saclay University, France, and Muhammad Ali Imran, University of Glasgow, UK.

Coordination with the steering chairs, Prof. Imrich Chlamtac, Prof. Carlo Cecati, and Prof. Song Guo, was essential for the success of the conference. We sincerely appreciate their constant support and guidance. It was also a great pleasure to work with such an excellent Organizing Committee team who worked hard in organizing and supporting the conference. In particular, we thank the Technical Program Committee, led by our TPC co-chairs, Dr. Quoc Tuan Vien and Dr. Nguyen-Son Vo, who completed the peer-review process of technical papers and compiled a high-quality technical program. We are also grateful to conference manager, Andrea Piekova, and all the authors who submitted their papers to INISCOM 2018.

A special thank is given to the Research Environment Links grant, ID 339568416, under the Newton Programme Vietnam partnership. The grant is funded by the UK Department of Business, Energy and Industrial Strategy (BEIS) and delivered by the British Council.

We strongly believe that INISCOM provides a good forum for all researchers, developers, and practitioners to discuss all science and technology aspects that are relevant to industrial networks and intelligent systems. We also expect that future INISCOM events will be as successful and stimulating as indicated by the contributions presented in this volume.

November 2018

Trung Q. Duong
Nguyen-Son Vo

Organization

Steering Committee

Imrich Chlamtac	University of Trento, Italy
Carlo Cecati	University of L'Aquila, Italy
Song Guo	The University of Aizu, Japan

Organizing Committee

General Chair

Trung Q. Duong — Queen's University Belfast, UK

General Co-chair

Le Nguyen Bao — Duy Tan University, Vietnam

TPC Chair and Co-chair

Quoc Tuan Vien	Middlesex University, UK
Nguyen-Son Vo	Duy Tan University, Vietnam

Web Chair

Dac-Binh Ha — Duy Tan University, Vietnam

Publicity and Social Media Chairs

Tomohiko Taniguchi	Fujitsu Laboratories, Japan
Al-Sakib Khan Pathan	Southeast University, Bangladesh
Yuanfang Chen	Hangzhou Dianzi University, China
Khaled M Rabie	Manchester Metropolitan University, UK

Workshop Chairs

Wei Liu	University of Sheffield, UK
Chinmoy Kundu	Queen's University Belfast, UK

Publications Chair

Nguyen-Son Vo — Duy Tan University, Vietnam

Panels Chairs

Nguyen Gia Nhu	Duy Tan University, Vietnam
Berk Canberk	Istanbul Technical University, Turkey

Tutorial Chairs

Dang Viet Hung   Duy Tan University, Vietnam
Tuan Le   Middlesex University, UK

Demos Chairs

Nguyen Quang Sang   Duy Tan University, Vietnam
Zoran Hadzi-Velkov   Ss. Cyril and Methodius University, Macedonia

Posters and PhD Track Chairs

Le-Nam Tran   University College Dublin, Ireland
Daniel Benevides da   Federal University of Cearas, Brazil
   Costa

Local Chairs

Duong Nguyen   Duy Tan University, Vietnam
Anand Nayyar   Duy Tan University, Vietnam

Conference Manager

Andrea Piekova   European Alliance for Innovation (EAI), Slovakia

Technical Program Committee

Abbas Kouzani   Deakin University, Australia
Abdel-Hamid Ayman   Arab Academy for Science, Technology, and Maritime
                                Transport, Egypt
Adriana Giret   Universidad Politécnica de Valencia, Spain
Aguiar Rui   University of Aveiro, Portugal
Alexandropoulos   Athens Information Technology, Greece
   George
Ali Hilal Al-Bayati   De Montfort University, UK
Al-Qahtani Fawaz   Texas A&M University at Qatar, Qatar
Al-Sakib Khan Pathan   Southeast University, Bangladesh
Anpalagan Alagan   Ryerson University, Canada
Atif Yacine   UAE University, UAE
Ban Nguyen   Posts and Telecommunications Institute of Technology,
                                Vietnam
Bang Giang Truong Vu   VNU, University of Engineering and Technology, Vietnam
Bao Vo Nguyen   Quoc Posts and Telecommunications Institute of
                                Technology, Vietnam
Bejaoui Tarek   University of Paris-Sud 11, France
Ben Horan   Deakin University, Australia
Ben-Othman Jalel   University of Paris 13, France
Binh Duong Nguyen   International University, Vietnam

Boggia Gennaro	Politecnico di Bari, Italy
Bonnie Law	Hong Kong Polytechnic University, SAR China
Bouras Christos	University of Patras and RACTI, Greece
Bouzguenda Lotfi	Université de Sfax, Tunisia
Bui Loc	Tan Tao University, Vietnam
Chau Yuen	Singapore University of Technology and Design, Singapore
Canberk Berk	Istanbul Technical University, Turkey
Chatzimisios Periklis Alexander	TEI of Thessaloniki, Greece
Che Anh Ho Chi Minh City	University of Technology, France
Chee Peng Lim	Deakin University, Australia
Chen Jiming	Zhejiang University, China
Chan Sammy	City University of Hong Kong, SAR China
Chen Thomas	City University London, UK
Chen Yuanfang	Institut Mines-Telecom, Telecom SudParis, France
Cheng Long	Singapore University of Technology and Design, Singapore
Chunsheng Zhu	University of British Columbia, Canada
Cho Sungrae	Chung-Ang University, South Korea
Cong-Kha Pham	The University of Electro-Communication, Japan
Constandinos Mavromoustakis	University of Nicosia, Cyprus
Da Costa Daniel Benevides	Federal University of Ceara, Brazil
Dang Ngoc	Posts and Telecommunications Institute of Technology, Vietnam
Dao Chien	Hanoi University of Science and Technology, Vietnam
Dezhong Peng	Sichuan University, China
Dinh-Duc Anh-Vu	University of Information Technology, Vietnam
Duong Trung Q.	Queen's University Belfast, UK
Duy Tran Trung	Posts and Telecommunications Institute of Technology, Vietnam
Dzung Nguyen	Hanoi University of Science and Technology, Vietnam
Edmundo Monteiro	University of Coimbra, Portugal
Edmund Lai	Massey University, New Zealand
El-Hajj Wassim	American University of Beirut, Lebanon
Elkashlan Maged	Queen Mary, University of London, UK
Fei Richard Yu	Carleton University, Canada
Feng-Tsun Chien	National Chiao Tung University, Taiwan
Fiedler Markus	Blekinge Institute of Technology, Sweden
George Grispos	Lero – The Irish Software Research Centre, Ireland
George Pallis	University of Cyprus, Cyprus
Gavrilovska Liljana	Ss Cyril and Methodius University – Skopje, Republic of Macedonia

Guan Zhangyu	State University of New York at Buffalo, USA
Guangjie Han	Hohai University, China
Guojun Wang	Central South University, China
Guizani Mohsen	University of Idaho, USA
Guizani Sghaier	Alfaisal University, Saudi Arabia
Hani Hamdan	SUPELEC, France
Hamid Abdi	Deakin University, Australia
Ha Vu INRS	University of Quebec, Canada
Han Guangjie	Hohai University, China
Hiep Pham	Thanh Yokohama National University, Japan
Hoang Thang	Hanoi University of Science and Technology, Vietnam
Hoang Van-Phuc	Le Quy Don Technical University, Vietnam
Honggang Wang	University of Massachusetts Dartmouth, USA
Ho-Van Khuong	Ho Chi Minh City University of Technology, Vietnam
Ho Seok Ahn	The University of Auckland, New Zealand
Hsi-Pin Ma	National Tsing Hua University, Taiwan
Huu Thanh Nguyen	Hanoi University of Science and Technology, Vietnam
Huynh Cuong	Ho Chi Minh City University of Technology, Vietnam
Ismail Mahamod	Universiti Kebangsaan Malaysia, Malaysia
Javier M. Aguiar	Universidad de Valladolid, Spain
Jianwei Niu	Beihang University, China
Josif Grabocka	University of Hildesheim, Germany
Joung Jingon	Institute for Infocomm Research, Singapore
Jucheng Yang	Tianjin University of Science and Technology, China
Karagiannidis George	Aristotle University of Thessaloniki, Greece
Kha Ha	University of Technology Sydney, Australia
Kim Dong In	Sungkyunkwan University, South Korea
Kim Dongwoo	Hanyang University, South Korea
Kim Kyeong Jin	Mitsubishi Electric Research Laboratories, USA
Korakis Thanasis	NYU Tandon School of Engineering, USA
Korpeoglu Ibrahim	Bilkent University, Turkey
Kostas Katsalis	Eurecom, France
Laddomada Massimiliano	Texas A&M-Texarkana, USA
Le Anh	Posts and Telecommunications Institute of Technology, Vietnam
Le Khoa	University of Western Sydney, Australia
Le Thi-Lan	MICA, HUST, Vietnam
Lee Jemin	Singapore University of Technology and Design, Massachusetts Institute of Technology, USA
Lei Xianfu	Utah State University, USA
Lesecq Suzanne	CEA-LETI, France
Le-Tien Thuong	Ho Chi Minh city University of Technology, Vietnam
Li Xinrong	University of North Texas, USA
Linh-Trung Nguyen	Vietnam National University, Hanoi, Vietnam
Liu Yang	Guangzhou University, China

Lloret Jaime	Universidad Politécnica de Valencia, Spain
Logothetis Michael	University of Patras, Greece
Lucia Lo Bello	University of Catania, Italy
Lu Chih-Wen	National Tsing Hua University, Taiwan
Lu Liu	University of Derby, UK
Md. Apel Mahmud	Deakin University, Australia
Maham Behrouz	University of Tehran, Iran
Manoj Panda	Swinburne University of Technology, Australia
Mai Linh	International University of Vietnam, Vietnam
Malone David	NUI Maynooth, Ireland
Matthaiou Michail	Queen's University Belfast, UK
Michalis Mavrovouniotis	De Montfort University, UK
Michele Minichino	ENEA, Italy
Milos Manic	University of Idaho, USA
Memon Qurban	United Arab Emirates University, UAE
Mojtaba Ahmadieh Khanesar	Semnan University, Iran
Morteza Biglari-Abhari	University of Auckland, New Zealand
Moussa Ouedraogo	Luxembourg Institute of Science and Technology, Luxembourg
Naccache David	ENS, France
Nallanathan Arumugam	King's College London, UK
Nam Pham	HUST, Vietnam
Nasser Nidal	Alfaisal University, Saudi Arabia
Ng Derrick Wing Kwan	University Erlangen-Nürnberg, Germany
Ngo Duc	Hanoi University of Science and Technology, Vietnam
Ngo Hien	Linkoping University, Sweden
Nguyen Dinh-Thong	University of Technology, Sydney, Australia
Nguyen Ha	University of Saskatchewan, Canada
Nguyen Huan	Middlesex University, UK
Nguyen Hung	University of Southampton, Vietnam
Nguyen Minh	International University, Vietnam
Nguyen Minh Son	International University at Ho Chi Minh, Vietnam
Nguyen Mui	Kyung Hee University, South Korea
Nguyen Nam Tran	University of Saskatchewan, Canada
Nguyen Tuan-Duc	International University, HCMC VNU, Vietnam
Nguyen-Le Hung	The University of Danang, Vietnam
Nguyen-Thanh Nhan	Telecom Paris Tech, France
Noël Crespi	Telecom SudParis, France
Panlong Yang	Institute of Communication Engineering, China
Panayotis Kikiras	AGT International, Switzerland
Panzieri Stefano	Università degli Studi Roma Tre, Italy
Paris Flegkas	CERTH-ITI, Volos, Greece
Philip Branch	Swinburne University of Technology, Australia
Pierluigi Siano	Università degli Studi di Salerno, Italy

Pavlina Fragkou	Technological Educational Institute of Athens, Greece
Peng Yuexing	Beijing University of Posts and Telecommunications, China
Pescapé Antonio	University of Naples Federico II, Italy
Pham Anh	The University of Aizu, Japan
Phung Phu	Chalmers University of Technology, Sweden
Pooneh Bagheri Zadeh	De Montfort University, UK
Qin Zhenquan	Dalian University of Technology, China
Rachedi Abderrezak	University of Paris-Est Marne-la-Vallée, France
Raveendran Paramesran	University of Malaya, Malaysia
Rodrigues Joel	Instituto de Telecomunicações, University of Beira Interior, Portugal
Said Youssef	Tunisie Telecom, Tunisia
Sarraf Mohsen	Applied Communication Sciences, USA
Shu Lei Guangdong	University of Petrochemical Technology, China
Skianis Harry	University of the Aegean, Greece
Simões Paulo	University of Coimbra, Portugal
Sotiris Moschoyiannis	University of Surrey, UK
Snape Richard	De Montfort University, UK
Spyros Lalis	University of Thessaly, Greece
Stempitsky Viktor Belarusian	State University of Informatics and Radioelectronics, Belarus
Strasser Thomas	Austrian Institute of Technology, Austria
Suraweera Himal	University of Peradeniya, Sri Lanka
Syed Pasha	University of Sydney, Australia
Ta Chien	Nitero, Australia
Taha Abd-Elhamid	Alfaisal University, Saudi Arabia
Yannis Manolopoulos	Aristotle University of Thessaloniki, Greece
Yong Li,	Tsinghua University, China
Yong Tang	South China Normal University, China
Yuanfang Chen	Institute Mines-Telecom and Pierre and Marie Curie University (Paris VI), France
Zhangbing Zhou	China University of Geosciences, China and Institute Mines-Telecom, France
Zuyuan Yang	Deakin University, Australia
Lars Schmidt-Thieme	University of Hildesheim, Germany
Thai Chan	IFSTTAR, France
Thang Truong Cong	The University of Aizu, Japan
Tran Cong Hung	Posts and Telecoms Institute of Technology, Vietnam
Tran Gia Khanh	Tokyo Institute of Technology, Japan
Tran Le-Nam	University of Oulu, Finland
Tran Nam	Le Quy Don University, Vietnam
Tran Quang	National Institute of Informatics, Japan
Tran Thien Thanh	Ho Chi Minh City University of Transport, Vietnam
Tran Xuan-Tu	Vietnam National University, Hanoi, Vietnam
Trong Tu Bui	Ho Chi Minh University of Science, Vietnam

Truong Kien	Posts and Telecommunications Institute of Technology, Vietnam
Tsiftsis Theodoros	Technological Educational Institute of Lamia, Greece
Tuan Pham	Da Nang University of Technology, Vietnam
Vien Quoc-Tuan	Middlesex University, UK
Vinel Alexey	Tampere University of Technology, Finland
Vinh Phan Cong	NTT University in Vietnam, Vietnam
Nguyen-Son Vo	Duy Tan University, Vietnam
Vollero Luca	Università Campus Bio-Medico (Roma), Italy
Vu Van Yem	Hanoi University of Science and Technology, Vietnam
Vu Xuan-Thang	LSS-SUPELEC, France
Wang Kun Nanjing	University of Posts and Telecommunications, China
Wang Wenwu	University of Surrey, UK
Wong Kai Kit	University College London, UK
Wu Xiaoling	Chinese Academy of Sciences, China
Wymeersch Henk	Chalmers University of Technology, Sweden
Xia Minghua	Institut National de la Recherche Scientifique, Canada
Yang Nan	The University of New South Wales, Australia
Ye Lu	Broadcom Corporation, USA
Yu Rong	Guangdong University of Technology, China
Yuan Jinhong	University of New South Wales, Australia
Yuen Chau	Singapore University of Technology and Design, Singapore
Zeadally Sherali	University of Kentucky, USA
Zhang Yan	Simula Research Laboratory and University of Oslo, Norway
Zhong Caijun	Zhejiang University, China
Zhou Zhangbing	Institute Telecom, France
Zhu Chunsheng	The University of British Columbia, Canada
Zhu Zuqing	University of Science and Technology of China, China

Contents

Optimal Beamforming for Multiuser Secure SWIPT Systems (Invited Paper)

Yuqing Su and Derrick Wing Kwan Ng[⊠]

School of Electrical Engineering and Telecommunications,
The University of New South Wales, Sydney, Australia
w.k.ng@unsw.edu.au

Abstract. In this paper, we study the beamforming design for simultaneous wireless information and power transfer (SWIPT) downlink systems. The design is formulated as a non-convex optimization problem which takes into account the quality of service (QoS) requirements of communication security and minimum harvested power. In particular, the proposed design advocates the dual use of energy signal to enable secure communication and efficient WPT. The globally optimal solution of the optimization problem is obtained via the semidefinite programming relaxation (SDR). Our simulation results show that there exists a non-trivial tradeoff between the achievable data rate and the total harvested power in the system. Besides, our proposed optimal scheme provides a substantial performance gain compared to a simple suboptimal scheme based on the maximum ratio transmission (MRT).

Keywords: SWIPT · Physical layer security · Energy beamforming

1 Introduction

With the rapidly growing number of wireless communication devices and applications such as automated control in smart cities, monitoring in e-health systems, and remote security sensing [1], it is expected that by 2020 there will be 50 billion wireless devices connected together via the Internet around the world [2]. As a result, the fifth-generation communication systems also target on enhancing mobile broadband (e.g. tactile internet), connecting massive Internet-of-Things (IoT) wireless devices, and enabling new mission-critical controls (e.g. autonomous vehicles) [3]. Among all 5G emerging technologies thriving to fulfill stringent quality of service (QoS) requirements, multiple-input multiple-output (MIMO) technology is a promising solution to reduce the energy consumption caused by the rocketing demand of ultra-fast data rate [4–6]. In addition, the extra degrees of freedom offered by multiple antennas enable an efficient interference management in wireless communication systems. However, due to high computational complexity at receivers, traditional MIMO architecture may not be suitable for portable devices. As an alternative solution, multi-user MIMO with a multiple-antenna transmitter serving multiple receivers equipped with single-antenna [7,8],

T. Q. Duong and N.-S. Vo (Eds.): INISCOM 2018, LNICST 257, pp. 1–14, 2019.
https://doi.org/10.1007/978-3-030-05873-9_1

has been proposed as it shifts the signal processing burden from receivers to the transmitter which allows simple designs and cheap receiver structure.

Recently, an increasing number of mobile devices such as wireless sensors for IoT applications is gaining their popularity among the industry and becoming essential parts of wireless communication networks [9]. In particular, most of these devices are battery-powered with finite energy storage capacity. Hence, the inconvenience/high-cost of battery charging or replacement are the major obstacles in realizing IoT implementation [9]. Consequently, energy harvesting technology is adopted as a viable solution to provide ubiquitous and self-sustainable networks. Although traditional energy harvesting technology, which collects energy from natural renewable energy sources (e.g. solar and tide), can enable self-sustainable communication networks to a certain extent, it is both climate-dependent and location-dependent, which makes renewable energy a perpetual but intermittent energy supply [10,11]. Therefore, directly integrating conventional energy harvesting technology into communication devices may result in unstable communication service. Instead of exploiting renewable energy for energy-limited systems, an emerging solution, RF-based energy harvesting communication [12,13], is considered as a practical approach and is served building block for sustainable wireless systems to unlock the potential of IoT networks.

1.1   Background

Proposed by Nikola Tesla back in the late nineteenth century, wireless power transfer (WPT) was implemented by a magnifying transmitter based on the Tesla coil transmitter [14]. Its ultimate goal was to broadcast wireless power to any location around the globe avoiding shortcomings of conventional cables such as being unaffordable and inconvenient to deploy [14]. Under massive impact of the industrial revolution in late 1800s, WPT was originally designed to apply on high-power machines. With public health concern about harmful electromagnetic radiation caused by large power emission from the transmit tower [1], progress on bringing WPT into practice was hindered. Besides, as antennas in reasonable size are required in practical systems to provide mobility for portable communication devices, the information signal is usually modulated at a high carrier frequency resulting in severe path loss, which leads to a small amount of received power at the receiver side. Therefore, low power transfer efficiency is one of the challenges in implementing WPT. Prevented by these two major challenges, WPT was not able to realize its further development until advancing silicon technology and wireless communication theory bring it back to life recently [1]. Therefore, collecting energy from background RF electromagnetic (EM) wave transmitted from ambient transmitters is feasible via advanced WPT technologies and the application of communication theory. In fact, various proof-of-concepts experiments and prototypes have been developed. For example, it has been shown that practical RF-based energy harvesting circuits are able to harvest microwatt to milliwatt of power over the range of several meters for a transmit power of 1 Watt at a carrier frequency of less than 1 GHz [15].

1.2 Communication Security

On the other hand, communication security is a critical issue that needed to be taken into account in wireless communication systems. Nowadays, most of the conventional cryptographic encryption methods are implemented in the application layer to ensure secure communication [16]. However, this technology relies on perfect secret key information protection and distribution which are not practical in some of wireless communication networks [1]. Furthermore, secret-key cryptography algorithms usually assume that potential eavesdroppers have bounded computational capabilities, which might pose a future threat on itself as computers with ultra-high computational capabilities would be the developing trend (e.g. quantum computers). In fact, the security issue is more prominent in RF-based energy harvesting communication systems. In particular, RF-based energy harvesting receivers are generally located closer to the transmitter than an information receiver for more efficient energy harvesting. Thus, such scenarios raise about communication security due to the broadcast nature of wireless channels and the relatively high transmit powers needed for SWIPT. Thus, the emerging and stringent QoS requirements on guaranteeing secure communication have drawn significant attention in SWIPT beamforming design [1]. For example, information-theoretic physical layer (PHY) security aims at protecting secure communication by making use of wireless communication channels' physical natures (e.g. channel fading, etc.) [17], has been proposed for SWIPT systems. In particular, in order to guarantee secure communication, an energy signal is generated deliberately not just for degrading the quality of eavesdroppers' channels, but also to help facilitate an efficient energy transfer to ERs, e.g. [18–25]. In this paper, we study the resource allocation algorithm design to enable secure SWIPT systems via multiple antennas.

The remainder of this paper is organized as follows: In Sect. 2, we introduce the SWIPT downlink system model. In Sect. 3, we formulate the resource allocation design as an optimization problem and solve it optimally. Section 4 presents the numerical performance results for the proposed optimal algorithm. In Sect. 5, we conclude the paper with a summary.

Notation

Key mathematical notations are given in Table 1. Boldface lower and capital case letters are used to denote vectors and matrices, respectively. $\text{Rank}(\mathbf{A})$, $\text{Tr}(\mathbf{A})$, and $\mathbf{A}^{\mathrm{H}}$ are the rank, the trace, and Hermitian transpose of matrix $\mathbf{A}$, respectively. $\mathbf{A} \succeq \mathbf{0}$ means $\mathbf{A}$ is a positive semi-definite matrix. $\mathbb{H}^{\mathrm{N}}$ represents all Hermitian matrix sets. $\mathbb{C}^{\mathrm{N} \times \mathrm{M}}$ and $\mathbb{R}^{\mathrm{N} \times \mathrm{M}}$ represent all $N \times M$ sets with complex and real entries, respectively. The circularly symmetric complex Gaussian (CSCG) distribution is denoted by $\mathcal{CN}(\mathbf{m}, \boldsymbol{\Sigma})$ with mean vector $\mathbf{m}$ and covariance matrix $\boldsymbol{\Sigma}$; $\sim$ indicates "distributed as"; $\mathcal{E}\{\cdot\}$ denotes statistical expectation; $|\cdot|$ represents the absolute value of a complex scalar. $[x]^{+}$ stands for $\max\{0, x\}$.

Table 1. Nomenclature adopted in this paper.

Notation	Description
$\mathbf{h}$	Channel vector between the information receiver (IR) and the transmitter
$\mathbf{g}_j$	Channel vector between energy harvesting receiver (ER) j and the transmitter
$\mathbf{w}$	Information beamforming vector
$\mathbf{w}_\mathrm{E}$	Energy signal beamforming vector
$\sigma_\mathrm{ant}^2,\ \sigma_\mathrm{s}^2$	Antenna and signal processing noise power
N_T	Number of transmit antennas
$R_\mathrm{ER}^\mathrm{Tol}$	Maximum tolerable data rate
P_max	Maximum transmit power of the transmitter
P_min	Minimum required power transfer to ERs

2   System Model

In this paper, we focus on a downlink SWIPT system. There is a transmitter equipped with N_T antennas, one information receiver (IR), and J energy harvesting receivers (ERs). In particular, both the IR and ERs are single-antenna devices [17], cf. Fig. 1. In the following, it is assumed that both the transmitter and IR know the perfect channel state information (CSI) for resource allocation. Table 1 shows the nomenclature adopted in this paper.

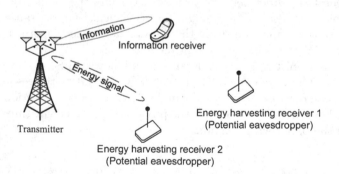

Fig. 1. A downlink SWIPT model with one IR and $J = 2$ ERs. For guaranteeing communication security, the ERs are also treated as potential eavesdroppers eavesdropping information transmitted from the transmitter to the IR.

In each time slot, the transmit signal vector $\mathbf{x}$ is given as

$$\mathbf{x} = \mathbf{w}\,s + \mathbf{w}_\mathrm{E}, \tag{1}$$

where $\mathbf{w}_\mathrm{E}$ is a pseudo-random energy signal and modeled as a complex Gaussian random vector with zero-mean and covariance matrix $\mathbf{W}_\mathrm{E}$, i.e., $\mathbf{w}_\mathrm{E} \sim$

$\mathcal{CN}(\mathbf{0}, \mathbf{W}_\mathrm{E})$, $\mathbf{w}$ is the beamforming vector of the information signal, and s is the information signal. Hence, the signals received at IR and ER $j \in \{1, \dots, J\}$ are given by

$$y = \mathbf{h}^H(\mathbf{w}s + \mathbf{w}_\mathrm{E}) + n \text{ and} \tag{2}$$

$$y_{\mathrm{ER}_j} = \mathbf{g}_j^H(\mathbf{w}s + \mathbf{w}_\mathrm{E}) + n_{\mathrm{ER}_j}, \forall j \in \{1, \dots, J\}, \tag{3}$$

respectively, where $\mathbf{h}$ and $\mathbf{g}_j$ are channel vectors between the transmitter and IR, transmitter and ER j, respectively, capturing the impact of small scale fading and large scale fading of the channels. Variables n, n_{ER_j} are additive white Gaussian noise (AWGN) of IR and ER j from the receiving antenna, respectively, with zero-mean and variance σ_s^2 and σ_j^2, respectively. We also assume without loss of generality that $\mathbb{E}\{|s|^2\} = 1$.

2.1 Energy Harvesting Model

In practice, an energy harvesting circuit consists of various hardware components, such as bandpass filter, receifying circuit, etc., cf. Fig. 2. In the SWIPT literature [26–32], for simplicity, the total harvested power at ER j is typically modelled by a linear equation[1]:

$$\Phi_{\mathrm{ER}_j}^{\mathrm{Linear}} = \eta_j P_{\mathrm{ER}_j} \quad \text{where} \tag{4}$$

$$P_{\mathrm{ER}_j} = \mathbb{E}\{|\mathbf{g}_j^H \mathbf{x}|^2\} = \mathrm{Tr}\Big((\mathbf{W}_\mathrm{E} + \mathbf{w}\mathbf{w}^H)\mathbf{G}_j\Big) \tag{5}$$

denotes the harvested power from the channel, $\eta_j \in [0, 1]$ is the RF-to-DC power conversion efficiency, and $\mathbf{G}_j = \mathbf{g}_j \mathbf{g}_j^H$.

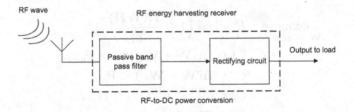

Fig. 2. Block diagram of an ER.

---

[1] Recently, various non-linear energy harvesting models have been proposed in the literature, e.g. [33,34]. However, in this paper, we adopt the conventional linear energy harvesting model for the ease of illustration.

2.2   Achievable Rate and Secrecy Rate

Given perfect CSI at the receiver, the achievable rate (bit/s/Hz) between the transmitter and the IR is given by

$$R = \log_2\left(1 + \frac{\mathbf{w}^H\mathbf{H}\mathbf{w}}{\sigma_s^2}\right), \tag{6}$$

where $\mathbf{H} = \mathbf{h}\mathbf{h}^H$. Note that since the pseudo-random energy signal, $\mathbf{w}_E$, is a deterministic sequence which is known by the IR, its impact on the achievable rate, i.e., $\mathrm{Tr}(\mathbf{H}\mathbf{W}_E)$, has been canceled with interference cancelation techniques before decoding the desired signal.

On the other hand, the capacity between the transmitter and ER j for decoding the signal of the IR can be expressed as

$$R_j^{\mathrm{ER}} = \log_2\left(1 + \frac{\mathrm{Tr}(\mathbf{W}\mathbf{G}_j)}{\mathrm{Tr}(\mathbf{G}_j\mathbf{W}_E) + \sigma_{s_j}^2}\right). \tag{7}$$

Thus, the achievable secrecy rate of IR is given by

$$R_{\mathrm{sec}} = \left[R - \max_{\forall j}\{R_{\mathrm{ER}_j}\}\right]^+. \tag{8}$$

3   Problem Formulation

The considered system design objective is to maximize the achievable rate of the IR while guaranteeing secure communication and facilitating efficient wireless power transfer. To this end, the resource allocation design is formulated as the following optimization problem:

Resource Allocation Design:

$$\underset{\mathbf{W},\mathbf{W}_E\in\mathbb{H}^{N_T}}{\text{maximize}} \quad R = \log_2\left(1 + \frac{\mathrm{Tr}(\mathbf{W}\mathbf{H})}{\sigma_s^2}\right) \tag{9}$$

$$\begin{aligned}
\text{s.t.} \quad & \mathrm{C1}: \ \mathrm{Tr}((\mathbf{W} + \mathbf{W}_E)\mathbf{G}) \geq P_{\min}, \\
& \mathrm{C2}: \ \mathrm{Tr}(\mathbf{W} + \mathbf{W}_E) \leq P_{\max}, \\
& \mathrm{C3}: \ R_j^{\mathrm{ER}} \leq R_{\mathrm{tol}}^{\mathrm{ER}}, \forall j \in \{1,\ldots,J\}, \\
& \mathrm{C4}: \ \mathbf{W} \succeq \mathbf{0}, \\
& \mathrm{C5}: \ \mathbf{W}_E \succeq \mathbf{0}, \\
& \mathrm{C6}: \ \mathrm{Rank}(\mathbf{W}) \leq 1.
\end{aligned}$$

where $\mathbf{G} = \sum_{j=1}^{J}\mathbf{G}_j\mathbf{G}_j^H$. Constants $P_{\min}$ and $P_{\max}$ in constraints C1 and C2 are the minimum required total harvested power and the maximum transmit power budget offered by transmitter, respectively. Constant $R_{\mathrm{ER}}^{\mathrm{Tol}} > 0$ in C3 is the maximum tolerable data rate which restricts the capacity of ER j if it attempts to decode the message of the IR. In practice, the transmitter sets $R_{\mathrm{ER}}^{\mathrm{Tol}} \to 0$, to ensure secure communication. If the above optimization problem is feasible, the adopted problem formulation guarantees that the achievable secrecy rate is bounded below by $R_{\mathrm{sec}} \geq R - R_{\mathrm{ER}}^{\mathrm{Tol}} > 0$.

3.1 Optimal Solution

The optimization problem in (9) is non-convex due to the constraint C3 and the rank-one matrix constraint C6. In general, there might be multiple local maximums [35] in a non-convex optimization problem. Although exhaustive search can be used to obtain the globally optimal solution, the incurred computational complexity increases exponentially with respect to the numbers of transmit antennas and the ERs. In order to design a computational efficient resource allocation, we apply the SDP relaxation to the considered problem. In particular, we remove constraint C6 and after some mathematical manipulation which yields:

Problem Transformation

$$\underset{\mathbf{W},\mathbf{W}_{\mathrm{E}}\in\mathbb{H}^{N_T}}{\text{minimize}} \quad -\mathrm{Tr}(\mathbf{W}\mathbf{H}) \tag{10}$$

$$\text{s.t.} \quad \mathrm{C1}: \; P_{\min} - \mathrm{Tr}((\mathbf{W} + \mathbf{W}_{\mathrm{E}})\mathbf{G}) \leq 0,$$

$$\mathrm{C2}: \; \mathrm{Tr}(\mathbf{W} + \mathbf{W}_{\mathrm{E}}) - P_{\max} \leq \mathbf{0},$$

$$\mathrm{C3}: \; \mathrm{Tr}(\mathbf{W}\mathbf{G}_j) - (2^{R_{\mathrm{tol}}^{\mathrm{ER}}} - 1)\,\mathrm{Tr}(\mathbf{G}_j\mathbf{W}_{\mathrm{E}} + \sigma_{\mathrm{s}_j}^2) \leq \mathbf{0}, \forall j,$$

$$\mathrm{C4}: \; -\mathbf{W} \preceq \mathbf{0},$$

$$\mathrm{C5}: \; -\mathbf{W}_{\mathrm{E}} \preceq \mathbf{0},$$

$$\mathrm{C6}: \; \cancel{\mathrm{Rank}(\mathbf{W}) \leq 1}.$$

Note that the transformed problem is convex and can be solved efficiently and optimally via standard convex program solvers, such as CVX. Yet, the rank constraint relaxation may not be tight $\mathrm{Rank}(\mathbf{W}) > 1$ occurs. Therefore, we reveal the tightness of the adopted SDP relaxation in (10) in the following theorem:

Theorem 1. *If channels $\mathbf{H}$ and $\mathbf{G}_j$ are statistically independent and the transformed problem in (10) is feasible, then the optimal information beamforming matrix $\mathbf{W}$ is at most rank-one with probability one, i.e., $\mathrm{Rank}(\mathbf{W}) \leq 1$.*

Proof: Please refer to the appendix.

Thus, the adopted SDP relaxation is tight as long as the channel conditions in Theorem 1 are fulfilled. Therefore, the considered information beamforming is optimal for the maximization of achievable data rate.

3.2 Suboptimal Solution

To further reduce the computational complexity, suboptimal scheme is proposed in this section. In particular, maximum ratio transmission (MRT) with respect to the direction of IR is adopted for information transmission, i.e., $\mathbf{w} = \frac{\mathbf{h}^*}{\|\mathbf{h}\|}$. Then, the transmit power of $\mathbf{w}$ and the covariance matrix of $\mathbf{W}_E$ are optimized to maximize the achievable data rate of the system subject to the same constraint set as in (10).

Table 2. Parameters in simulation.

Centre frequency of carrier signal	915 MHz
Bandwidth	200 kHz
Gain of transceiver antenna	10 dBi
Transmit antenna number N_T	$3, 6, 9$
Noise power σ^2	-95 dBm
Maximum transmit power P_{max}	1 W
$\mathbf{G}_j$ fading distribution	Ricean with Ricean factor 3 dB
$\mathbf{h}$ fading distribution	Rayleigh
R_{ER}^{Tol}	0.1375 bit/s/Hz
η_j RF-to-DC power conversion efficiency	0.5

4 Simulation Results

In this section, we demonstrate the performance of the proposed optimal resource allocation via simulations. Unless further specified, the important simulation parameters are listed in Table 2. Figure 3 shows the non-trivial trade-off between the total system data rate and the total system harvested power for the SWIPT model under proposed optimal beamforming scheme. In general, the area enclosed by the curve of a certain transmit antenna number is the achievable region. In other words, all the points lie inside or on the curve can be achieved by tuning the associated system parameters of the optimal scheme. By comparing intersecting points of both y and x-axis for different transmit antenna numbers, it can be observed that with increasing number of transmitter antennas, the maximum total system data rate of IR and the maximum total system harvested power of ERs increase due to extra spatial degrees of freedom supplied by multiple transmit antennas which improve the accuracy in beamforming. On the other hand, by comparing the two tradeoff regions (i.e., region between N_T = 3 and N_T = 6, region between N_T = 6 and N_T = 9), it is manifest that with increasing number of transmitter antennas, the increasing rate of the enlarged region decreases due to channel hardening [36]. For comparison, we also show the performance of the suboptimal MRT-based transmission scheme. It can be seen that the suboptimal scheme cannot achieve the maximum system data rate as communication security is taken into account. Also, the suboptimal scheme can only achieve a smaller tradeoff region compared to the proposed optimal scheme due to its less flexibility in beamforming.

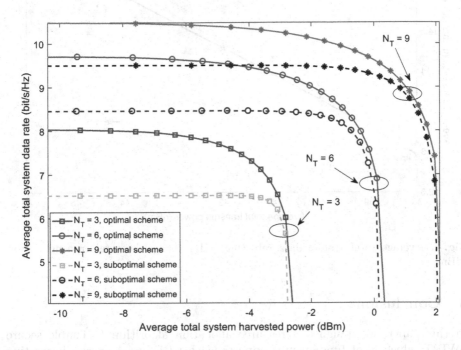

Fig. 3. Average total system data rate (bit/s/Hz) versus average total system harvested power (dBm).

Figure 4 shows the average data rate versus the total transmit power budget for different numbers of ERs of the proposed optimal scheme. It can be observed that for a given amount of total transmit power budget, the average total system data rate decreases when there are more ERs in the system. In fact, when there are more ERs in the system, both the QoS requirements on communication security and minimum required harvested power become more stringent. Particularly, the transmitter is forced to steer the direction of information signal towards the ERs. This will decrease the received signal strength of the desired signal at the IR. Besides, the transmitter would also increase the transmit power of energy signal to neutralize the higher potential of information leakage, which leads to a further reduction in the data rate. Furthermore, the proposed suboptimal scheme is able to guarantee both the QoS requirements of minimum required harvested power and communication security.

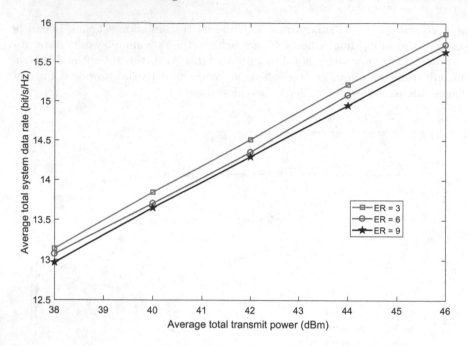

Fig. 4. Average total system data rate (bit/s/Hz) versus the total transmit power (dBm).

5 Conclusions

In this paper, we designed a resource allocation algorithm to enable secure SWIPT which is of fundamental importance for RF-based energy harvesting communication networks. In particular, the algorithm design was formulated as a non-convex optimization problem for the maximization of the achievable data rate of the IR while taking into account the QoS of efficient WPT and communication security. By exploiting the SDP relaxation, we solve the non-convex optimization problem optimally. Numerical results showed the potential gains in both data rate and harvested power enabled by the proposed optimization and its robustness against eavesdropping.

6 Appendix – Proof of Theorem 1

It can be verified that the transformed optimization problem in (10) is convex and satisfies the Slater's constraint qualification, hence, strong duality holds. In other words, solving its dual problem is equivalent to solving the primal problem. In this section, we intend to prove Theorem 1 via first defining the Lagrangian function:

$$
\begin{aligned}
L = \; & -\operatorname{Tr}(\mathbf{W}\mathbf{H}) - \lambda_{C1}\operatorname{Tr}((\mathbf{W} + \mathbf{W}_E)\mathbf{G}) + \lambda_{C2}\operatorname{Tr}(\mathbf{W} + \mathbf{W}_E) \\
& + \lambda_{C3}\operatorname{Tr}(\mathbf{W}\mathbf{G}_j) - \operatorname{Tr}(\mathbf{Y}\mathbf{W}) + \Delta,
\end{aligned} \tag{11}
$$

where Δ represents the variables and the constants that are independent of $\mathbf{W}$ and therefore irrelevant in the proof. $\mathbf{Y}$ and λ_{C1}, λ_{C2}, λ_{C3} are dual variables related to the constraints C4 and C1, C2, C3, respectively. Now, we can express the dual problem of (10) as

$$\max_{\mathbf{Y}, \lambda_{C1}, \lambda_{C2}, \lambda_{C3}} \quad \min_{\mathbf{W}, \mathbf{W}_E \in \mathbb{H}^{N_T}} \quad L. \tag{12}$$

Then, we study the structure of $\mathbf{W}$ via applying the Karush-Kuhn-Tucker (KKT) conditions:

$$\mathbf{Y} \succeq \mathbf{0}, \lambda_{C1}, \lambda_{C2}, \lambda_{C3} \geq 0, \tag{13}$$

$$\mathbf{YW} = \mathbf{0}, \tag{14}$$

$$\mathbf{Y} = -\mathbf{H} + \mathbf{B}, \tag{15}$$

where (14) is obtained by taking the derivative of the Lagrangian function with respect to $\mathbf{W}$ and $\mathbf{B} = -\lambda_{C1}\mathbf{G} + \lambda_{C2}\mathbf{I} + \lambda_{C3}\mathbf{G}_j$. Equation (13) is the complementary slackness property which implies that the columns of matrix $\mathbf{W}$ fall into the null-space spanned by $\mathbf{Y}$ for $\mathbf{W} \neq \mathbf{0}$. Hence, if we can prove that Rank$(\mathbf{Y}) \geq N_T - 1$, the optimal beamforming matrix $\mathbf{W}$ is a rank-one matrix or a zero matrix. Now, we study the structure of $\mathbf{Y}$ via examining (15). First, we prove by contradiction that $\mathbf{B}$ is a positive definite matrix with probability one. Suppose not, $\mathbf{B}$ is a positive semi-definite matrix. Then, there exist at least one zero eigenvalue and we denote the associated eigenvector as $\mathbf{v}$. Without loss of generality, we create a matrix $\mathbf{V} = \mathbf{vv}^H$ from the eigenvector. By multiplying both sides of (15) with $\mathbf{V}$ and applying the trace operation, we obtain

$$\mathrm{Tr}(\mathbf{YV}) = -\mathrm{Tr}(\mathbf{HV}) + \mathrm{Tr}(\mathbf{BV}) = -\mathrm{Tr}(\mathbf{HV}). \tag{16}$$

Since $\mathbf{H}$ and $\mathbf{G}_j$ are statistically independent, we have $\mathrm{Tr}(\mathbf{HV}) > 0$. This leads to contradiction as $\mathrm{Tr}(\mathbf{YV}) \geq 0$. Hence, matrix $\mathbf{B}$ is a positive definite matrix[2], i.e., Rank$(\mathbf{B}) = N_T$. To further proceed the proof, we introduce the following rank inequality:

Lemma 1. *Let $\mathbf{A}$ and $\mathbf{B}$ be two matrices with same dimension. The inequality of matrix* Rank$(\mathbf{A} + \mathbf{B}) \geq$ Rank$(\mathbf{A})$ − Rank$(\mathbf{B})$ *holds.*

Proof: By basic rule of inequality for the rank of matrix, Rank$(\mathbf{A})$ + Rank$(\mathbf{B}) \geq$ Rank$(\mathbf{A} + \mathbf{B})$ with both matrices of same dimension. Thus we have Rank$(\mathbf{A} + \mathbf{B})$ + Rank$(-\mathbf{B}) \geq$ Rank$(\mathbf{A})$. Since Rank$(\mathbf{B}) =$ Rank$(-\mathbf{B})$, the lemma is proved. $\square$

---

[2] It can be verified that matrix $\mathbf{B}$ is not a negative definite or a negative semi-definite matrix.

Now, we apply Lemma 1 on (14) which yields:

$$\text{Rank}(\mathbf{Y}) = \text{Rank}(-\mathbf{Y}) = \text{Rank}(-\mathbf{B} + \mathbf{H})$$
$$\geq \text{Rank}(-\mathbf{B}) - \text{Rank}(\mathbf{H}) = N_T - 1 \tag{17}$$

As $\text{Rank}(\mathbf{Y}) \geq N_T - 1$, we have $\text{Rank}(\mathbf{W}) \leq 1$ which completes the proof. $\square$

References

1. Ng, D.W.K., Leng, S., Schober, R.: Multiple Antennas and Beamforming for SWIPT Systems, pp. 170–216. Cambridge University Press, Cambridge (2016)
2. Ding, Z., et al.: Application of smart antenna technologies in simultaneous wireless information and power transfer. IEEE Commun. Mag. **53**(4), 86–93 (2015)
3. Wong, V., Schober, R., Ng, D.W.K., Wang, L.-C.: Key Technologies for 5G Wireless Systems. Cambridge University Press, Cambridge (2017)
4. Goldsmith, A.: Wireless Communications. Cambridge University Press, Cambridge (2005)
5. Ng, D.W.K., Lo, E.S., Schober, R.: Energy-efficient resource allocation in OFDMA systems with large numbers of base station antennas. IEEE Trans. Wirel. Commun. **11**(9), 3292–3304 (2012)
6. Wu, Q., Li, G.Y., Chen, W., Ng, D.W.K., Schober, R.: An overview of sustainable green 5G networks. IEEE Wirel. Commun. **24**(4), 72–80 (2017)
7. Marzetta, T.: Noncooperative cellular wireless with unlimited numbers of base station antennas. IEEE Trans. Wirel. Commun. **9**, 3590–3600 (2010)
8. Ng, D.W.K., Lo, E.S., Schober, R.: Energy-efficient resource allocation in multi-cell OFDMA systems with limited backhaul capacity. IEEE Trans. Wirel. Commun. **11**, 3618–3631 (2012)
9. Zorzi, M., Gluhak, A., Lange, S., Bassi, A.: From today's INTRAnet of things to a future INTERnet of Things: a wireless- and mobility-related view. IEEE Wirel. Commun. **17**, 44–51 (2010)
10. Ahmed, I., Ikhlef, A., Ng, D.W.K., Schober, R.: Power allocation for an energy harvesting transmitter with hybrid energy sources. IEEE Trans. Wirel. Commun. **12**, 6255–6267 (2013)
11. Ng, D.W.K., Lo, E.S., Schober, R.: Energy-efficient resource allocation in OFDMA systems with hybrid energy harvesting base station. IEEE Trans. Wirel. Commun. **12**, 3412–3427 (2013)
12. Chen, X., Zhang, Z., Chen, H.-H., Zhang, H.: Enhancing wireless information and power transfer by exploiting multi-antenna techniques. IEEE Commun. Mag. **4**, 133–141 (2015)
13. Varshney, L.: Transporting Information and Energy Simultaneously. In: Proceedings of IEEE International Symposium on Information Theory, pp. 1612–1616, July 2008
14. Tesla Memroial Society of New York, "Nikola Tesla's Idea of Wireless Transmission of Electrical Energy is a solution for World Energy Crisis" (2011). [Online]. http://www.teslasociety.com/tesla_tower.htm
15. Powercast Coporation: "RF Energy Harvesting and Wireless Power for Low-Power Applications" (2011). [Online]. http://www.mouser.com/pdfdocs/Powercast-Overview-2011-01-25.pdf

16. Ng, D., Schober, R.: Max-min fair wireless energy transfer for secure multiuser communication systems. In: IEEE Information Theory Workshop (ITW), pp. 326–330, November 2014
17. Ng, D.W.K., Schober, R., Alnuweiri, H.: Secure layered transmission in multicast systems with wireless information and power transfer. In: Proceedings of IEEE International Communication Conference, pp. 5389–5395, June 2014
18. Wyner, A.D.: The Wire-Tap Channel. Technical report, October 1975
19. Zhu, J., Schober, R., Bhargava, V.: Secure transmission in multicell massive MIMO systems. IEEE Trans. Wirel. Commun. **13**, 4766–4781 (2014)
20. Goel, S., Negi, R.: Guaranteeing secrecy using artificial noise. IEEE Trans. Wirel. Commun. **7**, 2180–2189 (2008)
21. Wang, H.M., Wang, C., Ng, D., Lee, M., Xiao, J.: Artificial noise assisted secure transmission for distributed antenna systems. IEEE Trans. Sig. Process. **64**(15), 4050–4064 (2016)
22. Chen, J., Chen, X., Gerstacker, W.H., Ng, D.W.K.: Resource allocation for a massive MIMO relay aided secure communication. IEEE Trans. Inf. Forensics Secur. **11**(8), 1700–1711 (2016)
23. Ng, D.W.K., Lo, E.S., Schober, R.: Efficient resource allocation for secure OFDMA systems. IEEE Trans. Veh. Technol. **61**, 2572–2585 (2012)
24. Wang, H.M., Wang, C., Ng, D.W.K.: Artificial noise assisted secure transmission under training and feedback. IEEE Trans. Sig. Process. **63**(23), 6285–6298 (2015)
25. Chen, X., Ng, D.W.K., Chen, H.H.: Secrecy wireless information and power transfer: challenges and opportunities. IEEE Wirel. Commun. **23**(2), 54–61 (2016)
26. Zhou, X., Zhang, R., Ho, C.K.: Wireless information and power transfer: architecture design and rate-energy tradeoff. In: Proceedings of IEEE Global Telecommunication Conference, December 2012
27. Ng, D.W.K., Lo, E.S., Schober, R.: Energy-efficient resource allocation in multiuser OFDM systems with wireless information and power transfer. In: Proceedings of IEEE Wireless Communication and Networking Conference (2013)
28. Leng, S., Ng, D.W.K., Schober, R.: Power efficient and secure multiuser communication systems with wireless information and power transfer. In: Proceedings of IEEE International Communication Conference, June 2014
29. Ng, D.W.K., Xiang, L., Schober, R.: Multi-objective beamforming for secure communication in systems with wireless information and power transfer. In: Proceedings of IEEE Personal Indoor and Mobile Radio Communication Symposium (2013)
30. Ng, D.W.K., Schober, R.: Resource allocation for coordinated multipoint networks with wireless information and power transfer. In: Proceedings of IEEE Global Telecommunication Conference, pp. 4281–4287, December 2014
31. Chynonova, M., Morsi, R., Ng, D.W.K., Schober, R.: Optimal multiuser scheduling schemes for simultaneous wireless information and power transfer. In: 23rd European Signal Processing Conference (EUSIPCO), August 2015
32. Wu, Q., Tao, M., Ng, D.W.K., Chen, W., Schober, R.: Energy-efficient transmission for wireless powered multiuser communication networks. In: Proceedings of IEEE International Communication Conference, June 2015
33. Boshkovska, E., Ng, D., Zlatanov, N., Schober, R.: Practical non-linear energy harvesting model and resource allocation for SWIPT systems. IEEE Commun. Lett. **19**, 2082–2085 (2015)
34. Boshkovska, E., Ng, D.W.K., Zlatanov, N., Koelpin, A., Schober, R.: Robust resource allocation for MIMO wireless powered communication networks based on a non-linear EH model. IEEE Trans. Commun. **65**(5), 1984–1999 (2017)

35. Floudas, C.A.: Nonlinear and Mixed-Integer Optimization: Fundamentals and Applications, 1st edn. Oxford University Press, Oxford (1995)
36. Tse, D., Viswanath, P.: Fundamentals of Wireless Communication, 1st edn. Cambridge University Press, Cambridge (2005)

Logarithmic Spiral Based Local Search in Artificial Bee Colony Algorithm

Sonal Sharma[1], Sandeep Kumar[2], and Anand Nayyar[3(✉)]

[1] Poornima College of Engineering, Jaipur, India
sonal.gold@gmail.com
[2] Amity University Rajasthan, Jaipur, India
sandpoonia@gmail.com
[3] Duy Tan University, Da Nang, Vietnam
anandnayyar@duytan.edu.vn

Abstract. Artificial bee colony (ABC) algorithm is recent swarm intelligence based meta-heuristic that is developed to solve complex real problems which are difficult to solve by the available deterministic strategies. It mimics the natural behaviour of real honey bees while searching for food sources. The performance of ABC depends on the size of step during position update process, that is a combination of the arbitrary component ϕ_{ij} and a difference vector between the current solution and an arbitrarily identified solution. The high value of ϕ_{ij} and high difference between the vectors in the step generation process may generate the large size step which may leads to the skipping of true solution. Therefore, to avoid this situation a logarithmic spiral based local search strategy, namely logarithmic spiral local search (LSLS) is planned and incorporated with the ABC. The proposed hybridized ABC is named as logarithmic spiral based ABC (LSABC). To demonstrate the efficiency and accurateness of the LSABC, it is tested over 10 popular benchmarks functions and outcomes are equated with ABC, Modified ABC, and Best-so-far ABC. The reported results showed that the proposed LSABC is a new viable variation of ABC algorithm.

Keywords: Nature Inspired Algorithms · Swarm intelligence
Population based optimization algorithm

1 Introduction

Nature continuously provide some of the well planned way to solve the real life problems. The nature has progressed for the period of thousands of year elevating with innovative models, techniques and tools and established into well-defined disciplines of scientific aspire. Humankind has been continuously annoying to apprehend the nature from the time when by evolving some innovative techniques and tools day by day. The area of nature-inspired computing is a combination of computing science with knowledge from different streams like mathematics,

T. Q. Duong and N.-S. Vo (Eds.): INISCOM 2018, LNICST 257, pp. 15–27, 2019.
https://doi.org/10.1007/978-3-030-05873-9_2

biology, chemistry physics and engineering. The algorithms simulating processes in nature or inspired from some natural phenomenon are called Nature Inspired Algorithms (NIAs). According to the source of inspiration they are classified in different classes. Most of the NIAs are motivated by working of natural biological systems, swarm intelligence, physical and chemical phenomenon. For that reason, the name biology based, swarm-intelligence based, physics and chemistry based, according to their sources of stimulation. Basically main source of inspiration is nature and thus they called nature-inspired. More than hundred algorithms inspired by nature are now a day available and ABC is one of them. The ABC algorithm is capable to realize better results very quickly however like other stochastic population based approaches, it is not better in terms of exploration of search space. The performance of ABC algorithm is critically analyzed and concluded that it is better than some other NIAs most of the times [1,2] but it has some drawback also. It is revealed in study that original ABC [3] sometimes halt while inching ahead in the direction of the global optimum despite the fact the population has not converged to a local optimum [4]. It happens due to it's position update equation that is really good in exploration but pitiable at exploitation [5]. Thus it can be concluded that ABC has no proper balancing between exploration and exploitation process. For that reason these weaknesses require an amendment in position update process in ABC. To boost the exploitation, in this paper, a local search strategy namely, logarithmic spiral local search (LSLS) is anticipated and assimilated with ABC to magnify the exploitation potential of the ABC algorithm. The proposed LSLS is inspired from the equation of logarithmic spiral. The hybridized ABC algorithm is named as logarithmic spiral based ABC (LSABC) algorithm.

Financial system and growth of a country is highly reliant on the agricultural products. A large number of people are concerned in agricultural production where various categories of plants are cropped on a variety of lands based on the ecological circumstances. However, maintaining these plants suffers with a number of problems which are faced by the farmers such as plant diseases, quality of soil, selection of crop and many more. The plant diseases must be detected and prevented well on time to enhance the production. For that reason, an automated plant disease identification system can be very helpful for monitoring the plants. Leaves are primary part of the plants where the effect of a disease is generally visualized and may be identified for required prevention. Similarly in order to enhance productivity a farmer must be aware about quality of soil, so that he/she can use suitable seeds and pesticides. These two things, identification of plant diseases and prediction of soil quality can be easily done using LSABC. The LSABC may be used for clustering of images (images of plant leafs and different soil) while identifying different classes of leafs and soil.

2   ABC Algorithm

The ABC algorithm is invented by Dervis Karaboga in 2005 [3] inspired by extraordinary conduct of honey bees while piercing for food sources with best characteristics. Analogous to natural honey bees the ABC algorithm divided all bees in three different groups according to their behavior and nature of task performed. The whole population composed of three types of bees: employed, onlooker and scout bee. The employed bees are accountable for searching for new food sources and providing useful information regarding food sources to bees that residing in hive (onlooker bees). Based on information received from employed bee, onlooker bees start exploiting these food sources. If a food sources exhausted due to exploitation it is considered as abandoned and replaces by scout bee. The bees are continuously trying to improve solutions using greedy search strategy till the termination criteria meet and memories the best solution established till now. The success of ABC algorithm depends on balance between these two processes. Initialization of swarm also play important role in deciding direction of solution search process. The ABC algorithm modified number of times to improve its performance like memetic search in ABC [6], levy flight ABC [7], modified gbest ABC [8], lbest gbest ABC [9], fitness based position update in ABC [10] and memetic search in ABC with fitness based position update [11]. The practical implementation of ABC is easy, and has only three parameters. The complete ABC algorithm spliced into three phases. The Algorithm 1 revealed the core steps of ABC algorithm.

Algorithm 1. ABC Algorithm

Parameter Initialization;
while Stopping criteria is not fulfilled **do**
    Phase 1: New food source engendered using employed bee phase
    Phase 2: Onlooker bee for apprising the food sources conditional to the quantity of nectar;
    Phase 3: Scout bee for determining the new food sources in place of rejected food sources;
    Phase 4: Remember the most feasible food source established till now;
end while
Yield the most favourable solution identified till now.

2.1 Steps of ABC Algorithm

Initialization: The first phase in ABC is initialization of parameters (Colony Size, Limit for scout bees and maximum number of cycles) and set up an initial population randomly using Eq. 1.

$$p_{ij} = LB_j + rand \times (UB_j - LB_j) \tag{1}$$

Where, $i = 1, 2, .., (Colonysize/2)$ and $j = 1, 2.., D$. Here D represent dimension of problem. p_{ij} denotes location of ith solution in j^{th} dimension. LB_j and UB_j denotes lower and upper boundary values of search region correspondingly. $rand$ is a randomly selected value in the range (0, 1).

Employed Bee Phase: This phase try to detect superior quality solutions in proximity of current solutions. If quality of fresh solution is enhanced than present solution, the position is updated. The position of employed bee updated using Eq. 2.

$$V_{ij} = p_{ij} + \overbrace{\phi_{ij} \times (p_{ij} - p_{kj})}^{s} \tag{2}$$

Where, $\phi_{ij} \in [-1, 1]$ is an arbitrary number, $k \in 1, 2, .(Colonysize/2)$ is a haphazardly identified index such that $k \neq i$. In this equation s denotes step size of position update equation. A larger step size leads to skipping of actual solution and convergence rate may degrade if step size is very small.

Onlooker Bee Phase: The selection of a food source depends on their probability of selection. The probability is computed using fitness of solution with the help of Eq. 3.

$$Prob_i = \frac{fitness_i}{\sum_{i=1}^{colonysize/2} Fitness_i} \tag{3}$$

Scout Bee Phase: An employed bee become scout bee when the solution value not updated till the predefined threshold limit. This scout bee engenders new solution instead of rejected solution using Eq. 1.

3   Motivation

In most of the swarm based stochastic meta-heuristics, the potential to explore the divergent unidentified areas in the solution search region to identify the global optimum is denoted as the process of exploration, whereas the capability to process the information about the earlier good solutions to find best feasible solutions is termed as exploitation. When it comes down to it these two process (exploration and exploitation) are contradictory to each other in nature. It is essential to maintain proper balancing between these two process so that it may attain excelling performance while performing optimization.

The swarm in ABC update its position by the means of two essential processes: the process of exploration, which legitimize discovering diverse areas of the search reason, and the identification of best solution, that make sure the exploitation of the previous experience. Though, some research pointed out that the ABC sometimes not able to continue for the global optimum despite the fact that the swarm has not congregated to a local optimum [4]. It can be perceived that the process of solution searching in ABC is very efficient in exploration but not up to the mark in exploitation [5].

One of the simplest way to embellish the exploration potential of the swarm based algorithm is to increase the size of swarm [12, 13]. In this way, to circumvent the condition of premature convergence annexation of certain peripheral budding solutions are useful in avoiding trapping into local optima and sometimes removal of several individuals may help to accelerate the rate of convergence, if it is taking excessively large time to converge.

Further, the step size in position update equation of ABC has significant impact on the superiority of the updated solution. The high absolute value of ϕ_{ij} and large difference between current solution and arbitrarily chosen solution leads to large step size and it results in skipping the true solution and very small step size results in low convergence rate. A balanced step size can maintain proper balancing between exploration and exploitation potential of the ABC meanwhile. This balancing is not possible by hand as this step size involves some random component. For that reason, a self-adaptive strategy, which can adaptively modify the step size of an individual may also help for balancing the exploration and exploitation competencies of the algorithm.

The size of step during position update process in ABC is the combination of an arbitrary element ϕ_{ij} and a difference vector between the current solution and an arbitrarily identified solution. The high value of ϕ_{ij} and high difference between the vectors in the step generation process may generate the large step size which may leads to the skipping of true solution. Therefore, to avoid this situation and to embellish the exploitation competence of ABC algorithm, logarithmic spiral based local search strategy, namely logarithmic spiral local search (LSLS) is anticipated and assimilated with the ABC. The anticipated hybridized ABC is named as logarithmic spiral based ABC (LSABC). To demonstrate the efficiency and trustworthiness of the LSABC, it is implemented for 10 established benchmarks problems and equated with ABC, Best-so-far ABC, and Modified ABC. The reported results showed that the proposed LSABC is a new worthy variation of ABC algorithm.

4   Logarithmic Spiral Based ABC

A local search strategy inspired by the equation of logarithmic spiral is incorporated in basic ABC. A logarithmic spiral is a self-similar spiral curve which over and over again give the impression in nature [14]. As the solution search process of ABC algorithm is highly depends on a combination of random component ϕ_{ij} and a difference vector (refer Eq. 2). Hence, the high value of random component ϕ_{ij} and the difference vector results in more chances to avoid the actual solution. For that reason, a new approach is projected, which helps the current best solution in the swarm to exploit the search reasons in its locality in this article. The anticipated new local search algorithm is described in Algorithm 2.

The equation of logarithmic spiral is shown in Eq. 4. Logarithmic spiral is the locus of points analogous to the positions with time of a point moving away from a static point with a fixed speed beside a line which revolves with fixed angular velocity. Unvaryingly, in polar coordinates (r, θ) as designated by the Eq. 4 [14].

$$r = a \times e^{b\theta} \tag{4}$$

Here, a and b denotes some randomly generated positive real numbers while e is the base of natural logarithms. The parameter a used for turning the spiral, whereas the distance between consecutive turnings controlled by b as depicted by Fig. 1.

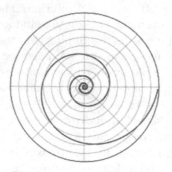

Fig. 1. Logarithmic spiral represented on a polar graph [14]

In the anticipated local search the most feasible solution in the current population is eligible for location update. The location update equation is developed by taking inspiration from the logarithmic spiral as shown in Eq. 5.

$$x_{new} = x_{best} + \text{step} \tag{5}$$

$$\text{where, step} = 2 \times sign \times U(0,1) \times (1 - \frac{Iter}{TI}) \times e^{\frac{sin(Iter)}{Iter}} \tag{6}$$

Here, x_{new} is the new location of the solution going to update, x_{best} denotes the highly fitted solution in the present population, $Iter$ is the local search iteration counter, TI denotes total iterations in local search, $sign$ is the addition or subtraction sign which is according to the fitness of the newly engendered solution. The step size is computed with the help of Eq. 6 which is developed by modification of logarithmic spiral equation. In this equation $a = 2 \times sign \times U(0,1) \times (1 - \frac{Iter}{TI})$ while $b = \frac{1}{Iter}$ and $\theta = sin(Iter)$. The step size is used to provide distance to the most suitable individual solution all along the local search process. The anticipated local search is elucidated in Algorithm 2.

In Algorithm 2, $U(0,1)$ denotes an arbitrary number that is evenly distributed in the range $(0,1)$ and D denotes the dimension of particular function.

Additional, the equation for position update in ABC algorithm is also improved. The original ABC updates the location of food sources using Eq. (2). The position update equation narrated by Eq. (2) is reformed as shown below with motivation from Gbest-guided ABC (GABC) [5] algorithm:

$$v_{ij} = p_{ij} + \phi_{ij}(p_{ij} - p_{kj}) + \psi_{ij}(p_{bestj} - p_{ij}),$$

here, ψ_{ij} is an arbitrarily generated evenly distributed number from 0 to C, with a positive constant C.

The pseudo-code of the newly anticipated hybridized LSABC algorithm is depicted in Algorithm 3.

Algorithm 2. Logarithmic Spiral Inspired Local Search Strategy

Input optimization function $Minf(x)$ and TI;
Identify the most feasible solution x_{best} in the population;
Initialize $Iter = 1$.
while $(Iter < TI)$ **do**
    Generate a new solution x_{new} as follows:
    Select a random dimension $j = U(1, D)$
    x_{new} is generated through equations 5 and 6;
    Compute $f(x_{new})$;
    if $f(x_{new}) < f(x_{best})$ **then**
        $x_{best} = x_{new}$;
    else
        $sign = -sign$;
    end if
    $Iter = Iter + 1$;
end while

Algorithm 3. Logarithmic Spiral Inspired Artificial Bee Colony (LSABC) Algorithm:

Parameter initialization;
while Stopping criteria **do**
    Step 1: Generation of new food sources using Employed bee phase.
    Step 2: Position of food sources updated using Onlooker bees phase depending on their nectar quality.
    Step 3: Scout bee phase to replace abandoned food sources with new food sources.
    Step 4: Identify the best solution;
    Step 5: Apply local search algorithm (LSLS) using Algorithm 2
end while
Print best solution.

5 Experiential Setup and Result Analysis

The performance of newly developed variant (LSABC) evaluated over ten unbiased standard problems and analysed here in terms of precision, efficacy and trustworthiness.

5.1 Considered Test Problems

A set of ten mathematical optimization problems (f_1 to f_{10}) with diverse degree of complexity are selected to confirm the efficiency of the planned LSABC algorithm. These functions are displayed in Table 1. All the selected functions are continuous in nature. These problems are minimization problems and solutions of the most of the functions does not exists on the origin, diagonal and axis.

Table 1. Test problems, AE: acceptable error

Objective function	Search range	Optimum value	D	AE
$f_1(x) = 1 + \frac{1}{4000}\sum_{i=1}^{D} x_i^2 - \prod_{i=1}^{n} \cos(\frac{x_i}{\sqrt{i}})$	$[-600\ 600]$	$f(0) = 0$	30	$1.0E - 05$
$f_2(x) = -\sum_{i=1}^{D-1}\left(\exp\left(\frac{-(x_i^2 + x_{i+1}^2 + 0.5x_i x_{i+1})}{8} \right) \times I \right)$ where, $I = \cos\left(4\sqrt{x_i^2 + x_{i+1}^2 + 0.5x_i x_{i+1}}\right)$	$[-5\ 5]$	$f(0) = -D + 1$	10	$1.0E - 05$
$f_3(x) = \sum_{i=1}^{D} (x_i - 1)^2 - \sum_{i=2}^{D} x_i x_{i-1}$	$[-D^2\ D^2]$	$f_{min} = -\frac{(D(D+4)(D-1))}{6}$	10	$1.0E - 01$
$f_4(x) = [1.5 - x_1(1 - x_2)]^2 + [2.25 - x_1(1 - x_2^2)]^2 + [2.625 - x_1(1 - x_2^3)]^2$	$[-4.5,\ 4.5]$	$f(3, 0.5) = 0$	2	$1.0E - 05$
$f_5(x) = 100[x_2 - x_1^2]^2 + (1 - x_1)^2 + 90(x_4 - x_3^2)^2 + (1 - x_3)^2 + 10.1[(x_2 - 1)^2 + (x_4 - 1)^2] + 19.8(x_2 - 1)(x_4 - 1)$	$[-10,\ 10]$	$f(1) = 0$	4	$1.0E - 05$
$f_6(x) = \sum_{i=1}^{11}[a_i - \frac{x_1(b_i^2 + b_i x_2)}{b_i^2 + b_i x_3 + x_4}]^2$	$[-5,\ 5]$	$f(0.192833, 0.190836, 0.123117, 0.135766) = 0.000307486$	4	$1.0E - 05$
$f_7(x) = \sum_{i=1}^{D-1}(100(z_i^2 - z_{i+1})^2 + (z_i - 1)^2) + f_{bias}$, $z = x - o + 1$, $x = [x_1, x_2, ...x_D]$, $o = [o_1, o_2, ...o_D]$	$[-100,\ 100]$	$f(o) = f_{bias} = 390$	10	$1.0E - 01$
$f_8(x) = \sum_{i=1}^{D} z_i^2 + f_{bias}$, $z = x - o$, $x = [x_1, x_2, ...x_D]$, $o = [o_1, o_2, ...o_D]$	$[-100,\ 100]$	$f(o) = f_{bias} = -450$	10	$1.0E - 05$
$f_9(x) = a(x_2 - bx_1^2 + cx_1 - d)^2 + e(1 - f)\cos x_1 + e$	$-5 \le x_1 \le 10, 0 \le x_2 \le 15$	$f(-\pi, 12.275) = 0.3979$	2	$1.0E - 05$
$f_{10}(x) = (1 + (x_1 + x_2 + 1)^2 \cdot (19 - 14x_1 + 3x_1^2 - 14x_2 + 6x_1x_2 + 3x_2^2)) \cdot (30 + (2x_1 - 3x_2)^2 \cdot (18 - 32x_1 + 12x_1^2 + 48x_2 - 36x_1x_2 + 27x_2^2))$	$[-2,\ 2]$	$f(0, -1) = 3$	2	$1.0E - 14$

5.2   Experimental Setting

The newly developed LSABC algorithm is compared with basic ABC and its two most recent variants, namely, Best-So-Far ABC (BSFABC) [15], and Modified ABC (MABC) [16] for the purpose of assessment to evaluate the efficiency, robustness, and reliability. Simulation results of the newly developed LSABC algorithm and the considered algorithms are presented in terms of average number of function evaluations (AFE), mean error (ME), and success rate (SR) as mentioned in Tables 2, 3, and 4 respectively. Standard deviation (SD) also measured for analysis. The considered algorithms are examined with the following experimental setting:

- Population size $NP = 50$ [17,18].
- Food source count $SN = NP/2$,
- Simulation count/run $=100$,
- $C = 1.5$ [5],
- Total local search iterations $TI = 11$ (It was set through empirical experiments).
- The termination for ABC,MABC and BSFABC is set to maximum number of function evaluations $= 200000$ or permissible error (Table 1) whichever meets first.
- Parameter settings for ABC [3] , MABC [16] and BSFABC [15] are taken from their base papers.

5.3   Analysis of Experimental Results

To evaluate the efficiency, an AFE based comparison is carried out in Table 2. It can be easily observed from this table that the AFE of LSABC is less for most of the functions i.e. the newly developed LSABC is converge to optima faster than the other considered algorithms. Further, the Table 3 shows the results of successful runs over 100 simulations. A simulation is considered successful, if the algorithm achieves optima at the level of acceptable error as mentioned in Table 1. The Table 3 elucidate that the newly developed LSABC is more reliable for most of the benchmark functions in terms of success rate as compared to the considered algorithms. The robustness and accuracy of the newly developed LSABC algorithm is measured by the ME as shown in Table 4. This table show competitiveness of the purposed algorithm in terms of accuracy for the considered Test Problems (TP).

5.4   Statistical Analysis

The comparison of LSABC with ABC, BSFABC and MABC is done on the basis of AFE, SR, and ME. The results in Tables 2, 3, and 4 shows that LSABC is very effective for 7 test problems (f_2, f_4 to f_8, and f_{10}) while these problems are of different nature, After observing these results it may be concluded that LSABC is able to balance the process of exploitation and exploration very effectively. MABC outperforms LSABC over test function f_3 while BSFABC outperforms LSABC over test functions f_1 and f_9.

The boxplot [19] analysis of AFE for LSABC, ABC, BSFABC, and MABC have been presented in Fig. 2 to denote the distribution of outcomes. It is clearly visible through the boxplots analyses of the results as shown in Fig. 2 that LSABC outperforms the considered algorithms. While observing the boxplots of success rate, the median of LSABC is high whereas interquartile range is low

Table 2. Comparison based on AFE.

TP	ABC	BSFABC	MABC	LSABC
f_1	75764	142277	199760.11	128080.67
f_2	87819.17	126256.06	69666.93	65283
f_3	197438.91	199276.24	126357.7	132566.87
f_4	16725.62	52087.08	10092.62	5862
f_5	199418.75	157990.69	135452.33	32653
f_6	186553.17	134165	180813.31	78254
f_7	177795.25	188112.09	161638.38	158374
f_8	120259.59	13107.29	14687.36	8556
f_9	189072.01	4524.29	197588.96	21596
f_{10}	31073.92	16246.31	8992.24	5312

Table 3. Comparison of success rate out of 100 runs.

TP	ABC	BSFABC	MABC	LSABC
f_1	100	96	1	95
f_2	93	81	100	100
f_3	5	2	99	79
f_4	100	92	100	100
f_5	1	42	63	99
f_6	17	57	18	95
f_7	20	13	38	56
f_8	55	100	100	100
f_9	13	100	3	100
f_{10}	100	100	100	100

Table 4. Comparison of mean error.

TP	ABC	BSFABC	MABC	LSABC
f_1	8.05E−06	8.03E−06	1.07E−03	1.34E−05
f_2	2.36E−02	8.77E−02	8.40E−06	7.45E−06
f_3	9.44E−01	4.10E+00	1.02E−01	1.03E−01
f_4	8.46E−06	2.42E−05	4.96E−06	4.51E−06
f_5	1.58E−01	2.98E−02	1.15E−02	6.59E−03
f_6	1.87E−04	1.28E−04	1.92E−04	4.78E−05
f_7	7.95E−01	2.85E+00	6.68E−01	7.80E+00
f_8	9.35E−07	6.19E−15	5.12E−15	6.12E−15
f_9	3.24E−05	4.49E−14	9.18E−04	4.32E−14
f_{10}	1.95E−03	1.95E−03	1.94E−03	1.77E−03

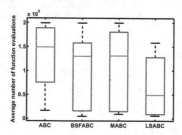

Fig. 2. Boxplots graph for AFE

as compared to the other considered algorithms which proves the reliability of the LSABC over the compared algorithms.

Further, some more statistical analyses, namely Mann-Whitney U (MWU) rank sum test [20] and Acceleration Rate (AR) test are done on the AFEs. As it is clear from the boxplots (refer Fig. 2) that the AFE's are not uniformly distributed, so the MWU Rank sum test is done at 5% level of significance ($\alpha = 0.05$) as shown in Table 5. The MWU rank sum test is applied for identification of the significant differences in function evaluations between LSABC-ABC, LSABC-MABC and LSABC-BSFABC. In Table 5, '+' symbol indicates the significant less function evaluations of LSABC, while '−' symbol indicates the significant high function evaluations of LSABC to the compared algorithm. The symbol '=' shows no significance difference of the function evaluations of the algorithms. The last row shows competitiveness of the proposed algorithm. As Table 5 contains 26 '+' signs out of 30 assessment. As a result, it can be declared that the outcomes of LSABC are significantly better than ABC, BSFABC and MABC over measured test problems.

Table 5. Comparison based on AFE and the MWU rank sum test [20].

TP	MWU rank sum test with LSABC		
	ABC	BSFABC	MABC
f_1	−	−	+
f_2	+	+	+
f_3	+	+	−
f_4	+	+	+
f_5	+	+	+
f_6	+	+	+
f_7	+	+	+
f_8	+	+	+
f_9	+	−	+
f_{10}	+	+	+

Further, a fair comparison in terms of the convergence speed is done through AR analyses on the AFEs of the considered algorithms. The AR is calculated by Eq. 7.

$$AR = \frac{AFE_{ALGO}}{AFE_{LSABC}}, \tag{7}$$

where, ALGO∈ {ABC, MABC, and BSFABC}. It is clear from Eq. 7 that the AR will be high for the algorithm having fewer AFEs and vice-versa. The calculated AR is presented in Table 6. While observing the Table 6, the value of AR is more

Table 6. Comparison of AR for LSABC with ABC, BSFABC, and MABC.

TP	ABC	BSFABC	MABC
f_1	0.591533445	1.110838974	1.559642919
f_2	1.345207328	1.933980669	1.067152704
f_3	1.489353335	1.503212982	0.953161978
f_4	2.853227567	8.885547595	1.721702491
f_5	6.10721067	4.838473953	4.148235384
f_6	2.383944207	1.714481049	2.310595113
f_7	1.12262903	1.187771288	1.020611843
f_8	14.05558555	1.531941328	1.716615241
f_9	8.754955084	0.209496666	9.149331358
f_{10}	5.849759036	3.058416792	1.692816265

than one for most of the function which shows that the newly developed LSABC is fast convergent algorithm as compared to the other considered algorithms.

6   Conclusion

Here a logarithmic spiral inspired local search algorithm assimilated in basic ABC algorithm to ameliorate its exploitation feature. The new local search is named as Logarithmic spiral based local search (LSLS) and variant of ABC is named as logarithmic spiral based ABC (LSABC). The performance of the newly anticipated LSABC is measured over 10 complex benchmark problems and results are compared with the ABC and its some state-of-art its variants. While statistical analyses of the results, it may be declared that the newly developed LSABC is an efficient variant of ABC and give a better trade of between exploration and exploitation abilities. In future LSABC may be implemented for multi variable function optimization. Further, LSABC may be implemented to classify the plant leaf and soil images into different categories.

Acknowledgement. This work was supported by Newton Prize 2017 and by a Research Environment Links grant, ID 339568416, under the Newton Programme Vietnam partnership. The grant is funded by the UK Department of Business, Energy and Industrial Strategy (BEIS) and delivered by the British Council. For further information, please visit www.newtonfund.ac.uk.

References

1. Karaboga, D., Basturk, B.: A powerful and efficient algorithm for numerical function optimization: artificial bee colony (ABC) algorithm. J. Glob. Optim. **39**(3), 459–471 (2007)

2. Karaboga, D., Basturk, B.: On the performance of artificial bee colony (ABC) algorithm. Appl. Soft Comput. **8**(1), 687–697 (2008)
3. Karaboga, D.: An idea based on honey bee swarm for numerical optimization. Technical report-TR06, Erciyes university, engineering faculty, computer engineering department (2005)
4. Karaboga, D., Akay, B.: A comparative study of artificial bee colony algorithm. Appl. Math. Comput. **214**(1), 108–132 (2009)
5. Zhu, G., Kwong, S.: Gbest-guided artificial bee colony algorithm for numerical function optimization. Appl. Math. Comput. **217**, 3166–3173 (2010)
6. Bansal, J.C., Sharma, H., Arya, K.V., Nagar, A.: Memetic search in artificial bee colony algorithm. Soft Comput. **17**(10), 1911–1928 (2013)
7. Sharma, H., Bansal, J.C., Arya, K.V., Yang, X.S.: Levy flight artificial bee colony algorithm. Int. J. Syst. Sci. **47**(11), 2652–2670 (2016)
8. Bhambu, P., Sharma, S., Kumar, S.: Modified Gbest artificial bee colony algorithm. In: Pant, M., Ray, K., Sharma, T.K., Rawat, S., Bandyopadhyay, A. (eds.) Soft Computing: Theories and Applications. AISC, vol. 583, pp. 665–677. Springer, Singapore (2018). https://doi.org/10.1007/978-981-10-5687-1_59
9. Sharma, H., Sharma, S., Kumar, S.: Lbest Gbest artificial bee colony algorithm. In: 2016 International Conference on Advances in Computing, Communications and Informatics (ICACCI), pp. 893–898. IEEE, September 2016
10. Kumar, A., Kumar, S., Dhayal, K., Swetank, D.K.: Fitness based position update in artificial bee colony algorithm. Int. J. Eng. Res. Technol. **3**(5), 636–641 (2014)
11. Kumar, S., Sharma, V.K., Kumari, R.: Memetic search in artificial bee colony algorithm with fitness based position update. In: Recent Advances and Innovations in Engineering (ICRAIE), pp. 1–6. IEEE, May 2014
12. Lanzarini, L., Leza, V., De Giusti, A.: Particle swarm optimization with variable population size. In: Rutkowski, L., Tadeusiewicz, R., Zadeh, L.A., Zurada, J.M. (eds.) ICAISC 2008. LNCS (LNAI), vol. 5097, pp. 438–449. Springer, Heidelberg (2008). https://doi.org/10.1007/978-3-540-69731-2_43
13. Wang, H., Rahnamayan, S., Wu, Z.: Adaptive differential evolution with variable population size for solving high-dimensional problems. In: Proceedings of IEEE Congress on Evolutionary Computation (CEC), pp. 2626–2632. IEEE (2011)
14. Logarithmicspiral. https://en.wikipedia.org/wiki/LogarithmicspiralAccessedon04/04/2018. Accessed 4 Apr 2018
15. Banharnsakun, A., Achalakul, T., Sirinaovakul, B.: The best-so-far selection in artificial bee colony algorithm. Appl. Soft Comput. **11**(2), 2888–2901 (2011)
16. Akay, B., Karaboga, D.: A modified artificial bee colony algorithm for real-parameter optimization. Inf. Sci. **192**(3), 120–142 (2010)
17. Diwold, K., Aderhold, A., Scheidler, A., Middendorf, M.: Performance evaluation of artificial bee colony optimization and new selection schemes. Memetic Comput. **1**(1), 1–14 (2011)
18. El-Abd, M.: Performance assessment of foraging algorithms vs. evolutionary algorithms. Inf. Sci. **182**(1), 243–263 (2011)
19. Williamson, D.F., Parker, R.A., Kendrick, J.S.: The box plot: a simple visual method to interpret data. Ann. Intern. Med. **110**(11), 916 (1989)
20. Mann, H.B., Whitney, D.R.: On a test of whether one of two random variables is stochastically larger than the other. Ann. Math. Stat. **18**(1), 50–60 (1947)

Outage Performance Analysis of Energy Harvesting DF Cooperative NOMA Networks over Nakagami-m Fading Channels

Van-Long Nguyen[1][✉] and Dac-Binh Ha[2]

[1] Graduate School, Duy Tan University, Danang, Vietnam
vanlong.itqn@gmail.com
[2] Faculty of Electrical and Electronics Engineering, Duy Tan University,
Danang, Vietnam
hadacbinh@duytan.edu.vn

Abstract. In this paper, we investigate the energy harvesting decode-and-forward cooperative non-orthogonal multiple access (NOMA) networks. We study the case of the better user play a role of relay to forward information to the worse user. Specifically, one source node wishes to transmit two symbols to two respective destinations directly and via the help of energy constraint node (better user) with applying the NOMA technique over Nakagami fading channels. In order to evaluate the performance of this considered system, we derive the closed-form expressions for the outage probability (OP) at each user based on the statistical characteristics of signal-to-noise ratio (SNR) and signal-to-interference-plus-noise ratio (SINR). Our analysis is confirmed by Monte Carlo simulation. Finally, in order to look insight this system, we also investigate the effect of several parameters, such as transmit power, power splitting ratio and the location of relay nodes to the outage performance of entire system.

Keywords: Energy harvesting · Non-orthogonal multiple access
Power splitting · Decode-and-forward · Cooperative network

1 Introduction

In recent years, a significant increase in the number of users and wireless devices, the future 5G networks are required to support low-latency, low-cost and more diverse services, yet at higher quality and a thousand-time faster data rate. In the search for new technologies, non-orthogonal multiple access (NOMA) technique has been viewed as one of the brightest candidates in meeting these requirements. The applying of NOMA can improve a significant spectral efficiency as it takes advantage of the power domain to serve multiple users at the same time/frequency/code that was not sufficiently utilized in previous generations.

© ICST Institute for Computer Sciences, Social Informatics and Telecommunications Engineering 2019
Published by Springer Nature Switzerland AG 2019. All Rights Reserved
T. Q. Duong and N.-S. Vo (Eds.): INISCOM 2018, LNICST 257, pp. 28–37, 2019.
https://doi.org/10.1007/978-3-030-05873-9_3

In addition, compared with conventional multiple access, NOMA can enhance citation the quality of user experience (QoE), especially at the cell edge [1].

In order to prolong the lifetime of energy-constrained networks as well as maintain network connectivity, radio frequency (RF) energy harvesting (EH) is a emerging solution that has been used and widely considered these days [2–5]. There are two information and power transfer architectures, namely time switching (TS) and power spliting (PS), to be used to extract the energy and information from RF signals [6,7]. With the TS architecture, the total received RF signal is used for both information processing and energy harvesting but in different time blocks; whereas in the PS architecture, the received signal is split into two parts: one for information detection and the other for energy harvesting. In this article, we combine these two architectures.

Cooperative relay scheme can improve the performance of NOMA energy harvesting networks [8–16]. NOMA cognitive radio networks with cooperative relay scheme has been considered in which the authors focused on cooperative communication protocols and performance analysis to confirm the advantage of this scheme [8–11]. The work in [12] presented a NOMA-based downlink cooperative cellular system, where the base station communicates with two paired mobile users through the help of a half-duplex amplify-and-forward (AF) relay. In order to evaluate the performance of the considered network, a closed-form expression of outage probability was derived and ergodic sum-rate was studied. By comparing to the conventional multiple access techniques, the authors showed that NOMA can offer better spectral efficiency and user fairness since more users are served at the same time/frequency/spreading code. Moreover, the exact and closed-form expressions of outage probability of NOMA cooperative networks were derived in [13] to investigate the performance of this network. One special point of their system model is that it consists of one base station (BS) and two users, in which user 1 communicates directly to the BS while user 2 communicates with the BS through the help of user 1. The investigated results showed that the system performance is improved significantly with NOMA technique. For NOMA network with RF-EH, the authors in [14] have studied data rate optimization and fairness of in NOMA systems with wireless energy harvesting based on time allocation. Moreover, NOMA scheme in simultaneous wireless information and power transfer (SWIPT) networks was investigated in [15]. Specifically, near NOMA users that are close to the source act as energy harvesting relays to help far NOMA users over small-scale Rayleigh fading channels. Far user combines the signals from BS and better user by using maximal-ratio combining (MRC) scheme. Furthermore, the authors investigated the performance of the considered systems by deriving the closed-form expressions for outage probability and system throughput under the random distribution of users' location. Analytical and simulation results showed that selecting users can reasonably reduce the outage probability.

Unlike above results, in this paper we focus on the energy harvesting decode-and-forward (DF) NOMA relaying networks, in which one source wants to transmit its two symbols to two destinations directly and via the help of better user with selection combining (SC) scheme at worse user over Nakagami fading channels using hybrid TS-PS energy harvesting architecture.

The rest of this paper is organized as follows. Section 2 describes network and channel model. Section 3 presents outage probability analysis at two the destinations. Section 4 discusses some numerical results. Finally, the conclusion is given in Sect. 5.

2   Network and Channel Models

The Fig. 1 depicts a system model of a M-user NOMA EH DF relaying network, where a source node S want to transmit its two symbols s_m and s_n ($1 \leq m < n \leq M$) to two destinaton nodes D_m and D_n, respectively, directly and via the help of better user. In this paper, we consider the following scenario

- The better user D_m harvests energy from the power transfer source S and helps the source convey information to the worse user D_n by using the power splitting relay protocol with power splitting ratio ρ ($0 < \rho < 1$) for energy harvesting and ($1 - \rho$) for decoding the source information.
- All nodes are equipped with a single antenna operating in half-duplex mode.
- All channels undergo Nakagami-m fading. Nakagami-m fading is more ganeral to describe the wireless channel than others.
- In each block time T, these channels are constant and independently and identically distributed (i.i.d).
- Without loss of generality, we assume that all the channel gains between S and users follow the order of $|h_{SD_1}|^2 \geq ... \geq |h_{SD_m}|^2 \geq ... \geq |h_{SD_n}|^2 \geq ... \geq |h_{SD_M}|^2$, where $|h_{SD_m}|^2$ and $|h_{SD_n}|^2$ are the channel power gain of the link from the source S to m^{th} user and the nth user, respectively.

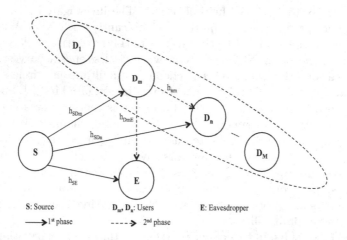

Fig. 1. Network model for NOMA energy constraint DF relaying.

The two-phase protocol of this SWIPT cooperative NOMA system is as follows

The First Phase. In this phase, the source node S broadcasts its signal containing two symbols s_m and s_n as a form $x = \sqrt{a_m P} s_m + \sqrt{a_n P} s_n$ in the time of αT, where $\mathbb{E}\left\{|s|^2 = 1\right\}$, P is a transmit power of source node, a_m and a_n respectively denote the power allocation coefficients for symbols s_m and s_n ($a_m + a_n = 1$, $a_m < a_n$), $0 < \alpha < 1$, and T is block time. Let $X_1 \triangleq |h_{SD_m}|^2$, $X_2 \triangleq |h_{SD_n}|^2$, where d_{SD_m}, d_{SD_n} are denoted as the Euclidean distances of $S - D_m$, $S - D_n$, respectively and θ returns the path-loss exponent. The energy at the mth user that harvests from S can be obtained as

$$E_{D_m} = \frac{\rho \eta P |h_{SD_m}|^2 \alpha T}{d_{SD_m}^\theta}, \tag{1}$$

where η ($0 < \eta < 1$) is depicted as energy convertion efficiency. Applying NOMA, D_m uses successive interference cancellation (SIC) to detect symbol s_n and subtracts this component from the received signal to obtain its own symbol s_m. We obtain the instantaneous SINR at the mth user to detect s_n as

$$\gamma_{SD_m}^{s_n} = \frac{a_n(1-\rho)\gamma|h_{SD_m}|^2}{a_m(1-\rho)\gamma|h_{SD_m}|^2 + d_{SD_m}^\theta} = \frac{b_4 X_1}{b_3 X_1 + 1}. \tag{2}$$

where $\gamma \triangleq \frac{P}{N_0}$ as the transmit SNR of S, $b_3 = \frac{a_m(1-\rho)\gamma}{d_{SD_m}^\theta}$, and $b_4 = \frac{a_n(1-\rho)\gamma}{d_{SD_m}^\theta}$. The received SNR for s_m at the mth user is given as

$$\gamma_{SD_m}^{s_m} = \frac{a_m(1-\rho)\gamma|h_{SD_m}|^2}{d_{SD_m}^\theta} = b_3 X_1. \tag{3}$$

Similarly, the instantaneous SINR at the nth user to detect s_n transmitted from S can be given as

$$\gamma_{SD_n} = \frac{a_n \gamma |h_{SD_n}|^2}{a_m \gamma |h_{SD_n}|^2 + d_{SD_n}^\theta} = \frac{b_2 X_2}{b_1 X_2 + 1}, \tag{4}$$

where $b_1 = \frac{a_m \gamma}{d_{SD_n}^\theta}$, and $b_2 = \frac{a_n \gamma}{d_{SD_n}^\theta}$.

The Second Phase. After D_m received signal from S, then harvested the energy and decoded the two symbols s_m and s_n, in this phase it forwards s_n to the worse user D_n in the time $(1 - \alpha)T$. Finally, D_n combines two received signals, i.e., the direct signal from S and the relaying signal from D_m, to decode its own symbol by using SC scheme. Let $\hat{X}_3 \triangleq |h_{mn}|^2$, where $|h_{mn}|^2$ and d_{mn} are denoted as the unordered channel gain of the link and the Euclidean distance of the mth user and the nth user, respectively. From (1), the transmit power of D_m is expressed as

$$P_{D_m} = \frac{\rho \eta \alpha P |h_{SD_m}|^2}{(1-\alpha)d_{SD_m}^\theta}. \tag{5}$$

The instantaneous SNR γ_{mn} at D_m in this phase is written as

$$\gamma_{mn} = \frac{\rho\eta\alpha\gamma}{(1-\alpha)d_{SD_m}^\theta d_{mn}^\theta}|h_{SD_m}|^2|h_{mn}|^2 = cX_1\hat{X}_3, \tag{6}$$

where $c = \frac{\rho\eta\alpha\gamma}{(1-\alpha)d_{SD_m}^\theta d_{mn}^\theta}$.

By using SC scheme, the end to end instantaneous SNR γ_{e2e} at the nth user to detect s_n can be given as follows

$$\gamma_{e2e} = \max\{\gamma_{SD_n}, \gamma_{mn}\} \tag{7}$$

The PDF and CDF of unordered variable X_i ($i \in \{1,2,3\}$), denoted as $\hat{X}_i$, are respectively expressed as

$$f_{\hat{X}_i}(x) = \frac{m_i^{m_i} x^{m_i-1}}{(m_i-1)!\lambda_i^{m_i}} e^{-\frac{m_i x}{\lambda_i}}, \tag{8}$$

$$F_{\hat{X}_i}(x) = 1 - \sum_{l=0}^{m_i-1} \frac{m_i^l}{l!\lambda_i^l} x^l e^{-\frac{m_i x}{\lambda_i}}, \tag{9}$$

where $\lambda_i = \mathbb{E}\left\{\hat{X}_i\right\}$ is interpreted as the average power gain of the S-D_m, S-D_n, and D_m-D_n channels, respectively; $m_i \geq 1/2$ is the fading severity factor. Note that the case of $m_i = 1$ corresponds to Rayleigh fading, whereas the case $m_i = (K_i + 1)^2/(2K_i + 1)$ approximates Rician fading with parameter K_i. In order statistics, PDF and CDF of X_j ($j \in \{1,2\}$) are respectively given by [16]

$$f_{X_j}(x) = \Phi_m \sum_{k=0}^{M-m} (-1)^k C_k^{M-m} f_{\hat{X}_j}(x) \left[F_{\hat{X}_j}(x)\right]^{m+k-1}, \tag{10}$$

$$F_{X_j}(x) = \Phi_m \sum_{k=0}^{M-m} \frac{(-1)^k}{m+k} C_k^{M-m} \left[F_{\hat{X}_j}(x)\right]^{m+k}, \tag{11}$$

where $\Phi_m = \frac{M!}{(M-m)!(m-1)!}$.

3 Outage Probability Analysis

In this paper, the receiver decodes successfully the information if its SINR or SNR satisfies the pre-defined threshold γ_t. In this section, we will derive the expressions of outage probability at D_m, D_n, and of whole system.

3.1 Outage Probability at D_m

An outage event happens when D_m unsuccessfully decodes both symbol s_m and s_n from S in the first phase. In this system, the desired symbol for user D_m is s_m, thus D_m has to successfully decode s_n first then using SIC to obtain s_m. Therefore, the outage probability at D_m can be formulated as

$$OP_{D_m} = \Pr(\gamma_{SD_m}^{s_n} < \gamma_t) + \Pr(\gamma_{SD_m}^{s_n} > \gamma_t, \gamma_{SD_m}^{s_m} < \gamma_t)$$

$$= \begin{cases} F_{X_1}\left(\frac{\gamma_t}{b_3}\right), & \gamma_t < \frac{a_n}{a_m} - 1 \\ F_{X_1}\left(\frac{\gamma_t}{b_4 - b_3 \gamma_t}\right), & \frac{a_n}{a_m} - 1 < \gamma_t < \frac{a_n}{a_m} \end{cases}$$

$$OP_{D_m} \overset{(a)}{=} \Phi_m \sum_{k=0}^{M-m} \frac{(-1)^k}{m+k} C_k^{M-m} \left[1 - \sum_{l=0}^{m_1-1} \frac{m_1^l}{l!\lambda_1^l}\left(\frac{\gamma_t}{b_3}\right)^l e^{-\frac{m_1}{\lambda_1}\left(\frac{\gamma_t}{b_3}\right)}\right]^{m+k},$$

$$OP_{D_m} \overset{(b)}{=} \Phi_m \sum_{k=0}^{M-m} \frac{(-1)^k}{m+k} C_k^{M-m} \left[1 - \sum_{l=0}^{m_1-1} \frac{m_1^l}{l!\lambda_1^l}\left(\frac{\gamma_t}{b_4 - b_3 \gamma_t}\right)^l e^{-\frac{m_1}{\lambda_1}\left(\frac{\gamma_t}{b_4 - b_3 \gamma_t}\right)}\right]^{m+k}, \quad (12)$$

Note that step (a) is obtained by assuming the following condition holds $\gamma_t < \frac{a_n}{a_m} - 1$ and result (b) is obtained by assuming the following condition holds $\frac{a_n}{a_m} - 1 < \gamma_t < \frac{a_n}{a_m}$.

3.2   Outage Probability at D_n

An outage event happens when D_n unsuccessfully decodes the symbol s_n both from S in the first phase and from near user D_m in the second phase. The outage probability at D_n can be formulated by

$$OP_{D_n} = \Pr(\gamma_{SD_m}^{s_n} < \gamma_t, \gamma_{Sn} < \gamma_t) + \Pr(\gamma_{SD_m}^{s_n} \geq \gamma_t, \gamma_{e2e} < \gamma_t)$$

$$\overset{(c)}{=} F_{Y_1}\left(\frac{\gamma_t}{b_2 - b_1 \gamma_t}\right)\left[F_{X_1}\left(\frac{\gamma_t}{b_4 - b_3 \gamma_t}\right) + I\right], \quad (13)$$

where $I = \Pr(\gamma_{SD_m}^{s_n} \geq \gamma_t, \gamma_{mn} < \gamma_t)$. Note that step (c) is obtained by assuming the following condition holds $\gamma_t < \frac{a_n}{a_m}$. We calculate I as follows

$$I = \Pr\left(X_1 \geq \frac{\gamma_t}{b_4 - b_3 \gamma_t}, \hat{X}_3 < \frac{\gamma_t}{cX_1}\right) = \int_\beta^\infty F_{\hat{X}_3}\left(\frac{\gamma_t}{cy}\right) f_{X_1}(y) dy$$

$$\overset{(d)}{=} 1 - F_{X_1}(\beta) - \frac{\Phi_m \mu_1^{m_1}}{(m_1 - 1)!} \sum_{k=0}^{M-m} (-1)^k C_k^{M-m} \sum_{l=0}^{m_3-1} \frac{1}{l!}\left(\frac{\mu_3 \gamma_t}{c}\right)^l$$

$$\times \left[\int_\beta^\infty y^{m_1-l-1} e^{-\mu_1 y - \frac{\mu_3 \gamma_t}{cy}} dy + \sum_{j=1}^{m+k-1} \sum_j \widetilde{(-1)^j U_j V_j} \int_\beta^\infty y^{m_1+\tilde{j}-l-1} e^{-\mu_1(j+1)y - \frac{\mu_3 \gamma_t}{cy}} dy\right]$$

$$\overset{(e)}{=} 1 - \Phi_m \sum_{k=0}^{M-m} \frac{(-1)^k}{m+k} C_k^{M-m} \left(1 - \sum_{l=0}^{m_1-1} \frac{\mu_1^l}{l!} \beta^l e^{-\mu_1 \beta}\right)^{m+k}$$

$$- \frac{\Phi_m \mu_1^{m_1}}{(m_1 - 1)!} \sum_{k=0}^{M-m} (-1)^k C_k^{M-m} \sum_{l=0}^{m_3-1} \frac{1}{l!}\left(\frac{\mu_3 \gamma_t}{c}\right)^l \left[\int_\beta^\infty y^{m_1-l-1} e^{-\mu_1 y}\left(1 - \frac{\mu_3 \gamma_t}{cy}\right) dy\right.$$

$$\left. + \sum_{j=1}^{m+k-1} \sum_j \widetilde{(-1)^j U_j V_j} \int_\beta^\infty y^{m_1+\tilde{j}-l-1} e^{-\mu_1(j+1)y}\left(1 - \frac{\mu_3 \gamma_t}{cy}\right) dy\right] \quad (14)$$

$$
\overset{(f)}{=} 1 - \Phi_m \sum_{k=0}^{M-m} \frac{(-1)^k}{m+k} C_k^{M-m} \left(1 - \sum_{l=0}^{m_1-1} \frac{\mu_1^l}{l!} \beta^l e^{-\mu_1 \beta} \right)^{m+k}
$$

$$
- \frac{\Phi_m \mu_1^{m_1}}{(m_1-1)!} \sum_{k=0}^{M-m} (-1)^k C_k^{M-m} \sum_{l=0}^{m_3-1} \frac{1}{l!} \left(\frac{\mu_3 \gamma_t}{c} \right)^l
$$

$$
\times \left\{ \left[\mu_1^{l-m_1} \Gamma(m_1-l, \mu_1\beta) - \frac{\mu_3\gamma_t}{c} \mu_1^{l-m_1+1} \Gamma(m_1-l-1, \mu_1\beta) \right] \right.
$$

$$
+ \sum_{j=1}^{m+k-1} \overset{\sim}{\sum_j} (-1)^j U_j V_j \left[(\mu_1(j+1))^{l-m_1-j} \Gamma(m_1+j-l, \mu_1(j+1)\beta) \right.
$$

$$
\left. \left. - \frac{\mu_3\gamma_t}{c} (\mu_1(j+1))^{l-m_1-j+1} \Gamma(m_1+j-l-1, \mu_1(j+1)\beta) \right] \right\}.
$$

$$
(15)
$$

where $\beta \triangleq \frac{\gamma_t}{b_4 - b_3 \gamma_t}$, $\overset{\sim}{\sum_j} \triangleq \sum_{j_1=0}^{j} \sum_{j_2=0}^{j-j_1} \cdots \sum_{j_{m_1-1}=0}^{j-j_1-j_2-\cdots j_{m_1-2}}$, $U_j \triangleq C_j^{m+k} C_{j_1}^{j}$

$C_{j_2}^{j-j_1} \cdots C_{j_{m_1-1}}^{j-j_1-j_2-\cdots j_{m_1-2}}$, $V_j \triangleq \left(\frac{\mu^{m_1-1}}{(m_1-1)!} \right)^{j-j_1-j_2-\cdots j_{m_1-2}} \prod_{l=0}^{m_1-2} \left(\frac{\mu^l}{l!} \right)^j_{l+1}$, $\bar{j} \triangleq$

$(m_1-1)(j-j_1) - (m_1-2)j_2 - (m_1-3)j_3 - \cdots j_{m_1-1}$, $\mu_1 = \frac{m_1}{\lambda_1}$, $\mu_3 = \frac{m_3}{\lambda_3}$.
Note that step (d) is obtained by using the Eq. (17) in the work [16], step (e) is
obtained by using the following approximation $e^{1-\alpha x} \approx 1 - \alpha x$ for small value
of $|x|$ and step (f) is obtained by using the equation (3.381-3) [17].

4 Nummerical Results and Disscussion

This section provide nummerical results and disccussion of the outage performance at both D_m and D_n via Monte Carlo simulation and theoretical results.

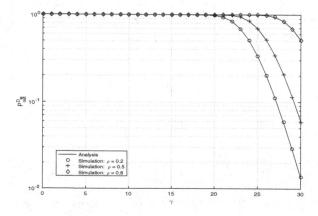

Fig. 2. OP at D_m with $\gamma_T = 8.5$ dB, $\alpha = 0.4$, $m_1 = 2$, $a_m = 0.1$, $a_n = 0.9$.

4.1   Results and Disccussion of the Outage Performance at both D_m

The theoretical results for OP at D_m are compared to the simulation results in Fig. 2. In this figure, we see that the OP at D_m increases when the power allocation factor increases. The power used to decode the signal at D_m decreases because the power splitting ratio increases, so the OP at D_m increases. In addition, we also see that the OP at D_m decreases when the transmit power increases.

4.2   Results and Disccussion of the Outage Performance at D_n

Figure 3 plots OP at D_n versus γ. It is seen from Fig. 3 that when ρ increases, OP at D_n decreases. When ρ increases, the power used to forward the signal at D_m increases, Therefore, the signal acquisition capacity at D_n increases. In other words, the OP at D_n decreased.

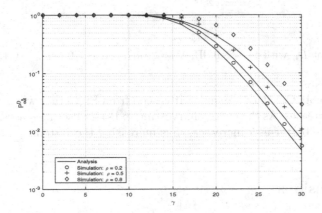

Fig. 3. OP at D_n with $\gamma_T = 8.9$ dB, M $= 2$, $m_1 = 1$, $\alpha = 0.2$, $a_m = 0.1$, $a_n = 0.9$

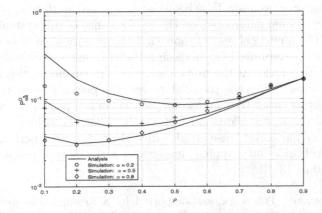

Fig. 4. OP at D_n with $\gamma_T = 8.9$ dB, M $= 3$, $m_1 = 2$, $a_m = 0.1$, $a_n = 0.9$.

Figure 4 plots OP at D_n versus ρ. It is seen from Fig. 4 that when α increases, OP at D_n decreases. This is expected since more time spent for energy harvesting as α grows leads to higher transmission power, hence better OP results.

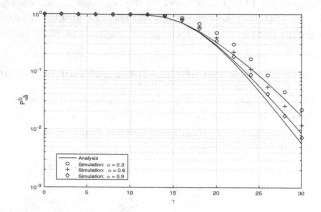

Fig. 5. OP at D_n with $\gamma_T = 8.9$ dB, M $= 2$, $m_1 = 1$, $\rho = 0.3$, $a_m = 0.1$, $a_n = 0.9$.

Figure 5 plots OP at D_n versus γ. It is seen from Fig. 5 that when α increases, OP at D_n decreases. As the energy harvesting in better user increases, the energy received in the better user increases, thus increasing the systems performance at bad user.

5   Conclusions

In this paper, we study energy harvesting technique in the NOMA relaying networks. To study the performance of this system, the closed-form expressions of outage probability of two users have been derived. Based on these expressions the affection of key system parameters on the performance of this systems has been investigated. The numerical results show that the performance at better user of this system can be improved by increasing the transmit power and decreasing the power splitting ratio. At the worse user, the performance of this system can be improved by increasing the transmit power, the power allocation factor and the power splitting ratio. We can also see that there is a pair of the power splitting ratio and the power allocation factor value that can achieve the optimal performance. In particular, case studies investigated system performance versus γ, analytical lines and mapping simulations with the transmit power values ranging from 0 to 13 dB.

Acknowledgement. This work was supported by a Research Environment Links grant, ID 339568416,under the Newton Progamme Vietnam partnership. The grant is funded by theUK department of Business, Energy and Industrial Strategy (BEIS) and deliveredby the British Council. This research was also funded by Vietnam National

Foundation for Science and Technology Development (NAFOSTED) under Grant Number 102.04-2017.301.

References

1. Raghunathan, V., Ganeriwal, S., Srivastava, M.: Emerging techniques for long lived wireless sensor networks. IEEE Commun. Mag. **44**(4), 108–114 (2006)
2. Paradiso, J.A., Starner, T.: Energy scavenging for mobile and wireless electronics. IEEE Pervasive Comput. **4**(1), 18–27 (2005)
3. Ulukus, S., et al.: Energy harvesting wireless communications: a review of recent advances. IEEE J. Sel. Areas Commun. **33**(3), 360–381 (2015)
4. Medepally, B., Mehta, N.B.: Voluntary energy harvesting relays and selection in cooperative wireless networks. IEEE Trans. Wirel. Commun. **8**(11), 3543–3553 (2010)
5. Zhou, X., Zhang, R., Ho, C.K.: Wireless information and power transfer: architecture design and rate-energy tradeoff. IEEE Trans. Commun. **61**(11), 4754–4767 (2013)
6. Zhang, R., Ho, C.K.: MIMO broadcasting for simultaneous wireless information and power transfer. IEEE Trans. Wirel. Commun. **12**(5), 1989–2001 (2013)
7. Nasir, A.A., Zhou, X., Durrani, S., Kennedy, R.A.: Wireless-powered relays in cooperative communications: time-switching relaying protocols and throughput analysis. IEEE Trans. Commun. **63**(5), 1607–1622 (2015)
8. Ding, Z., Perlaza, S.M., Esnaola, I., Poor, H.V.: Power allocation strategies in energy harvesting wireless cooperative networks. IEEE Trans. Wirel. Commun. **13**(2), 846–860 (2014)
9. Ding, Z., Peng, M., Poor, H.V.: Cooperative non-orthogonal multiple access in 5G systems. IEEE Commun. Lett. **19**(8), 1462–1465 (2015)
10. Do, N.T., Costa, D.B.D., Duong, T.Q., An, B.: BNBF user selection scheme for NOMA-Based cooperative relaying systems With SWIPT. IEEE Commun. Lett. **21**(3), 664–667 (2017)
11. Lee, S., Costa, D.B., Vien, Q.T., Duong, T.Q., de Sousa Jr, R.T.: Non-orthogonal multiple access schemes with partial relay selection. IET Commun. **11**, 846–854 (2016). https://doi.org/10.1049/iet-com.2016.0836
12. Men, J., Ge, J.: Performance analysis of non-orthogonal multiple access in downlink cooperative network. IET Commun. **9**(18), 2267–2273 (2015)
13. Kim, J.-B., Lee, I.-H.: Non-orthogonal multiple access in coordinated direct and relay transmission. IEEE Commun. Lett. **19**(11), 2037–2040 (2015)
14. Diamantoulakis, P.D., Pappi, K.N., Ding, Z., Karagiannidis, G.K.: Optimal design of non-orthogonal multiple access with wireless power transfer. In: IEEE International Conference on Communications (ICC), Kuala Lumpur, Malaysia, pp. 1–6 (2016)
15. Liu, Y., Ding, Z., Elkashlan, M., Poor, H.V.: Coperative non-orthogonal multiple access with simultaneous wireless information and power transfer. IEEE J. Sel. Areas Commun. **33**(4), 938–953 (2016)
16. Han, W., Ge, J., Men, J.: Performance analysis for NOMA energy harvesting relaying networks with transmit antenna selection and maximal-ratio combining over Nakagami-m fading. IET Commun. **10**(18), 2687–2693 (2016)
17. Gradshteyn, I.S., Ryzhik, I.M.: Table of Integrals, Series and Products, 7th edn. Academic Press, London (2007)

Multi-path Routing for Mission Critical Applications in Software-Defined Networks

Ramon Carreras Ramirez[1], Quoc-Tuan Vien[1], Ramona Trestian[1],
Leonardo Mostarda[2(✉)], and Purav Shah[1]

[1] School of Science and Technology, Middlesex University, London NW4 4BT, UK
RC934@live.mdx.ac.uk, {q.vien,r.trestian,p.shah}@mdx.ac.uk
[2] Scuola di Scienze e Tecnologie, Universita' degli Studi di, Camerino,
Camerino, Italy
leonardo.mostarda@unicam.it

Abstract. Mission critical applications depends on the communication among other systems and/or users and thus, the traffic/flows generated by these applications could bring profound consequences in sectors such as military, hospital, automotive safety and air-traffic control systems. These critical flows require stringent QoS requirements on parameters such as throughput, packet loss, latency, jitter and redundancy. Network operators must have tools that allow them to provide special treatment to such mission-critical flows based on specific application requirements. Due to the constraints of traditional networks, we should seek for solutions supported by de-centralised approaches offered by SDN.

In this paper, we propose a solution to achieve the stringent QoS requirement of such mission critical flows in multi-path environments based on SDN. This solution allows the network operator to prioritise traffic between specific end points. Also, using the overall view of the network, the solution allows evaluation of the path loads between two endpoints and to opt for the less congested path. Moreover, this paper tries to demonstrate a satisfactory network performance by presenting trade-offs between throughput and the number of hops within a multi-path network. The proposed solution is implemented in the application and control layer of the OpenDaylight Controller. The networking devices were simulated using Mininet simulator and background traffic was generated using Iperf.

1 Introduction and Related Works

Mission-critical networks provide applications to control and monitor energy, railways, public-safety, aviation management, etc. The network thus must be reliable and resilient in order to support such applications where by in case of a network failure, people's lives and companies can be at risk. Load balancing has been an existing problem in providing reliable network by assuring the system is

© ICST Institute for Computer Sciences, Social Informatics and Telecommunications Engineering 2019
Published by Springer Nature Switzerland AG 2019. All Rights Reserved
T. Q. Duong and N.-S. Vo (Eds.): INISCOM 2018, LNICST 257, pp. 38–48, 2019.
https://doi.org/10.1007/978-3-030-05873-9_4

not overloaded and also by high resource utilisation [1,2]. With the emergence of the Software Defined Networking (SDN) paradigm which has been triumphed as a reliable and resilient solution for future Internet, mission-critical networks can be improved in reliability, resilience and security by leveraging SDN's features. SDN promises to make the existing networks more easier to supervise, configure, deploy and monitor. The traditional networks are getting more complex by having a combined control and forwarding planes on the same device. Thus, the simplification of the SDN architecture by separating these planes and thus simplifying the communication using the standardised OpenFlow (OF) protocol [6] is a key to improving the network performance. A SDN controller configures the network elements by distributing the forwarding rules to the switches using low-level language.

In SDN, OpenFlow is a widely used protocol that enables the controller to communicate with OF-switches [6]. OpenFlow enables the controller to remotely install, modify, and even delete forwarding rules in OF-switches flow table via the OpenFlow channel [7]. When a new flow enters the network, the OF-switch will encapsulate a new packet of the flow into a Packet-In message and send to the controller to ask for an appropriate behaviour. The controller then selects a forwarding path for the new flow based on the current global network status and sets up the forwarding path by using Flow Mod messages to reactively install rules on related OF-switches.

Several recent studies have focused on load balancing in the control plane. BalanceFlow [3] introduces an extension for OF-switch known as controller X. However, it comes at a cost of more complexity on the OF-switches and increased communication overhead at the control plane from using a periodical approach. In this scheme, the controllers need to publish their load information periodically via a cross-controller communication channel. In [5], a dynamic load re-balancing method based on switch migration mechanism for clustered controllers was introduced. However, the problem of cross-controller communication was not resolved. Thus, its implementation on large-scale networks is very limited. Furthermore, another dynamic and adaptive algorithm named DALB was proposed by Zhou et al. [8], that forces the controllers to collect load from each other in case load on each controller exceeds a certain load threshold at the same time. This approach also causes high communication overhead in the control plane.

Motivated from these above challenges, this paper proposes an efficient scheme to balance the load on the network when considering mission-critical flows. The main contributions of this paper can be summarised as follows:

1. Firstly, by using multi-path routing, the proposed approach eliminates the single point of failure problem in the data plane and provides a reliable network, which is essential for SDN-based mission-critical applications.
2. Simulation results show that the proposed scheme achieves good load-balancing performance in the network as compared with previous schemes based on shortest path based routing.

The rest of this paper is organised as follows. Section 2 presents the proposed solution for multi-path routing and prioritisation. Section 3 analyses the proposed solution and presents the a discussion on the results achieved from the simulation setup and finally, Sect. 4 finishes the paper with a brief conclusion and remarks for future work.

2 Proposed Multi-path Routing and Prioritisation Solution

The proposed solution is implemented using the OpenDaylight SDN controller to meet three primary objectives:

1. Provide prioritisation to selected traffic in a multi-path network environment
2. Using the SDN central management, choose the less congested path between two endpoints for the prioritised traffic
3. Demonstrate that in a multi-path environment, by considering path(s) with more hops but less congestion, it can achieve a satisfactory performance of throughput and packet loss.

2.1 Proposed Solution

As shown in Fig. 1a, the proposed framework architecture consists of three layers: infrastructure layer, control layer, the application layer. The infrastructure layer or data plane consists of a set of OpenFlow Switch connected to each other to build a network topology. These switches are connected to the SDN controller; therefore, they are dynamically programmable by using southbound APIs. The control layer consisted of some internal modules of the OpenDaylight controller, those are Statistic Collector, Device Tracking, Flow programming. The top layer of the diagram is the application layer, this is the core of our proposed framework. In this layer, there are some modules developed for providing distinct roles, such as Topology Service, Path Generation, Optimal Path Selection Service, Flow Installation Service, and Management.

Control Layer. In this layer, the internal modules of the OpenDayLight controller are utilised to collect all the network states (statistics collector). The traffic load information of each link is utilised to select the less congested path. The controller sends an FLOW STATS REQUEST to the switches to send information for collection. This message is replied with an FLOW STATS REPLY along with the information requested. This information is maintained in the controller and it will be consulted using REST API by the application layer. SDN allows three kinds of operational modes to setup a new flow rule (flow entry) namely proactive, reactive and hybrid modes [4], which makes it flexible to add/forward flow entries into the OF-switches.

OpenFlow controllers are able to collect statistics of a switch at different aggregation levels such as port, table, flow, and queue. In the proposed framework, we are using the port level statistics. Port-level statistics provides bytes counter (transmitted and received), packets counter (transmitted and received),

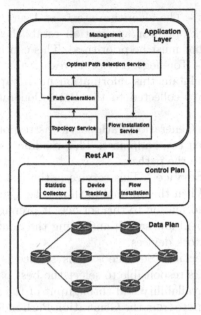

(a) Proposed Solution Framework

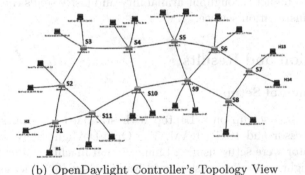

(b) OpenDaylight Controller's Topology View

Fig. 1. Proposed solution and topology view

packets dropped or errors occurred. The Device tracking module consists on discovery each device, ports, and links of the whole network. The information collected by this module helps the application plane to keep a global view of the network. The Flow Programming module is responsible for pushing the flow rules to the OpenFlow device by means of Southbound API.

Application Layer. The main functionality of our proposed solution is located in the Application Layer.

User/Management - The algorithm starts at User/ Management service. Basically, the network operator/application establishes the source and the target of the traffic that must be prioritised. This can be done at port-level.

Topology Service - Topology service is responsible to obtain the entire network view and maintain this information in the application layer. This module sends the data collected to the Path Generation service for path computations.

Generation Path - Generation Path service is responsible to compute all possible paths between two endpoints. This module uses a modified depth-first search to generate the paths.

Flow Installation Service - This service is the last step of the proposed algorithm solution. When the optimal path selection service selects the best path and send the information to this service, it prepares the flow entries to be inserted in the OpenFlow switch and using the Southbound API installs new flows on the network devices.

Optimal Path Selection Service - This service is implemented in the application layer and is responsible to select the best path between two endpoints based on the availability and throughput of the paths. It receives the distinct paths to compute from the Generation Path module. Using the information from the statistics collector, it evaluate the load of each path, chooses the path with higher throughput availability and lastly sends the path elected to the flow installation service.

3   Simulation and Results

3.1   Experimental Setup

The simulation was set up on a Laptop with standard physical specifications of Intel i7 processor and 16 GB RAM. The OpenDaylight SDN controller and Mininet simulator were setup using a Linux virtual machine named ODL-MIN. The OpenDaylight controller version used was Carbon (6th release). The additional features that were required by the controller in order to run the simulation were customised according to the proposed algorithm and statistics monitor. For simplicity, we have not considered any background traffic, however, considering background traffic would significantly impact the network performance.

Topology. We took the Abilene topology from TopologyZoo. This network was a high-performance backbone network created by the Internet2 community in the late 1990s. It comprises eleven nodes within USA. Abilene was selected since it is an ideal multi-path topology to test our proposed solution.

The network topology was simulated using Mininet simulator. This network topology as shown in Fig. 1b consists of one SDN Controller node, eleven (11) OpenFlow Switches and two (2) hosts connected to each switch. We used the host H1 and H2 as source nodes and H13 and H14 as destination nodes. The bandwidth of each link is limited to 100 Mbps from switch to switch and switch

Table 1. List of paths

Path no.	Path (Source to Dest)	Hop count	Load/Traffic (Availability)
1	S7, S8, S9, S10, S11, S1	6	25 Mbps
2	S7, S6, S9, S10, S11, S1	6	25 Mbps
3	S7, S6, S5, S4, S3, S2, S1	7	100 Mbps
4	S7, S6, S5, S9, S10, S11, S1	7	25 Mbps
5	S7, S6, S9, S10, S11, S2, S1	7	25 Mbps
6	S7, S8, S9, S10, S11, S2, S1	7	25 Mbps
7	S7, S6, S5, S9, S10, S11, S1	7	25 Mbps

to host. From the sources and destination, we have multiples paths. The shortest path is Path 1, which has six hops. But as our solution considers all possible paths, we have multiple paths to get the destination with more hops. For our scenarios, we are going to consider the following paths as shown in Table 1. These paths have between six and seven hops. Some links are part of more than one path; for example, paths 1 and 2 share the links between S10, S11, and S1. Iperf is used to generate traffic/load from sources to destinations. Iperf provides statistical information about the packet loss, throughput, and jitter from the traffic generated. This information was used to create the statistics report of the simulation. This tool allows generating traffic UDP or TCP. For our scenarios, we are using UDP traffic only for ease of purposes and since TCP traffic cannot provide packet loss information. The parameters used in the simulation are: Throughput = 75 Mbps, Simulation Time = 60 s and Monitoring iteration time = 2 s.

3.2 Test Case 1

This test case has two main objectives: first, prioritise selected traffic in a multi-path network environment and second, selecting the less congested path between two endpoints using the SDN central management for the prioritised traffic. As shown in Table 2, we prepared two new flows from source to destination. Every new flow generates the same predetermined amount of load/traffic from source to destination using Iperf generation tool for the time shown in Table 2. The Host H1 started to send traffic in the first second, when the switch S1 received the first packet sent by H1, the switch did not have any flow entry defined, therefore encapsulated the packet and sent it to the controller. The OpenDaylight controller took the packet and evaluated the header of the packet and determined the best path based on the shortest path algorithm. The path selected is Path 1 since it has six hops. The controller prepared the flow entries and installed them to each switch of the path 1. So, the traffic from **H1 to H13** took the **Path 1**. During the simulation, at 20 s, the Host H2 started to send traffic to the Host H14. Basically, the switch and the controller executed the same procedure of H1 and H13 because the S1 did not have any flow entry about how to deal with this traffic. Based on the shortest path, it selected the path 1. At the same instance, we executed our proposed solution for the traffic from H2 to H14. Our algorithm evaluated the load of every path taking into account the bottleneck link. As shown in Table 1, we are considering seven paths, if we compare the table with the topology, we realise that some links from the path 1 are used on the other paths, except in the path 3. Each link from the path 1 has a load of 75 Mbps. Therefore, the throughput availability of each path that share at least one link with the path 1 is 25 Mbps. The proposed solution selects the path 3 as it has the higher availability and redirects the traffic from **H2 to H14 via path 3**.

Table 2. Test case 1: simulation parameters

Flow no.	Source/Dest	Traffic/Load	Duration (sec)	Start at
1	H1 to H13	75 Mbps	60	0 s
2	H2 to H14	75 Mbps	40	20 s

3.3 Test Case 2

The second test case is realized with the objective to compare our proposed solution (PS) with algorithm based on the shortest path (SP). First, give a special treatment to a selected traffic in a multi-path network environment; and to prove that the shortest path is not always the best option. In this scenario, we sent traffic from H1 to H13 and from H2 from H14. The throughput used in this scenario is 75 Mbps. The controller was taking routing decision based on the shortest path. Therefore, both flows are using the same path (1). We captured the statistic information for 60 s. After that, we executed the proposed solution for the traffic from H2 to H14 and started to send traffic again from both sources (H1 and H2). The Flow 1 kept the same path (1) and the flow 2 was redirected to the path 3. We captured the statistic information for 60 s. In this test case, both the flows 1 and 2 of Table 2 have same characteristics, except both are for 60 s duration and both start at 0 s.

3.4 Results

In Fig. 2, throughput and packet loss are compared for both the flows for test case 1. As seen in Fig. 2a, when the flow 2 started to send traffic, it had a direct impact on the throughput of Flow1 since both flows were taking the same path and those links have a bandwidth of 100 Mbps. Both flows are trying to send traffic with a throughput of 75 Mbps, but the most that they can achieve before the execution of the proposed algorithm is around 40 Mbps. After the PS was executed, we can see that both flows achieve a throughput of about 75 Mbps. We can clearly observe that our solution increases the overall throughput of the network by using multiple paths and exceeding 100 Mbps link in total. We are prioritising flow 2 and by using our algorithm, this flow is achieving the throughput that the host is sending by taking the less congested path.

Figure 2b shows the results of Packet Loss that was monitored in the test case 1. Packet loss is typically caused by network congestion. When the Flow 2 started to send traffic, the links of path 1 get congested. The percentage of packet loss reaches a peak of around 50% in both flows. After our solution inserted the new flows entries in the switches, the packet loss percentage returned back to 0%.

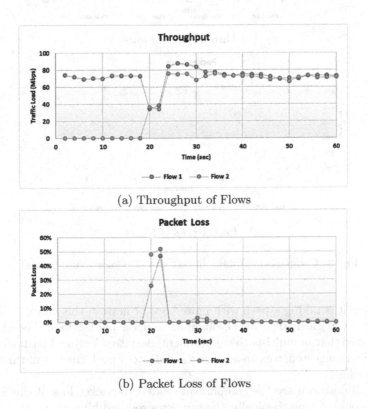

(a) Throughput of Flows

(b) Packet Loss of Flows

Fig. 2. Comparison of flows in test case 1

The throughput results of test case 2 are shown in Fig. 3a. Here, we compare the resulting flows of our proposed solution (PS) with the existing shortest path (SP) solution. The bottom flows peaking at maximum around 47 Mbps are the results of trying to transmit 75 Mbps each flow into links limited to a 100 Mbps bandwidth. The problem with this routing is that once the algorithm selected the shortest path (SP), it keeps sending traffic by this path even it is overloaded. Using the PS, we can see clearly the difference in the throughput. The top flows peaking at maximum of 75 Mbps are taking the less congested path, the grey is taking the path 1 and the yellow is taking the path 3. We demonstrated that

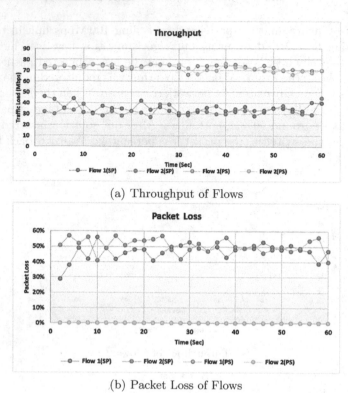

(a) Throughput of Flows

(b) Packet Loss of Flows

Fig. 3. Comparison of flows in test case 2 (Color figure online)

even though it is taking a path with more hops (7 hops in this case), it achieved a satisfactory performance on throughput and on packet loss. Therefore, we demonstrated that in multi-path environment, despite selecting a path with more hops but less congested, we can achieve a satisfactory performance of throughput and packet loss.

In Fig. 3b, we can see the comparison related to packet loss of the SP algorithm and our solution. Basically, the top flows red and blue are overloading the links of path 1 trying to send 75 Mbps each one into links limited to 100 Mbps. Using the PS, the prioritised traffic is sent by a different path which is not congested. Therefore, the packet loss is maintained at 0% as we can see in the flows grey and yellow (bottom). In this test, we managed to treat mission critical flows with the stringent QoS requirements of zero packet loss.

4 Conclusions and Future Work

This paper proposes a system to provide special treatment to mission-critical applications in multi-path environments based in SDN. The system was implemented in the OpenDaylight controller and using the OpenFlow protocol. It was developed in the Application layer taking advantage of the rich northbound API provided by the control plane. We can conclude that our solution achieved the three proposed objectives. First, give a special treatment to a selected traffic in a multi-path network environment, second, using the SDN central management choose the less congested path between two endpoints for the prioritised traffic, and third, demonstrate that in multi-path environment, consider paths with more hops but less congested, it can achieve a satisfactory performance of throughput and packet loss. The results of the simulation validate that our solution increases the overall throughput of the entire network. Furthermore, observing the comparison of our solution versus shortest path algorithm, we can conclude that our solution offers better performance than the shortest path algorithm and single path routing. However, it was not possible to monitor the latency and the delay of these flow paths. Generally, those values may increase while more hops there are into a path. For future work, latency and delay should be considered in the evaluation of choosing the less congested path and establishing a threshold for this value. Also, it is recommended to keep monitoring the QoS parameters and handle policies and events to assure continuous QoS on selected traffic. To realise these possible improvements, it is necessary to develop a statistic module within the controller. Unlike to our solution which processes the information in the application layer to obtain the throughput, there will be an increase in the response time due to policy execution to prioritise the traffic. A statistic module in the control plane will allow creating monitors for distinct QoS parameter such as delay, throughput, packet loss, among others in real time. The efficiency of this module is the base for any QoS system that offers solutions to the relentless growth of traffic volume, network size, and diversity in QoS requirements.

References

1. Dilmaghani, R., Kwon, D.: Evaluation of OpenFlow load balancing for navy. In: 2015 IEEE Military Communications Conference, MILCOM 2015, pp. 133–138, October 2015. https://doi.org/10.1109/MILCOM.2015.7357431
2. Hojiev, S.Q., Kim, D.S.: Dynamic load balancing algorithm based on users immigration in wireless LANs. J. Adv. Comput. Netw. 3(2), 114–118 (2015)
3. Hu, Y., Wang, W., Gong, X., Que, X., Cheng, S.: BalanceFlow: controller load balancing for OpenFlow networks. In: 2012 IEEE 2nd International Conference on Cloud Computing and Intelligence Systems, vol. 02, pp. 780–785, October 2012. https://doi.org/10.1109/CCIS.2012.6664282
4. Karakus, M., Durresi, A.: Quality of service (QOS) in software defined networking (SDN): a survey. J. Netw. Comput. Appl. 80, 200–218 (2017)
5. Liang, C., Kawashima, R., Matsuo, H.: Scalable and crash-tolerant load balancing based on switch migration for multiple open flow controllers. In: 2014 Second International Symposium on Computing and Networking, pp. 171–177, December 2014
6. McKeown, N., et al.: OpenFlow: enabling innovation in campus networks. SIGCOMM Comput. Commun. Rev. 38(2), 69–74 (2008)

7. Verma, D.C.: Simplifying network administration using policy-based management. IEEE Netw. **16**(2), 20–26 (2002)
8. Zhou, Y., et al.: A load balancing strategy of SDN controller based on distributed decision. In: 2014 IEEE 13th International Conference on Trust, Security and Privacy in Computing and Communications, pp. 851–856, September 2014. https://doi.org/10.1109/TrustCom.2014.112

Natural Disaster and Environmental Threat Monitoring System: Design and Implementation

Ba-Cuong Huynh[1], Thanh-Hieu Nguyen[2], Thanh-Duong Vu[3], Nguyen-Son Vo[1(✉)], and Trung Q. Duong[4]

[1] Duy Tan University, Da Nang, Vietnam
cuonghuynh18590@gmail.com, vonguyenson@gmail.com
[2] Ho Chi Minh City University of Transport, Ho Chi Minh City, Vietnam
thanhhieu.nguyen@ut.edu.vn
[3] Quang Nam University, Tam Kỳ, Vietnam
thanhduong.dhqn@gmail.com
[4] Queen's University Belfast, Belfast, UK
trung.q.duong@qub.ac.uk

Abstract. Nowadays, climate change is mainly caused by human activities. As a consequence, natural disasters such as flooding, storm, and drought are attacking people at high frequency and extreme damage. In addition, many megacities have been facing the rapid urbanization problem of the increase in carbon emission, noise, dust, and temperature that seriously impacts on the living conditions of the people. In this paper, we design and implement a monitoring system for early detecting and warning the natural disasters and the environmental threats of the rapid urbanization in two typical provinces in the Central Vietnam, i.e., Quang Nam and Da Nang. The system will sense, communicate, store, process, and display the important information including precipitation, wind speed and direction, water level, and landslide/earthquake in Quang Nam and CO_2 emission, temperature, dust, and noise in Da Nang. The experimental results can help the local government and citizens with better management of natural disasters and environmental threats in the future.

Keywords: Climate change
Natural disaster and environmental management
Wireless monitoring system · Wireless sensor network

1 Introduction

Climate change is one of the most challenges to the world. On the one hand, it has really made human life suffering from many unknown natural and environmental disasters, e.g., storm, flooding, and drought, etc. As reported by the government [1], Vietnam is one of the countries affected by climate change most

T. Q. Duong and N.-S. Vo (Eds.): INISCOM 2018, LNICST 257, pp. 49–62, 2019.
https://doi.org/10.1007/978-3-030-05873-9_5

seriously. About 10–12% of Vietnam population will be directly under the impact of climate change, causing a loss of 10% GDP. Efforts have been dedicated to enhancing capacity of risk management, for instance, improving infrastructures for dyke systems, storm shelters, and flood control works. Yet, 500 people are dying of natural disasters annually, not to mention economic loss. Due to geographic conditions, the Central Vietnam provinces, e.g., Quang Nam, Quang Tri, Quang Binh, are most severely affected by natural disasters.

On the other hand, several provinces such as Da Nang and Quang Ngai, in the Central Vietnam, have developed quickly. The rapid urbanization introduces high environmental threats of carbon emission, noise, dust, and temperature. Together with the problem of natural disasters, to ensure sustainable development, these provinces should have a good policy and a system to track, measure, and estimate the effect of rapid urbanization on the human life so that all the problems of natural and environmental disasters are under controlled.

As communication networks play an important role in grasping dynamic changes of disasters, monitoring environment, and providing advanced applications and services, many wireless communication architectures, i.e., wireless mesh networks [2], mobile ad hoc networks [3], wireless multimedia sensor networks [4], and 5G networks [5] have been studied. Although these works are potential to be applied, the proposed architectures have not been developed in particular areas for practical evaluations.

In efforts to reduce the risk and strengthen the capacity for natural and environmental disaster management, the nation has deployed many monitoring stations of air (26 stations), water (56 stations), and sea water (6 stations). These stations are effectively integrated with the meteorological, hydrographic and navigational monitoring network to send data to the environmental monitoring center of the Vietnam Environment Administration [6]. In addition, some big cities, particularly Ho Chi Minh city, the traffic pollution monitoring program has been operated to grasp the environment changes [7]. Importantly, after the marine environmental disaster in Ha Tinh, Quang Binh, Quang Tri, and Thua Thien Hue provinces, the local governments have installed a seawater monitoring station and an air monitoring station around Vung Ang economic zone and its vicinity. It is expected that a whole marine environment monitoring system including automatic and continuous seawater and air monitoring stations will be developed in these provinces in the future [8].

In Da Nang, in January 2018, the Center of Integrated Circuits (CENTIC) initially deployed four solar energy-based monitoring stations for monitoring the water quality and warning the pollution level in the local rivers and lakes. These stations also monitor and evaluate the quality of industrial wastewater for the purpose of protecting the water environment and preventing the pollution incidents. They automatically send sensed data to the data center. The data is then uploaded to the Environmental Monitoring Center's website of Danang Department of Natural Resources and Environment. In Quang Nam, the authors have deployed a project entitled "Research on the dynamic changes in natural disasters (flooding and drought) in Quang Nam in the context of climate change".

The objectives of the project are to evaluate the existing problems and build a decision support system with capability of calculating, updating, and warning the dynamic changes of natural disasters to minimize the damage caused by flooding and drought [9].

The aforementioned monitoring systems, especially in Quang Nam and Da Nang, can provide the natural and environmental disaster management agents and policy makers with useful information. This way, it is possible to set a better disaster management and more reasonable urbanization road map for sustainable development and thus provide a higher quality of citizens' life. However, there is a lack of monitoring storm and landside/earthquake in Quang Nam and temperature, noise, dust, and CO_2 concentration in Da Nang. In this paper, we therefore design and implement an integrated wireless system (IWS) for sensing, monitoring, detecting, and warning the natural and environmental disasters. Particularly, the IWS will sense, communicate, store, process, and display important information including precipitation, wind speed and direction, water level, landslide/earthquake in Quang Nam and temperature, noise, dust, and CO_2 concentration in Da Nang.

The rest of this paper is organized as follows. In Sect. 2, we present the requirements of the system and the proper device vendor. We design the IWS in Sect. 3. Section 4 is dedicated to experimental results and system performance evaluation. Finally, we conclude the paper in Sect. 5.

2 System Requirements and Vendors Selection

2.1 System Requirements

Feasibly, the IWS is designed by integrating wireless sensor networks (WSNs) and mobile cellular networks (MCN). The WSNs are equipped with sensor nodes including weather, horizontal and vertical liquid level, vibration, CO_2, temperature, dust, and microphone sensors. The WSN in Quang Nam grasps the natural disasters, i.e., storm, flooding, and landslide/earthquake by using weather, liquid level, and vibration sensors. The weather sensors are in charge of tracking the precipitation and direction and speed of wind. The horizontal and vertical liquid level sensors are used to observe the flooding. And, the vibration sensors are to measure the ground vibration for detecting landslide/earthquake. In Da Nang, the CO_2, temperature, dust, and microphone (noise) sensors in WSN grasps the environmental threats caused by rapid urbanization.

All the sensors are connected to their corresponding sinks integrated with general packet radio service (GPRS) modules. The sensed data from the sinks is sent to a server via the GPRS of the MCN. The server stores, analyze, and process the sensed data and timely warn the local governments or the citizens (if needed) about the natural disasters and environmental threats in Quang Nam and Da Nang. Furthermore, a web portal is designed at the server for explicitly monitoring the sensed data in chart form.

2.2 Vendors Selection

Based on the aforementioned requirements of the IWS, we take into account three sensor device vendors that are Libelium (http://www.libelium.com), Device Modern (http://moderndevice.com), and Seeed (http://www.seeedstudio.com). In particular, Libelium provides full requirements of weather, horizontal and vertical liquid level, vibration, CO_2, temperature, dust, and microphone sensors, integrated boards (i.e., smart city, event, gas, and agriculture), GPRS modules, and Li-Ion rechargeable batteries. Meanwhile, Modern Device offers only wind and temperature sensors and there is no information about integrated wireless communication solutions. Seeed supports more sensors than Modern Device including temperature, dust, CO_2, water flow, wind, and integrated wireless communication solutions, but not good enough compared to Libelium. Furthermore, with regard to technical support, deployment, and troubleshooting, Libelium is superior to both Device Modern and Seeed. Obviously, the IWS is at higher cost of using Libelium devices than Device Modern and Seeed. In this paper, Libelium is the best choice thanks to its full solution and technical support. The detailed evaluation of the three vendors is summarized in Table 1.

Table 1. Vendors evaluation

No.	Functions/devices	Vendors		
		Libelium	Modern device	Seeed
1	Sensors	*****	**	***
2	Integration	*****		****
3	GPRS communications	*****		*****
4	Power supply	*****	*****	*****
5	Catalogs and technical support	*****	**	**

3 System Design

3.1 Installation Locations

Choosing the installation locations of the sensors plays an important role in deploying the IWS. A selected location must satisfy the following three factors: (1) highly vulnerable to natural disasters and environmental threats, (2) high signal-to-noise ratio for communicating between the sinks in WSN and the base stations in MCN, and (3) the support of citizens, e.g., in using their power supply, notifying when having problems, etc. To do so, we have considered some survey sites given as follows:

- In Quang Nam, we investigated Song Tranh Hydropower, Phu Ninh District, Dai Dong Commune - Dai Loc District, Hoi An, Ha Lam - Thang Binh, and Dien Hoa Commune - Dien Ban District.

– In Da Nang, we investigated Hoa Tien Commune - Cam Le District, Khue Trung Ward - Cam Le District, Man Thai Ward - Son Tra District, Hoa Khe Ward - Thanh Khe District, and Hoa Khanh Nam Ward - Lien Chieu District.

After carefully investigating, the final installation locations selected that meet the three requirements above are Hoi An, Ha Lam - Thang Binh, and Dien Hoa Commune - Dien Ban District in Quang Nam and Khue Trung Ward - Cam Le District, Man Thai Ward - Son Tra District, and Hoa Khe Ward - Thanh Khe District in Da Nang.

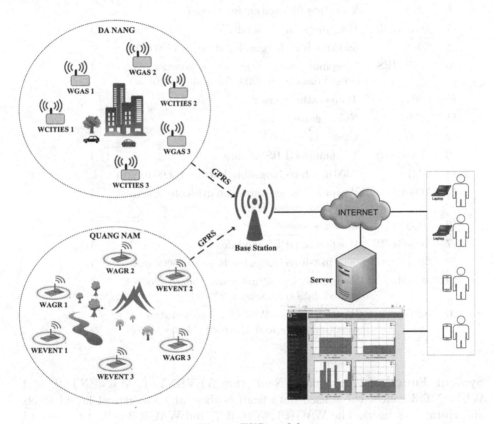

Fig. 1. IWS model

3.2 Integrated Wireless System

System Model: The proposed IWS is shown in Fig. 1. It consists of the sensors and main boards (WEVENT, WCITIES, WGAS, and WAGR) with integrated GPRS modules in WSNs, the base station in MCN, and the server that allows web portal deployment. The detailed Libelium devices used for the IWS are listed in Table 2.

Table 2. Libelium devices

No.	Device code	Device description	Quantity
1	WA-GPRS	Waspmote GPRS/GSM (GPRS module)	3
2	6041	6600mA-h rechargeable battery (6600 mA-h)	3
3	WEVENTS	Waspmote events sensor board (collect sensed data from 9242, 9240, and 9222)	3
4	9242	Vertical liquid level sensor	3
5	9240	Horizontal liquid level sensor	3
6	9222	Vibration film sensor for thread	3
7	WA-GPRS	Waspmote GPRS module	3
8	6041	6600mA-h rechargeable battery (6600 mA-h)	3
9	WCITIES	Waspmote smart cities sensor board (collect sensed data from 9203, 9259 and 9320)	3
10	9203	Temperature sensor	3
11	9259	Noise sensor	3
12	9320	Dust sensor	3
13	WA-GPRS	Waspmote GPRS module	3
14	6041	6600 mA-h rechargeable battery (6600 mA-h)	3
15	WGAS	Waspmote gases sensor board (collect sensed data from sensor 9230)	3
16	9230	CO_2 gas sensor	3
17	WA-GPRS	Waspmote GPRS module	3
18	6041	6600 mA-h rechargeable battery (6600 mA-h)	3
19	WAGR	Waspmote agriculture sensor board (collect sensed data from sensor 9256)	3
20	9256	Weather station WS-3000 (precipitation, direction and speed of wind)	3

System Functions: In Quang Nam, the WEVENT 1, WEVENT 2, and WEVENT 3 collect the sensed data from horizontal and vertical liquid levels and vibration sensors. The WAGR 1, WAGR 2, and WAGR 3 collect the sensed data from precipitation and direction and speed of wind sensors of the Weather Station WS-3000. Similarly, in Da Nang, the WCITIES 1, WCITIES 2, and WCITIES 3 gather the sensed data from dust, temperature, and noise sensors, while the WGAS 1, WGAS 2, and WGAS 3 gather the sensed data from CO_2 sensors. The main boards, i.e., WEVENT, WCITIES, WGAS, and WAGR, are equipped with GPRS modules and rechargeable batteries to send the sensed data to the server via the base station in MCN. All the main boards, GPRS modules, and sensors can work in the power saving mode when not transmitting and thus they are able to work in months. In some convenient cases, the main power is supplied by suitable adapters.

System Operation: The main boards are embedded with codes so that they can send the sensed data to the server via GPRS in POST form. The server receives the data and compares them to the pre-defined thresholds for making decision, then warns if needed and displays on the web portal. The web portal designed meets the following requirements.

- Account management to authorize the administrators and the users of the web portal;
- Display interface to show the continuous sensed data on the graph by date, week, month, and year;
- Excel export to convert the sensed data to excel format for the ease of processing and reporting;
- Adding and removing sensor/device function;
- Warning function to send warnings to the designated email list if the sensed values exceed the pre-defined thresholds. The warnings are also displayed on the web portal;
- And regular data backup.

4   Experimental Results and System Performance Evaluation

4.1   System Performance

To evaluate the system performance, it is necessary to determine the ratio of valid samples to the total number of predefined samples sent to the server. After getting the statistical data from the server, we evaluate the system performance as shown in Table 3. It can be observed that the system performance is high, i.e., 94% of the total samples are valid on average.

Table 3. Performance evaluation

No.	Monitoring parameters	Total samples	Valid samples	Performance (%)
1	Vertical liquid level	20,000	18,400	92
2	Horizontal liquid level	20,000	19,000	95
3	Vibration film	20,000	17,800	89
4	Temperature	20,000	18,000	90
5	Noise	20,000	19,200	96
6	Dust	20,000	19,400	97
7	CO_2	20,000	18,400	92
8	Anemometer	20,000	18,200	91
9	Wind vane	20,000	19,600	98
10	Pluviometer	20,000	10,600	98
System performance				**94**

4.2   Experimental Results

In this paper, the experimental results are almost unchanged on weekdays, therefore, we do not discuss here. It should be noted that the results presented below are the ones exactly done during the survey period, i.e., from March 2016 to March 2017, at the selected installation locations in Da Nang and Quang Nam.

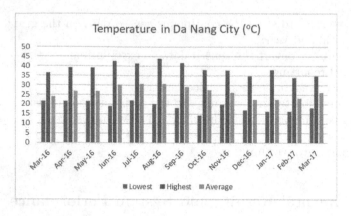

Fig. 2. Temperature in Da Nang

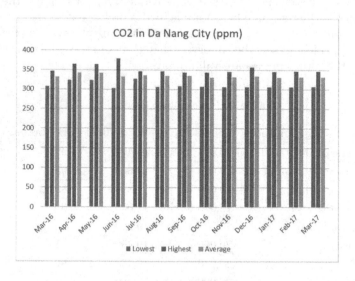

Fig. 3. CO_2 in Da Nang

Results in Da Nang: In Da Nang, the sensed data includes temperature, CO_2, dust, and noise. The sensed data is shown per month by three values: lowest, highest, and average. The detailed results are presented in the sequel.

Temperature: As we can see in Fig. 2, the high-temperature points (close to and greater than 40°) appeared from April to September. The average highest temperature was at 30° occurring between June and September. The lowest temperature period fell from October to February at around 15°. It is important to note that the hot weather frequency was relatively high and serious at 44° occurring in August. The gap between the highest and lowest temperatures in each month was also very high, i.e., more than 20°, especially the difference was up to 24° in August, September, and October.

CO_2: Figure 3 shows the CO_2 concentration measured in parts per million (ppm) every month. The average CO_2 concentration was below 350 ppm and the highest was 380 ppm, which is relatively low compared to the Earth's CO_2 emission threshold, i.e., 408.48 ppm reported on April 1, 2017 (https://www.co2.earth) [10]. The CO_2 emission in Da Nang was well controlled as there was no sign of getting increased in 2016 and the first quarter of 2017.

Dust: Vietnam has set a limit on the safety of dust at $300 \, \mu g/m^3$ [11]. The results in Fig. 4 show that the average dust concentration in Da Nang increased from March 2016 to March 2017. There were many alarms about dust levels in most of the months, even two times higher than the limit ($600 \, \mu g/m^3$).

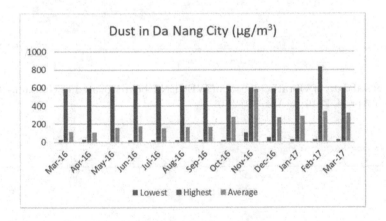

Fig. 4. Dust in Da Nang

Noise: While the noise limit in Vietnam is 70 dB [11], the average noise level in Da Nang was close to the limit as shown in Fig. 5. The alarms occurred at high intensity and frequency every month, i.e., from 80 dB to 100 dB. In addition, the lowest noise level of 50 dB is also an early alarm to the local government and citizens.

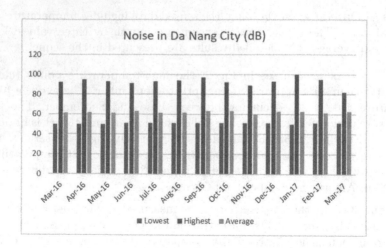

Fig. 5. Noise in Da Nang

Graph with day to 04/08/2016 - 04/10/2016

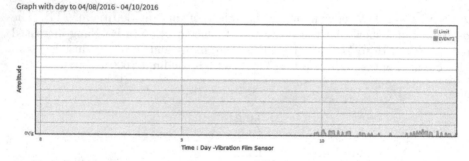

Fig. 6. Sensed data of vibration sensor of WEVENT board located in Ha Lam - Thang Binh from Apr. 8, 2016 to Apr. 10, 2016

Graph with day to 04/12/2016 - 04/14/2016

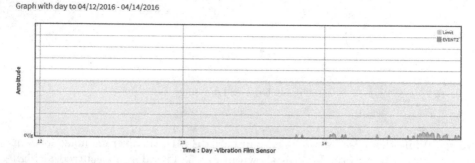

Fig. 7. Sensed data of vibration sensor of WEVENT board located in Ha Lam - Thang Binh from Apr. 12, 2016 to Apr. 14, 2016

Results in Quang Nam: In Quang Nam, the sensed data includes vertical and horizontal liquid levels, vibration, precipitation, and direction and speed of wind. The detailed results are presented in the sequel.

- Vertical and horizontal liquid levels: There was no important sensed data from March 2016 to March 2017. In fact, there was no flooding event and thus we do not discuss in this paper.
- Vibration: There was no significant sensed data occurred from March 2016 and March 2017, excepted that the two consecutive earthquakes were on April 9 and April 13, 2016.
- Precipitation and speed and direction of wind: They are shown per month by three values: lowest, highest, and average.

Earthquake: In 2016, two consecutive earthquakes stuck at the site of the Song Tranh 2 hydropower plant in Bac Tra My district, Quang Nam province. In particular, the Earthquake Information and Tsunami Warning Center (EITWC) of the Vietnam Global Physics Institute said the first earthquake, which was measured at magnitude 3.2 in Richter scale with its epicenter located at 7 km deep, hit the Song Tranh 2 hydropower plant at 22h12'50", on Apr. 9, 2016. Four days after that, at 14h46'21", on Apr. 13, 2016, the second earthquake, which was measured at magnitude 3.0 in Richter scale with its epicenter located at 5 km deep, recorded by the EITWC in the same area.

Most of the vibration sensors of the WEVENT boards detected these two earthquakes. For example, the vibration sensors, which located near these epicenters, i.e., located in Ha Lam - Thang Binh (55 km) and Hoi An (90 km), recorded the earthquake events. However, because of the relatively long distances, the detected signals were not strong enough as shown in Figs. 6, 7, 8, and 9. In addition, some of them did not detect because the detected signals were quite weak. The reason is that these earthquakes happened in the short time at a not very high in Richter scale, meanwhile, the epicenters were 100 km far from the installation locations of the vibration sensors in Dien Hoa Commune - Dien Ban District. As a result, the sensors did not detect the earthquake events from 8 to 14 Apr. 2016. There were the other earthquakes occurred on February 26 and March 1, 2017. But the distances from the epicenters were too far from the vibration sensors, the sensors were not able to detect them.

Precipitation: As shown in Fig. 10, the precipitation in Quang Nam was light, i.e., less than 1 mm/h. However, the precipitation was very diversity between the lowest and highest values, especially from September to December 2016.

Speed and Direction of Wind: The direction of wind in Quang Nam was mainly south-east, with average speed ranging from 1 to 5 km/h (Fig. 11). The wind speed varied from month to month, especially in September 2016, from 0.8 km/h (lowest) to 30 km/h (highest). But it was still harmless even at the highest speed of 30 km/h.

Graph with day to 04/08/2016 - 04/10/2016

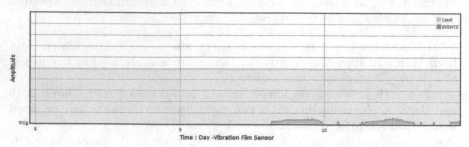

Fig. 8. Sensed data of vibration sensor of WEVENT board located in Hoi An from Apr. 8, 2016 to Apr. 10, 2016

Graph with day to 04/12/2016 - 04/14/2016

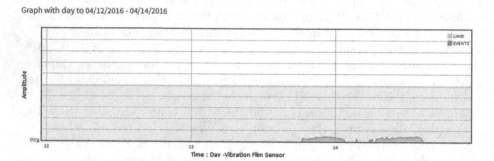

Fig. 9. Sensed data of vibration sensor of WEVENT board located in Hoi An from Apr. 12, 2016 to Apr. 14, 2016

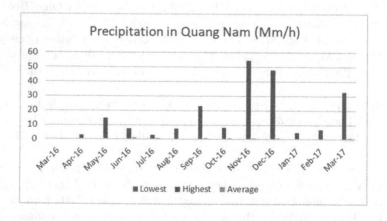

Fig. 10. Precipitation in Quang Nam

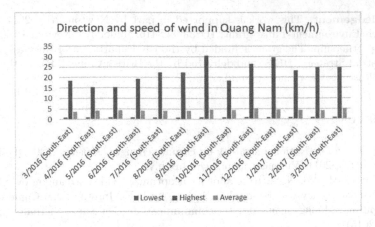

Fig. 11. Wind in Quang Nam

Table 4. Sensed data evaluation in Da Nang and Quang Nam

No.	Sensor type	Evaluation					Location
		1	2	3	4	5	
1	Temperature	x					Da Nang
2	CO_2			x			Da Nang
3	Dust	x					Da Nang
4	Noise	x					Da Nang
5	Precipitation				x		Quang Nam
6	Earthquake			x			Quang Nam
7	Storm				x		Quang Nam

Summary: In summary, the sensed data in Da Nang and Quang Nam can be comprehensively evaluated in Table 4, ranked from 1 for the worst to 5 for the best. It should be noted that the experimental results will be more accurate if the duration of collecting sensed data is longer, e.g., more than two years.

5 Conclusion

In this paper, we have designed and implemented an integrated wireless system (IWS) for monitoring the natural disasters and environmental threats in Quang Nam and Da Nang provinces in the Central Vietnam. The IWS can provide many important parameters of temperature, CO_2, dust, and noise in Da Nang and vertical and horizontal liquid levels, vibration, precipitation, and direction and speed of wind in Quang Nam. The experimental results show reasonable evaluations that can help the local government and citizens manage the natural disasters and environmental threats better in the future.

Acknowledgement. This work is supported in part by Newton Prize 2017 and by a Research Environment Links grant, ID 339568416, under the Newton Programme Vietnam partnership. The grant is funded by the UK Department of Business, Energy and Industrial Strategy (BEIS) and delivered by the British Council. For further information, please visit www.newtonfund.ac.uk.

References

1. Dung, N.T.: The National strategy on climate change. Decision 2139/QD-TTg on December 5, 2011, Vietnam. http://chinhphu.vn
2. Nguyen, T.-H., Vo, N.-S.: Rate allocation optimized video streaming over heterogeneous emergency wireless mesh networks. In: IEEE International Conference on Computing, Managements and Telecommunications, Da Nang, Vietnam, pp. 178–183 (2015)
3. Vo, N.-S., Duong, T.Q., Guizani, M.: Quality of sustainability optimization design for mobile ad hoc networks in disaster areas. In: IEEE Global Communications Conference, San Diego, CA, pp. 1–6 (2015)
4. Vo, N.-S., Nguyen, T.-H., Nguyen, H.K.: Joint active duty scheduling and encoding rate allocation optimized performance of wireless multimedia sensor networks in smart cities. Mobile Netw. Appl. 1–11 (2017)
5. Vo, N.-S., Duong, T.Q., Guizani, M., Kortun, A.: 5G optimized caching and downlink resource sharing for smart cities. IEEE Access **6**, 31457–31468 (2018)
6. The environmental monitoring network serves to assess, forecast timely and accurately environmental changes. Integrated Portal Of The Vietnam Environment Administration, 9 September 2016
7. Monitoring station region III. Decision 1211-QD/MTg on October 22, 1994, Minister of Natural Resources and Environment, Vietnam. http://www.hcmier.edu.vn
8. Approving the investment strategy for a project to building a marine environmental monitoring and warning system in 04 provinces of Central Vietnam. Decision No. 1307/QD-TTg of the Prime Minister, 3 September 2017
9. Lan, V.T., Son, H.T.: Research on the dynamic changes in natural disasters (flooding and drought) in Quang Nam in the context of climate change. Vietnam J. Earth Sci. **35**(1), 55–74 (2013)
10. QCVN 05:2009/BTNMT written by Drafting Committee of national technical regulations on air quality; submitted by Vietnam Environment Administration, Department of Science and Technology, Department of Legal Affairs; and promulgated in line with the Circular No. 16/2009/TT-BTNMT dated October 7, 2009, of the Minister of Natural Resources and Environment
11. QCVN 26:2010/BTNMT written by the Compilation Board of the National Technical Regulation on noise and vibration; submitted by the Vietnam Environment Administration, the Department of Science and technology, and the Legal Department; and promulgated together with the Circular No. 39/2010/TT-BTNMT dated December 16, 2010, of the Minister of Natural Resources and Environment

Sport News Semantic Search with Natural Language Questions

Quang-Minh Nguyen[1(✉)], Son-Hong Ngo[2], and Tuan-Dung Cao[2]

[1] School of Electronics and Telecommunications,
Hanoi University of Science and Technology, Hanoi, Vietnam
minh.nguyenquang@hust.edu.vn
[2] School of Information and Communication Technology,
Hanoi University of Science and Technology, Hanoi, Vietnam
{sonnh, dungct}@soict.hust.edu.vn

Abstract. An increasingly huge amount of sport news published from a number of heterogeneous sources on the Web brings challenges to the traditional searching method using keywords. Providing an expressive way to retrieve news items and exploiting the advantages of semantic search technique in the development of Web-based sports news aggregation system is within our consideration. This paper presents a method to translate natural language questions into queries in SPARQL, the standard query language recommended by W3C for semantic data. Our contribution consists mainly of the construction of a semantic model representing a question, the detection of ontology vocabularies and knowledge base elements in question, and their mapping to generate a query. We evaluate the method based on a set of natural language questions and the results show that the proposed method achieves good performance with respect to precision.

Keywords: Natural language interface · Semantic web · Question answering
Sport news aggregator · SPARQL

1 Introduction

The Web is considered one of the most popular sports news sources that is quickly updated about worldwide events but the readers encounter an enormous amount of news items including coincident and redundant information or information which is outside their interest area. Therefore, the rapid, convenient and effective retrieval of information for readers is always a challenging issue with regard to the development of Web-based news systems [9]. However, keyword search often returns news that contains the terms in questions, but without understanding the meaning of the desired questions. The semantic search promises to solve above problems by returning results relevant to the readers' desire. Furthermore, besides reading news, readers are also interested in information related to entities appearing in news, such as certain characters or organizations. This makes features for sport news. For instance, readers are likely to view news about Lionel Messi, Christiano Ronaldo together with "the classic matches"

T. Q. Duong and N.-S. Vo (Eds.): INISCOM 2018, LNICST 257, pp. 63–73, 2019.
https://doi.org/10.1007/978-3-030-05873-9_6

In previous work, we proposed a sports news system based on semantics and discussed benefits that the system could bring and introduced methods for generating RDF semantic annotation for news items [7, 8]. Our next research topic involves the building of a news semantic search system which can return both precise and suitable news in accordance with readers' requirement. Moreover, this search system should be friendly to users who are just normal readers with limited knowledge about technology. In this paper, we present a method for translating questions from natural language form, for example *"Which team defeated Chelsea this season?"* or *"Who transferred to Barcelona this year?"* into SPARQL queries. These queries will be sent to semantic search engine [4] to find related news associated with the answers to questions.

The remainder of this paper is organized as follows. We start with the presentation of some existing works on information retrieval, especially the systems support natural language questions in searching for information. Section 3 describes the method for translating a natural language question in sport domain to SPARQL query. Experimental scenario and results are presented in Sect. 4, followed by the conclusion.

2   Related Works

There were many researches on information retrieval from semantic data store, and some of researches using directly SPARQL statements to query information from data of semantic knowledge store [12]. However, the use of SPARQL syntax has some weaknesses such as complex query language syntax. In addition, it requires users to understand inside structure of semantic knowledge store. Several other researches enhance the friendliness to users by providing graphical user interface based on ontology to formulate SPARQL queries [2]. Although these graphical interfaces make the query formulation work easier, they are still not perfectly suited to end-users as they still requires users to basically understand ontology.

Providing an intuitive way to express complicated user needs, natural language interfaces such as question answering (QA) systems have received increasing attention. Guo and Zhang [6] raised the importance and feasibility of a Question-Answering system in the Chinese language. Their QA system was established relying on three models, i.e. the question's semantic comprehension model based on Ontology and Semantic Web, FAQ-based question similarity matching model, and document warehouse-based automatic answer fetching model. Certain works based on controlled natural languages, such as Squall2Sparql [5] typically consider an unambiguous restricted subset of natural language that can be translated directly into SPARQL.

PANTO [11] is a Portable nAtural language interface to ontologies allowing users to represent "information needs" by means of natural language without knowledge of RDF, OWL syntax, query languages, or vocabularies of the ontologies. It uses WordNet and String metric algorithm to map words in the user's query to entities in the ontology (concepts, entities, relations). However, the system still has limitations in processing negative questions, and is unable to deal with the question of quantity.

QuestIO [3] (Question-based Interface to Ontologies) is a domain-independent tool which uses natural language to serve query of large knowledge stores in ontology. The drawback of this tool comes from the detection of relations in the question as it is based on rules and not on syntax analysis of questions at deep level. Therefore, it cannot deal with questions with complex semantic. The test was carried out on data set of 22 questions from GATE user mailing list, where users enquired about various GATE modules and plug-ins, the precision obtained was 71.88%.

ORAKEL [1] is a nature language interface which accepts questions and returns answers on the basis of a given ontology. This task is performed thanks to the Query Interpreter, which interprets the input question and transforms it to a first order logic representation and the Query Converter, which is in charge of transforming the logical representation of the question to SPARQL query. This system can only work with the type wh-question type and not the yes/no question type.

The above works show the significance of semantic search function in the form of natural language question. However, they focus more on general domain than specific one. In addition, porting Natural Language Interfaces to other domains [1] is never an easy task. The question types are identified by above systems usually having simple structures, which cannot describe all of the readers' demand of information.

Our research aims to develop a search system using natural language, friendly to users, allowing users without knowledge about complex query language to use the system effectively. In order to do so, the precision of transforming query into its semantic form needs to be enhanced. We propose a novel method consisting of several stages in order to do this in the domain of sports news. Question modeling, analysis of grammar structure recognition and transforming to corresponding semantic representation play an important role in our method.

3   Translating Natural Language Question into SPARQL

The proposed process of translating natural language question to SPARQL query comprises 5 main phases as presented in Fig. 1. The procedure is as follows. The input question in the form of natural language is normalized by the Question Preprocessing component, which helps for others components to function effectively and precisely. After that, the question is sent to Syntax Analysis to analyze grammar elements and its relations. Then the question is represented in the form of semantic model by the component of Semantic Representation of Question. From the semantic model, the

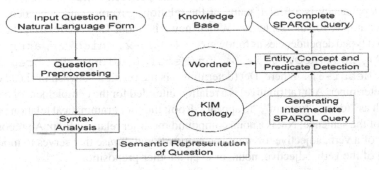

Fig. 1. Natural language to SPARQL query translation process

intermediate SPARQL query is generated. Component of named entity, concept and predicate detection then annotates variables in the intermediate SPARQL query by URIs in ontology and knowledge base of the system. And at last, a complete SPARQL query is generated.

3.1 Question Preprocessing

Preprocessing component is responsible for normalizing abnormal tokens and determining the temporal property of question. Readers are accustomed to use abbreviations when they write. Therefore, we make statistics of common abbreviations and build a normalization table. Each token in questions is examined, and then abbreviations are replaced. In parallel, statistics of temporal labels were classified into the following types: one-day period (i.e. "today", "yesterday", etc.), one-week, one-month and one-year period. Based on the time readers make a query, a specific temporal value corresponding to each temporal label is computed, then replaces temporal label in questions with this specific value.

The system accepts even input as imperative sentences and reduced ones. In order for syntax analyzer to function properly and to simplify next steps processing, the system transforms such sentences into one of two standard question forms with equivalent meaning. The former is wh-question and the latter is yes/no question. For example, the question *"news about Lionel Messi"* is transformed into better standard grammar question as *"Which news is about Lionel Messi?"*.

3.2 Syntax Analysis

Syntax Analysis is in charge of identifying the form of question, grammar elements and its relation in questions, thanks to the POSTagged, Phrase Structure Tree and Typed Dependencies analysis. The results are used in tasks of next phases such as identification of question form, building triple relations, annotation of entity, class and property.

Phrase Structure Tree is a visual way to represent the output of the syntax analysis process. Phrase Structure Tree shows 3 aspects of sentence structure: linear order of words in the sentence, groups of words to establish a phrase and hierarchical structure of the phrase. There, the root node of tree identifies form of question. The Typed Dependencies represent the grammatical relations between words in a sentence. Each Typed Dependency is a triplet: name of the relation, governor and dependent.

For example, for the sentence "Which news is about Lionel Messi?", the system figures out typed dependencies as follows: det(news-2, Which-1), attr(is-3, news-2), root(ROOT-0, is-3), nn(Messi-6, Lionel-5), prep_about (is-3, Messi-6). Where, Det (determiner) is the relation between the head of a NP and its determiner; Attr (attributive) is a relation intended for the complement of a copular verb such as "to be", "to seem", "to appear"; Root: the root grammatical relation points to the root of the sentence; Nn: is a noun compound modifier relation; Prep: A prepositional modifier of a verb, adjective, or noun is a prepositional phrase that serves to modify the meaning of the verb, adjective, noun, or even another preposition.

In this work, only certain Typed Dependencies are concerned, which show constraints between elements of question, such as subject-verb, verb-object, passive verb-agent, word-accompanied preposition, noun-adjective modifier, etc. They help to identify important words in the sentence and relations between them. On that basis, constraints by triple patterns are formulated in the SPARQL query.

3.3 Building Semantic Representation of Question

We define a semantic representation model covering two basic question forms: wh-question and yes/no question as the aggregation of the following elements: variable list, constraints of variables and constraints of dependent relations. This model plays the role of the intermediate representation to generate SPARQL query.

Each variable in the *Variable list* represents a token in the question and is named in the format "symbol string + ID (for example: ?x1, ?x2, etc.). Variable label is the word it represents. Variables are divided into two types, in which *query variables* are hidden variables containing information to be queried for the question and *ordinary variables* are the rests. For wh-question, at least one query variable is required to exist in the variable list while for yes/no question, query variable does not exist.

Two types of query variable are proposed, quantity query variable with wh-question of "how many", represented in variables list as COUNT(?variable_-name) and object query variable with wh-question of "who/what/which/where", represented in variables list as ?variable_name.

Constraints of variables include the constraint on the label of variables, constraints on relationship between variables and quantity constraints. We consider the quantity constraint of certain variable by a specific value through relationships as: "=" (equal), ">" (morethan), "> = (moreORequal)", "<" (lessthan), "<=" (lessORequal).

Constraints of dependent relationship include AND/OR constraints indicating the dependent relationships occurring at any time and temporal constraints.

Transforming Grammar Structure into Semantic Representation
This task consists of four main steps: identification of query variable, identification of ordinary variables and constraint of dependent relationship between variables and identification of quantity and temporal constraints.

Identification of Query Variable
The existence of a query variable in the variable list depends on the form of input questions. If the input is a yes/no question, query variable does not exist in the variables list. On the contrary, if the input is a wh-question, then subject is identified corresponding to query words. For "who/what/where", we determine subject as the query word, while with "how many/which", it is determined as the noun following a query word, by which the query variable is identified.

Identification of Ordinary Variables and Constraint of Dependent Relationship Between Variables
Each Typed Dependency is a triplet, i.e. name of the relation, governor and dependent. From Typed Dependencies, related words and relations among them (name of Typed

Dependency) are inferred mutually. These words are represented by variables, i.e. query variables and ordinary variables).

An important part of a SPARQL query is triple set what defines the constraints of the question. However, each dependent relationship between variables which are identified from Typed Dependency is just in the form of double. These relationships belong to one of two types, subject - predicate or predicate – object. Therefore, we propose to combine typed dependencies to generate constraint by triple patterns. Two typed dependencies representing two types of different dependence which share predicate are to determine to create the triple relationship in the form (subject, predicate, object). The following examples illustrate how to carry out the method for questions with simple semantics S-V-O. For the question "Did Barcelona defeat Chelsea?", the syntax analysis phase finds out two main typed dependencies which are nsubject(?x-defeat, ?y-Barcelona) and dobject(?x-defeat, ?z-Chelsea). Where, nsubject represents the subject – predicate relationship and dobject represents the predicate – object relationship. By combining these two typed dependencies, we obtain a triple pattern (?y-Barcelona, ?x-defeat, ?z-Chelsea).

Identification of Quantity Constraint
Two types of quantity constraints are considered:

(1) Constraint based on comparing the number of certain objects with a specific value, i.e. "Who scored more than 3 goals?" and
(2) Superlative quantity constraint of a certain object, i.e. "Who scored the most goals?" "Which team conceded the least goals?"

For the type (1), the type dependency num(?object, ?quantvalue) shows the existence of a quantity constraint for the ?object object based on the relationship with the ?quantvalue value. In order to identify the relationship between an object and this quantity value, we consider the existence of another typed dependency which is quantmod(?quantvalue, "than").

The detection of "the most/least" + noun structure can be based on two typed dependencies det(?object, "the") and amod(?object, ?dep). For example, if the value of ?dep in the typed dependency admod is "most", it means the ?object object has the most occurrences, the module generates themost(?object) constraint in the semantic model.

Identification of Temporal Constraint
A raised problem is to identify elements of time-related question and to transform it into a semantic model. To do that, in the semantic model, we define one "Interval" including two fields: Interval (BEGIN, END). Interval type shows the constraint that time of events must be in the form of from BEGIN to END.

In the scope of this work, we only concern minimum time unit in day (day, week, month, year). If the question mentions temporal context, then the syntax analyzer generates Typed Dependency prep_in (object, time_label). In which, time_label is the temporal label to show the temporal context of question. If time_label is a certain day, BEGIN and END attain the same value. If time_label is a certain week, month, year or season, we base on time user making queries to compute BEGIN and END value.

3.4 Generation of Intermediate SPARQL Query

The SPARQL intermediate queries just have frames containing variables, which consist of two main elements, namely question clause and condition clause. In addition, for special questions such as questions with quantity constraint and question with temporal constraint, there are additional constraint clauses. Figure 2 shows the process of generating these queries from semantic model of question and main principle of each step as well.

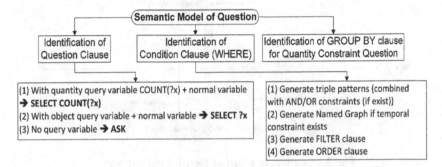

Fig. 2. Generation of intermediate SPARQL query.

3.5 Entity, Class and Predicate Detection and Complete SPARQL Generating

Type and value for these variables in the semantic representation model are identified in the phase of ontology's elements detection. Each label of variable is mapped to knowledge base and ontology to detect corresponding entity, class or property.

Named Entity Detection
KIM [10] is a platform that provides a knowledge and information management infrastructure, services for automatic semantic annotation, indexing and document extraction. However, KIM is built to serve in a public domain and it is equipped with a high level ontology named PROTON and a knowledge base with a large number of entities of general importance. In order to identify name entities in news item, we extend KIM for serving a specific domain such as sport. An ontology named BKSport with classes and properties at a lower and more detailed level is integrated into KIM' ontology, as shown in Fig. 3.

By enriching the KIM's knowledge base, named entities appearing in input questions are detected as instance of detail class in sport ontology rather than the one of KIM's general types. Each variable in the question's semantic model has a corresponding private label which is used to semantically annotate variables. Most of variables have been labeled with the name of proper noun corresponding to an entity in the knowledge base. If a variable is recognized as an entity in the knowledge base, it is replaced with a URI of the entity. In case that a label is proper noun but not recognized, a FILTER statement is added into intermediate SPARQL query to bind label value to variable.

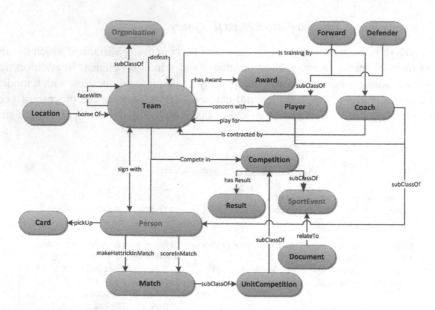

Fig. 3. Snapshot of a part of the BKSport ontology

Classes and Properties Detection

For variables act as subject or object in the constraint by triple patterns but not entities, class constraint for such variables is built. To recognize class of variables, firstly a list with two fields is built. The former is URIs of all classes in ontology and the latter is corresponding labels of those classes. Secondly, WordNet is applied to search synonyms of labels of each above URI, generating a set of representing words for each URI. Finally, the system examines which set of representing words the labels of each variable belong to. From that, we identifies corresponding URIs to variables and adds a triple pattern with the syntax of `<variable_name> <rdf:type> <class_URI>` into SPARQL query to identify type of variables.

Property recognition of variables acting as predicate in triples is also carried out like class recognition.

Generation of Complete SPARQL Query

As results of the detection phase, all variables in semantic model are identified. Generation of complete SPARQL query is simply accomplished by replacing variables in the intermediate SPARQL query by corresponding URIs. For example: a SPARQL query is generated as follows for the question "Which team defeated Chelsea in 08/08/2015?":

```
SELECT ?x WHERE {
?graph { ?x <http://bk.sport.owl#defeat> <http://bk.sport.owl#Chelsea>.    }
?x rdf:type <http://bk.sport.owl#team>.
?g <http://bk.sport.owl#hasTime> ?t.
?t rdf:type time:Instant.
?t time:inXSDDateTime ?instantDate.
FILTER (?instantDate >= "2015-05-08"^^<xsd:dateTime> && ?instantDate <= "2015-05-08"^^<xsd:dateTime>).
```

4 Experiments

To evaluate the proposed method, we setup an experiment dataset consisting of 41 questions belong to frequently used question in sports domain. These questions are sent to the system to be transformed into SPARQL queries automatically. Evaluators then consider whether the generated SPARQL query describes information shown in the natural language input question accurately and fully or not.

However, "correct level" evaluation of SPARQL query is difficult. We determine a SPARQL query including three main clause types including *question clause, WHERE clause, other constraint clauses* (for example temporal constraint clause, quantity constraint clause, etc.). In order to measure the precision of query, we firstly measure the precision of each type of clauses. In order to do it, we base on unit elements. We define a "correct unit element" which is a variable satisfying one of the following conditions: *recognized, corresponding to one URI, type is identified clearly and constraints on label value are identified clearly.*

We evaluate precision of a query generated by the system based on not only satisfied number of elements but also importance of each element. Weight is assigned for each type of clause in query based on our opinion about its importance. Considering n_i as the number of correct unit elements of clause i, N_i is the number of unit elements of clause i needed to identify in a query written by an expert, then $\frac{n_i}{N_i}$ is the precision of clause i. In special case, there is no specific question variable in "SELECT *" question clause, we assign 0.5 to the precision of question clause.

Finally, we determine the general formula to measure the precision of a query q generated by the system as follows:

$$Precision(q) = b \times \frac{\sum_{i=1}^{M}\left(a_i * w_i * \frac{n_i}{N_i}\right)}{\sum_{i=1}^{M} a_i w_i}$$

In which:

- b gets the value of 0 or 1:
  - $b = 0$, if identification of question clause (SELECT or ASK) is wrong, or wrong identification of all question variables.
  - $b = 1$ in the other cases.
- M is the number of clause types in the query written by experts.
  - $a_i = 1$ if the clause exists in the query generated by the system, $a_i = 0$ otherwise.

Table 1 illustrates a part of the experiment dataset of questions and the corresponding precision computed using proposed formula. For the whole set of questions, the average formula in measurement of all experiment questions is applied and the overall precision obtained is 91.89%.

Table 1. Certain input questions and corresponding precision of transforming process

ID	Question	Precision
	*** *Definition question*	
1	Who is Lionel Messi?	1
	*** *Yes/no question*	
2	Was Chelsea defeated by Barcelona last year?	1
3	Did Barcelona defeat Chelsea?	1
4	Did Wayne Rooney dispute with Alex Ferguson yesterday?	1
	*** *Predicative question*	
5	Which team defeated Chelsea?	1
6	Which team defeated Chelsea this season?	0.83
7	Which event relates to Lionel Messi?	1
8	Which team did Lionel Messi transfer to?	1
	*** *Opinion question*	
12	What did Lionel Messi say about Chelsea?	1
	*** *Phrase-verb*	
13	News about Chelsea	1
	*** *Quantity question*	
14	How many clubs defeated Chelsea?	1
	*** *Comparative, superlative question*	
15	Who won the most games this year?	1
16	Who won more than 1 game this year?	1
	*** *Association question*	
20	What is the result of the match between Chelsea and Barcelona?	1
21	What happened between Chelsea and Barcelona?	1
	*** *Multi-subject, multi-object question*	
22	Which team defeated Chelsea and Barcelona?	1
23	Did Manchester United and Chelsea defeat Barcelona in 2014?	1
24	Was Barcelona defeated by Manchester United and Chelsea this year?	1

5 Conclusion

We have presented a method of transforming natural language query into SPARQL query. Based on grammar structure preprocessing and deep analysis of questions, the proposed system is able to process some complex forms of questions such as comparative question, superlative question, question with many subjects and objects, question with abnormal grammar structure, etc. Although the proposed method handles improperly in some cases due to the complexity of generated syntax analysis results or the complexity of question semantics, such complex questions were handled partly correctly. By experiment and evaluation with question set including various questions, the proposed system obtains high precision and competitive with related approach. In future works, we focus on improving cases that the current system cannot handle properly. "The tense" of relationship which is verb should be studied to capture semantics of question more correctly.

References

1. Cimiano, P., Haase P., Heizmann, J.: Porting natural language interfaces between domains: an experimental user study with the orakel system. In: Proceedings of the 12th International Conference on Intelligent User Interfaces, pp. 180–189 (2007)
2. Clemmer, A., Davies, S.: Smeagol: a "specific-to-general" semantic web query interface paradigm for novices. In: Hameurlain, A., Liddle, Stephen W., Schewe, K.-D., Zhou, X. (eds.) DEXA 2011. LNCS, vol. 6860, pp. 288–302. Springer, Heidelberg (2011). https://doi.org/10.1007/978-3-642-23088-2_21
3. Damljanovic, D., Tablan, V., Bontcheva, K.: A Text-based query interface to OWL ontologies. In: Proceedings of the 6th Language Resources and Evaluation Conference (LREC), Morocco, ELRA, pp. 205–212 (2008)
4. Erling, O., Mikhailov, I.: RDF support in the virtuoso DBMS. In: Pellegrini, T., Auer, S., Tochtermann, K., Schaffert, S. (eds.) Networked Knowledge—Networked Media, vol. 221, pp. 7–24. Springer, Heidelberg (2009). https://doi.org/10.1007/978-3-642-02184-8_2
5. Ferré, S.: SQUALL: a controlled natural language for querying and updating RDF graphs. In: Kuhn, T., Fuchs, N.E. (eds.) CNL 2012. LNCS (LNAI), vol. 7427, pp. 11–25. Springer, Heidelberg (2012). https://doi.org/10.1007/978-3-642-32612-7_2
6. Guo, Q., Zhang, M.: Question answering system based on ontology and semantic web. In: Wang, G., Li, T., Grzymala-Busse, J.W., Miao, D., Skowron, A., Yao, Y. (eds.) RSKT 2008. LNCS (LNAI), vol. 5009, pp. 652–659. Springer, Heidelberg (2008). https://doi.org/10.1007/978-3-540-79721-0_87
7. Nguyen, Q.M., Cao, T.D.: A novel approach for automatic extraction of semantic data about football transfer in sport news. J. Pervasive Comput. Commun. 11(2), 233–252 (2015)
8. Nguyen, Q.M., Cao, T.D., Nguyen, H.C., Hagino, T.: Towards efficient sport data integration through semantic annotation. In: Proceeding of The Fourth International Conference on Knowledge and Systems Engineering, KSE 2012, pp. 99–106 (2012)
9. Paliouras, G., Mouzakidis, A., Moustakas, V., Skourlas, C.: PNS: a personalized news aggregator on the web. In: Virvou, M., Jain, L.C. (eds.) Intelligent Interactive Systems in Knowledge-Based Environments, vol. 104, pp. 175–197. Springer, Heidelberg (2008). https://doi.org/10.1007/978-3-540-77471-6_10
10. Popov, B., Kirayakov, A., Ognyanoff, D., Manov, D., Kirilov, A.: KIM—a semantic platform for information extraction and retrieval. Nat. Lang. Eng. 10(3/4), 375–392 (2004)
11. Wang, C., Xiong, M., Zhou, Q., Yu, Y.: PANTO: a portable natural language interface to ontologies. In: Franconi, E., Kifer, M., May, W. (eds.) ESWC 2007. LNCS, vol. 4519, pp. 473–487. Springer, Heidelberg (2007). https://doi.org/10.1007/978-3-540-72667-8_34
12. Yamaguchi, A., Kozaki, K., Lenz, K., Wu, H., Kobayashi, N.: An intelligent SPARQL query builder for exploration of various life-science databases. In: Proceedings of the 3rd International Conference on Intelligent Exploration of Semantic Data, pp. 83–94 (2014)

Smart City Total Transport-Managing System

(A Vision Including the Cooperating, Contract-Based and Priority Transport Management)

Nguyen Dinh Dung[(✉)] and Jozsef Rohacs

Budapest University of Technology and Economics, Budapest 1111, Hungary
{ddnguyen, jrohacs}@vrht.bme.hu

Abstract. Today's nations are facing numerous challenges in transforming living environments in a way better-serving people's demands of the future. The principal point in this transformation is reinventing cities as smart cities that combine their data, their resources, their infrastructure and their people to continually focus on improving livability while minimizing the use of resources. The usage data and sensor network is the primary characteristics of any smart city. However, just having data is not enough, data points themselves are only information. It is good to have, but hardly useful by themselves.

This paper gives a short overview of the concepts for transport management system in the smart city and proposes a new transport management approach that is contract-based and priority transport management. These methods allow to estimate and control traffic efficiently. Based on these concepts, the authors propose a new transport management system that is working as a single system. This proposed system has three layers: physical, info-communication and control generation. The system deals with four different classes of tasks: (i) handling the non-cooperative vehicle, (ii) traffic management based on the cooperative vehicle information, (iii) contract-based traffic management, (iv) priority transport management. Some benefits of implementing this system are also expected in this paper.

Keywords: Smart city · Transport system · Intelligent transport management
Contract-based transport management · Priority transport management

1 Introduction

There are various definitions of a smart city, which have been given over the years. The original concept is the "information city", and then evolving into an idea of the information and communication technology (ICT) – centered smart city. There are six main dimensions of the concept: (1) a smart economy, (2) smart mobility, (3) a smart environment, (4) smart people, (5) smart living, and (6) smart governance [1, 2]. The smart city prefers to focus on factors such as human capital and education than the digital city or intelligent city do.

Taking everything into account, the functions of a smart city are (i) reducing the amount of resources used to provide features to citizens, (ii) focusing on services – not

T. Q. Duong and N.-S. Vo (Eds.): INISCOM 2018, LNICST 257, pp. 74–85, 2019.
https://doi.org/10.1007/978-3-030-05873-9_7

on infrastructure – to provide these functions, and (iii) integration of services, respectively service providers to gain a suitable and pleasant user experience [3].

In this paper, transportation management will be determined by the authors due to it is the primary public function of a smart city. Drivers and service providers (SPs) have different goals that each strives to satisfy. SPs try to optimize overall system-wide task. Drivers discovery trip plans whose performance will complete their individual travel preferences subject to current network conditions. They also want their trip performance to be equivalent to others who are making similar trips through the network. A driver might interpret this trips because of the need to optimize the perceived 'quality of service'. Therefore, the roadway routing problem is solved by global solutions that must be acceptable on two levels: (i) satisfying individual drivers, and (ii) the network as a whole.

The successes of the intelligent transportation system (ITS) technologies, such as route guidance systems [4], were measured by improvements in system performance as a function of market penetration and levels of user compliance. The issue of individual driver quality of service is overlooked or down-played as supply-side traffic management systems strive to optimize network performance. Traffic assignment approaches are based on assumptions regarding driver preferences and system-wide objectives [5]. While "equilibrium" solutions often satisfy supply-side objectives [6, 7], there is little evidence that expectations of individual drivers are met.

Approaches to traffic management that will be perceived by drivers as increasing quality of service and resulting in increased compliance will require the more active involvement of drivers themselves. Directly involving drivers in the assignment process can help achieve this goal. A traffic management system that is a contract-based method has the potential to overcome the problems of compliance and encouraging drivers to work with the system operators to generate assignments that are mutually beneficial.

Contract-based transport management (CTM) proposes an innovative traffic management solution able to face the challenge of traffic growth and improves the efficiency of the European transportation systems.

There is only a few research on the effect of the transport management approach, which is based on the contract between management centers and drivers, or priority management. All most of them consider the cooperative system or full priority such as emergency and police vehicles. In this study, we propose novels contract-based and priority transportation management, which is likely to the cooperative type but the users have to pay for the contract to get service.

The objective of this paper is introducing the transport management depending on the possible cooperativeness and needs of drivers.

This paper is organized as the following: Sect. 2 provides an overview of the related literature and determines the research questions. The supporting systems are presented in Sect. 3. The proposed transport management system is given in Sect. 4. There are some conclusions in Sect. 5.

2 Literature Review

While giving a better service to the inhabitants by fulfilling personal demands in an easy to use, flexible and cost-efficient way, the smart transport expects to reduce the pollution and the related health risks to improve quality of life in urban areas [8]. To allow active traffic management the current traffic situation and operating grade has to be captured to enable predictions on future conditions to hedge decisions made by actors and decision engines.

The significant issue while implementing those kinds of systems is providing a reasonable basis for a decision. The best foundation for decisions is knowing about the current state of the overall system. Therefore, the autonomic transport management systems (ATMS) has an essential aspect that is the data acquisition and the networks transporting the information about the current situation in the field. At this point, it was evident that the reliability of data is essential and an essential requirement to the ATMS. Establishing an autonomic system gives some significant advantages in increasing the safety of the overall operations.

Traffic planning is powered by online timetables calculating the optimal way to use public transport regarding time and costs. The travel planner can use a typical graph for all requests and generate lookup tables to guaranty short answering times. Since the upcoming mega trends smart cities, personalized medicine and personalized production by industry 4.0, the ATMS used to establish smart transportation that solves a wide range of new requirements for transport. Moreover, the smart transit provides personalized mobility and goods transportation.

However, users prefer to share contents than to interconnect themselves to remote devices. Thus, the trend of imminent possibility of the future smart cities will affect enhancing services that will be given and employed [9].

Contract-based and priority transportation management approaches are relatively new ones. Consequently, very little academic literature deals directly with such managing types applied in city transport. In the only relevant research paper found on the topic, [10] presented a case study based on blockchains. This research is a case study for blockchain-based real-time ride-sharing services, in which the newly developed parallel transportation management systems are built based on blockchain that is considered as one of the secured and trusted architectures.

In the level of priority transportation management, an adaptive control algorithm for pre-signals tailored to real-time private and public transportation demands was developed as well as established the necessary infrastructure to operate an adaptive pre-signal [11]. They suggested that the travel times of buses in cities can be reduced by using the pre-signal approach to public transport. The primary purpose of this method is to allow buses to jump the car queues upstream of the intersection while cars can still use all the lanes at the primary signal to fully utilize the capacity of the intersection. By this way, the bus delays are reduced, while the capacity loss at the intersection is minimized.

The operation buses on signal-controlled using special lanes was also evaluated [12]. The results showed that bus lanes with intermittent priority did not significantly reduce street capacity, but increased the average traffic density at which the demand is served, and therefore, increased traffic delay. Thus, they proposed the homogeneous system which is estimated travel and arrival times reduction.

Above literature suggests that the real-time dynamic roadway routing problem on instrumented networks is unique in that communication between drivers and system managers (or service providers) is needed to promote optimal solutions. The real-time information is necessary to choose an effective route. Although drivers can aid with routing decisions, they are not capable of handling the significant data needs and must rely on the knowledge network to gather needed information from the system. Thus, it is necessary to establish the contract between demand (drivers) and supply-side (service providers) because of both attain efficient capacity allocation network-wide and satisfy each driver's routing needs and preferences.

The overall idea of contract based transport management in a smart city has been born by studying the air traffic management.

The air traffic is observed and managed by air traffic controllers. The primary surveillance system uses radar for observing all the flying objects. The secondary radar system detects the – so-called – cooperative targets, namely aircraft identifying themselves by sending a response on the radar signal [13]. The contract-based air transport concept introduced a new principle in air traffic management by moving it to a market-driven air transportation system [14].

Principally connected transport principle [15] is on somehow analogue to the cooperating transport idea.

3 Supporting Services

3.1 Internet of Thing Devices

Everything in a smart city is tied together by the Internet of things (IoT) that is one of the essential components in any smart cities. The smart cities work based on the data created by sensor networks that gather and share useful information. The cities can be managed in real-time with this information, and minimize unintended consequences with sufficient integration data. Because of dependence on sensors grows, it is the need that sensors are reliable and that the systems to which sensors are connected will be able to put up with the inevitable failures.

The transfer usable and potentially could be challenged since cities improve from millions to billions, and the trillions of devices. It is noticed that the need for a user-selected can be fulfilled nearby, by which the convenience will be presented without tying up some of the bandwidth of the carrier data networks.

Using data and sensor technologies to present insight is at the core of what it means to be an ITS. However, just having data is not enough; data points themselves are merely information. It is good to have, but hardly useful by themselves. So, a transport system that can connect data points can build a picture of its users. A knowledgeable transport system combines enough data to make unique practices better. Because more and more objects become joined through the Internet of Things (IoT), so they accumulate more data that can be used to inform planning tailored to individuals and populaces. It is estimated that 50 billion objects will be connected by 2020 for an added economic value of US$19 billion [16, 17].

3.2   Information and Communication Technology – Cooperating Transport

Information and Communication Technology (ICT) platforms became the ground floor of the Smart City foundation, thanks to their capability to offer advanced services in ITS, environmental and energy monitoring, building management, healthcare, public safety and security, remote working, and e-commerce domains [18]. Therefore, ICT plays a vital role by interconnecting all the actors of a Smart City [19] and by supporting the provision of seamless, everywhere services [20].

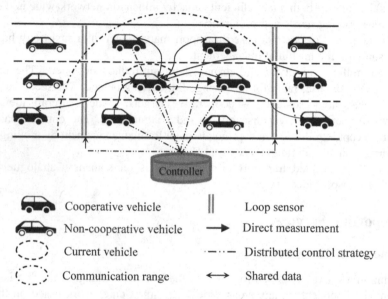

<span></span>	Cooperative vehicle	‖	Loop sensor
	Non-cooperative vehicle	→	Direct measurement
	Current vehicle	·—··—	Distributed control strategy
	Communication range	↔	Shared data

Fig. 1. The general framework of cooperative systems

Advances in ICT enable the transportation community to anticipate dramatic improvements concerning a more efficient, environmentally friendly and safe traffic management. One of potential traffic management systems is a cooperative system with the methodological concerns following the development of such systems [15]. The cooperative traffic models were used as a framework to simulate. The simulation results have shown that (i) with a high amount of cooperative vehicle, only fewer lane changes occur because the cars are organized around an equilibrium state; (ii) the defined cooperative strategy presents a higher impact than adaptive cruise control (ACC) for low penetration rates; (iii) the transition between congested and free flow states appears smoother with the cooperative strategy than the ACC system; (iv) the trust layer reinforces the homogenization effect and this approach provides the best performances.

Furthermore, Information-Centric Networking (ICN) is a concept which has risen as a hopeful candidate for the design of the future Internet [21]. The ICN provides the use of in-network caching and multicast mechanisms by indexing information at the network layer. Therefore, the facilitation is efficiency, and the delivery of the data to the users is opportune. They provided seven ICN approaches to provide a unified view of the alternative proposals by defining a set of core ICN functionalities, e.g., name resolution and data routing, mobility and security.

However, despite the fact that the current Internet architecture corroborates the communication among all of these technologies, it appears a set of conditions related to the decoupling of materials from the knowledge of their location, security aspects, and services scalability.

The future Internet can contribute solutions to many requests that transport management system has to face, but, contrariwise, also this system can give an outstanding experimental environment for probing advantages and disadvantages of future Internet architectures in a variety of application domains.

Each of these technologies works together to make a transport system even more improvement. Because the world's population grows, and more people move into urban areas, the need for more improved transport system will increase to make the best use of available sources.

In our terminology, the non-connected vehicles are the non-cooperative vehicles. The connected ones are the cooperating vehicles. The cooperation might be realized on the local basis (as is shown in Fig. 1) or globally when the transport management is harmonized from the control center. Generally, the cooperating vehicles publish and allow to use their vehicle performance, travel goal and (GPS) positioning data.

4 The Transport Management System

4.1 System Description

The developing traffic-managing system (TMS) is a net-centric system uses the military strategic command concept that in the highest level called as C4ISR (Command, Control, Communications, Computers and Intelligence, Surveillance, and Reconnaissance). The system uses the vast distributed network of sensors for surveillance and recognition of the different cooperating and non-cooperating vehicles, extreme traffic situations (Fig. 2). The sensors are mechanical, optical, electromagnetic, biological, etc. sensors. Extensive wireless communication transfers the sensed data to the system center (working as a command point). The intelligent system generates the controls for avoiding the extreme and dangerous situations as well as for managing the more useful, greener traffic and supporting the contracting vehicles and priority traffic. The controls are realized through the traffic controls (control lights, control signalization, and actuators integrated into the infrastructure). There is no principal difference in cases when the vehicles are moving autonomously, or driver controlled. A driver screen may show the position of the other vehicles, obstacles around the vehicle.

The system has three layers: physical, info-communication and control generation. The physical part including all the vehicles, the available infrastructure and the sensor network, traffic controls integrated into the infrastructure. The infrastructure takes part in the system entirely. That means, for example, a series of signal lights are built into the line dividing the lanes. The communication is based on the wireless system, partly on using the Internet. The control layer is a hierarchically organized software set. It is

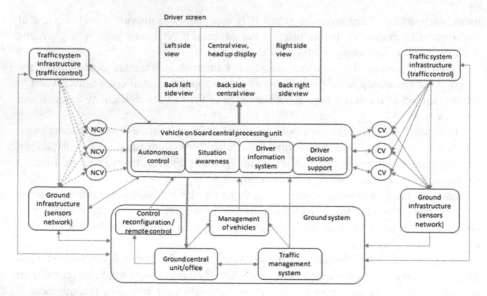

Fig. 2. The traffic-managing system architecture (NCV - non-cooperative vehicle, CV - cooperative vehicle)

used for recognition and classification of the vehicles, traffic situation awareness, conflict detection and resolution including the sense and avoidance of the obstacles, other vehicles, people, etc. The system uses the simulation evaluation of the systems and developing the required traffic and vehicle controls.

4.2   Concept of Operation

The system is working as a single system, while it deals with four different classes of tasks.

Handling the Non-cooperative Vehicles

The system collects all the available and measured information about the traffic infrastructure condition, traffic intensity and complexity, and about all the vehicles regardless of they are cooperative or non-cooperative vehicles, here vehicles mean all the types of vehicles including the road or railway, water transport vehicles, that may be individual, personal vehicles, or vehicles of mass transport systems. This information is the primary input (data of primary surveillance).

The system identifies the non-cooperative vehicles and classifies them depending on their size, mass, predicted performances (as acceleration, turning radius) and predictable goal of trips. The optical, infrared, ultrasonic, radar, etc. sensors built into the traffic infrastructure, into the road, lampposts, traffic lights, nearby buildings, etc. as the elements of the first surveillance, provide the inputs. The system applies this information in short time forecast of the traffic intensity and complexity together with information provided by the cooperating vehicles. The goal is to evaluate where, which direction will increase the traffic, where the traffic jam might appear. With managing such traffic situation, traffic jam, the developing system will support even the drivers of the non-cooperative vehicles. For example, the 4 lanes road might be dynamically controlled: two lanes supporting the traffic into the more intensive traffic direction and one, only, for the other direction, while one lane will dedicate to the contract-based and priority traffic.

Traffic Management Based on the Cooperative Vehicle Information

In air traffic management, the aircraft have transponders that reply to each interrogation signal by transmitting a response containing encoded data identifying the given aircraft. This is the secondary surveillance. In a smart city, net-centric transport managing system, the cooperating vehicles continually provide information about the type of vehicle, motion condition (velocity, changing in velocity, direction, etc.) and actual (GPS) position. This is simple, first level cooperation. The connected vehicles provide this information to the nearby vehicles, too, and harmonize their motions.

The second level of cooperation is characterized by sending the information about the goal and target (geographical positions) of trips to the traffic-managing center. The vehicles applying the third level cooperation send data to the traffic managing center about the nearby vehicles, the condition of infrastructure, traffic situations, etc.

The inputs from primary and secondary surveillance allow introducing total traffic management. Of course, the traffic management may support the cooperating vehicles directly.

Contract-Based Traffic Management

Nowadays, many researchers, cities are working on developing special support the smarter transport. These activities deal with developing the information systems for mass transport, supporting the mode choices, developing optimal transport modal systems, developing the control for connected vehicles, organization, and management of information control for groups of vehicles, supporting the car sharing, information on parking availability, autonomous control of vehicles, etc.

The contract-based traffic management (CTM) introduces a new service opening new market segments for people would like to use it for reducing their travel time. That possible services may start from the dedicated parking areas at P + R systems, through the special small buses transport from parking place to the city business centers, drop off car system. When the driver stops the car anywhere in the city and the traffic-managing system will park it at the nearest parking place. Later the system will transfer the car to the driver defined place. The system may use a remote control or the car may have the required information from the transport-managing system and may pay autonomous.

The CTM is a supporting service that brings together drivers and service providers (SPs) of transportation services to increase the efficiency of both drivers and SPs operation. One way this service benefits driver is that they are the first people who have essential information, thereby they increasing their satisfaction and obtaining their goals. SPs, too, may benefit from contract-based by gaining profit from this contract, allowing for efficient operation increasing supply competition.

The CTM model uses goals rather than positions to achieve an efficient reallocation of network-capacity over time and space without seriously violating any individual driver's preferences for routing, departure, and arrival time. The goal is to achieve more efficient metering of scarce roadway capacity by steering drivers toward paths that will satisfy their individual needs while also improving overall network execution. A good solution derived from CTM will result in drivers being satisfied that their needs and preferences were achieved by their resultant trip itinerary and the SPs being satisfied with the improved system-wide performance.

Figure 3 presents the CTM for roadway routing. It is proposed that if drivers and SPs pursue a collaborative, problem-solving approach to negotiate trip planning, better results will be realized on both sides. Drivers will be able to follow a trip itinerary that better meets their travel objectives.

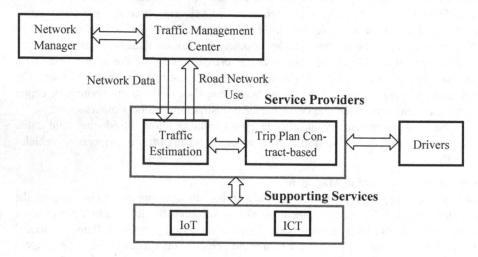

Fig. 3. Conceptual view of contract-based transport management

The SP receives traffic data and driver information from the network managers and in return provides the network managers with information on current and anticipated roadway use gathered from its driver base. Drivers rely on the SP for giving traffic advisories and seek to negotiate trip plans. However, since the SP has a cooperative arrangement with the system, and the system-side traffic management objectives may differ from those of the individual driver, there is always an element of distrust among drivers.

The top level of contract-based traffic may include a "semi priority" system. By using this principle, the contract-based vehicles may use for example bus lane opening provisionally (for a short time) if it was not disturbing the bus transport. In such cases, the drivers will have information from the transport-managing centers about the recommended shortest ways and possible shorting the traveling time and the same time they will see special commanding signals on the road (appearing for the short time and the given vehicles).

The contract-based transport may work on the basis first come – first served without disturbing the existing transport concept and system and even – in case of extra service – it may apply as the personalized taxi.

Priority Transport Management

The developing transport management system uses all the available information about the transport infrastructure, vehicles, traffic complexity, appeared in traffic situations and may simulate the transport and determine the future optimal, more efficient transportation. Therefore, it may manage the priority transport, too.

Generally, the transport-managing system uses passive methods for monitoring the non-cooperative transport, semi-active methods (for partly controlling) the cooperating transport, active method for supporting the contract-based transportation and proactive approach for managing with priority transport.

The priority transport (as police, fire machines, ambulances, traveling the protected persons, etc.) might be supported by opening them the free lanes, freeways by the total transport-managing system.

4.3   Benefits

At a conceptual level, the CTM and priority transport management (PTM) can be regarded as operational ways of achieving the satisfaction of users and service providers. They represent a possible means by which all the operators can share a unique and impartial view of each other's priorities. Thus, they ensure a standard translation and representation of the performance targets to be achieved by the overall transport chain.

At a second more operational level, CTM unequivocally identifies the transfer of responsibility areas between partners, and at the same time, they organize a way of controlling doubtfully and monitoring disruptions. Measurement of compliance with CTM established during the negotiation process could represent a new and reliable metric for assessing the quality of a provided service.

CTM and PTM concepts are expected to directly bring the following substantial benefits to the transport management system:

Firstly, more punctuality at the destination: The CTM and PTM are designed to achieve an ultimate goal, namely arrival on time at the destination places. Through the CTM and PTM the users, drivers, and services providers share the same goal for the vehicles represented by an agreed contract. The synergy between the service providers and users is thus reinforced. Users will reduce delay-related costs and optimize their cars. Service providers will be able to optimize their operations and maximal profits. Even though the efficiency design target identified by IoT and ICT services applies to on-time, a strong correlation between punctuality at departure and the destination exists. It would be interesting to evaluate this correlation during the assessment process, especially in the real-time process.

Secondly, improved predictability: The CTM and PTM are designed taking into account vehicle technical constraints, with built-in scope for disruption management aiming to achieve the ultimate target of the objectives, which is arriving on time at the destination. Each operator knows it's part of the contract, i.e., those contract it must fulfill. Drivers will be able to build on their programs, as predictability will be improved, and they should get a better payoff from their fleet. The management centers will also be able to rely on their schedules, and so optimization of their operations will be possible, which will not only enhance the quality of service delivered to users but also improve the infrastructure pay-off.

Thirdly, reduced overall costs: Drivers will be able to place more trust in scheduling, which will allow them to mitigate delay-related costs, and thus improve the operational costs of vehicles. Providers will get a better approach and better scheduling of their operations, and will, therefore, be able to dedicate the right number of resources to service equipment, which will lead to cost-efficiency.

Finally, reduced environmental impacts: like cost reductions, environmental benefits are mostly linked to the better use of resources (e.g., real-time information, capacities of traffic system) and improved predictability. Drivers will state their preferred routes by economic business models, and thus minimize fuel/energy consumption and improve the "distance/fuel consumption" ratio.

5 Conclusion

The primary aim of implementing the contract-based and priority transportation management is to work towards real-time operations in the traffic system. As these CTM and PTM are consensual trade-offs issued during negotiation between all the actors, even if the economic models of this different actors change, this proposed concepts will be still viable. These concepts also allow all traffic management operational methods, which are bound up with the specific characteristics of the various types of traffic and local areas, to become genuinely adaptable.

This study also describes the transport management system including three layers: physical, info-communication and control generation. The system uses the vast distributed network of sensors for surveillance and recognition of the different cooperating and non-cooperating vehicles. The operational system based on four different classes of tasks. Some benefits of this system are also estimated as the results of this research.

Our future research considers the assessments regarding the systemic and operational issues. The traffic management system works in these ways for establishing fitness for the proposed ideas, based on pieces of evidence.

References

1. Lee, J.H., Phaal, R., Lee, S.-H.: An integrated service-device-technology roadmap for smart city development. Technol. Forecast. Soc. Change **80**(2), 286–306 (2013)
2. Giffinger, R., Gudrun, H.: Smart cities ranking: an effective instrument for the positioning of the cities? ACE Arch. City Environ. **4**(12), 7–26 (2010)
3. Schlingensiepen, J., Mehmood, R., Nemtanu, F.C.: Framework for an autonomic transport system in smart cities. Cybern. Inf. Technol. **15**(5), 50–62 (2015)
4. He, Z., Guan, W., Ma, S.: A traffic-condition-based route guidance strategy for a single destination road network. Transp. Res. Part C Emerg. Technol. **32**, 89–102 (2013)
5. Peeta, S., Ziliaskopoulos, A.K.: Foundations of dynamic traffic assignment: the past, the present and the future. Netw. Spat. Econ. **1**(3–4), 233–265 (2001)
6. Yang, H., Bell, M.G.: Traffic restraint, road pricing and network equilibrium. Transp. Res. Part B Methodol. **31**(4), 303–314 (1997)
7. Yang, H., Huang, H.J.: The multi-class, multi-criteria traffic network equilibrium and systems optimum problem. Transp. Res. Part C Emerg. Technol. **38**(1), 1–15 (2004)

8. Accenture: Building and managing an intelligent city. Technical report (2011). http://www.accenturecom/SiteCollectionDocuments/PDF/Accenture-Building-Managing-Intelligent-City.pdf
9. Ahlgren, B., Dannewitz, C., Imbrenda, C., Kutscher, D., Ohlman, B.: A survey of information-centric networking. IEEE Commun. Mag. **50**(7), 26–36 (2012)
10. Yuan, Y., Wang, F.-Y.: Towards blockchain-based intelligent transportation systems. In: 19th International Conference on IEEE Intelligent Transportation Systems (ITSC) (2016)
11. Guler, S.I., Gayah, V.V., Menendez, M.: Bus priority at signalized intersections with single-lane approaches: a novel pre-signal strategy. Transp. Res. Part C Emerg. Technol. **63**, 51–70 (2016)
12. Eichler, M., Daganzo, C.F.: Bus lanes with intermittent priority: strategy formulae and an evaluation. Transp. Res. Part B Methodol. **40**(9), 731–744 (2006)
13. Galotti, V.: The Future Air Navigation System (FANS): Communications, Navigation, Surveillance – Air Traffic Management (CNS/ATM), p. 362. Aldershot, Avebury Aviation (1997)
14. Guibert, S., Guichard, L., Rihacek, C., Castelli, L., Ranieri, A.: Contract-based air transportation system to fulfil the arriving–on-time objectives. In: 26th International Congress of th International Council of aeronautical sciences, (ICAS), p. 8 (2008). https://www.eurocontrol.int/eec/gallery/content/public/document/eec/conference/paper/2008/004_CATS.pdf
15. Guériau, M., Billot, R., El Faouzi, N.E., Monteil, J., Armetta, F., Hassas, S.: How to assess the benefits of connected vehicles? A simulation framework for the design of cooperative traffic management strategies. Transp. Res. Part C Emerg. Technol. **67**, 266–279 (2016)
16. Minch, R.P.: Location privacy in the Era of the Internet of Things and Big Data analytics. In: 48th Hawaii International Conference on IEEE System Sciences (HICSS), pp. 1521–1530 (2015)
17. Coppa, I., Woodgate, P., Mohamed-Ghouse, Z.: Global outlook 2016: spatial information industry (2016)
18. Piro, G., Cianci, I., Grieco, L.A., Boggia, G., Camarda, P.: Information centric services in smart cities. J. Syst. Softw. **88**, 169–188 (2014)
19. Hernández-Muñoz, J.M., et al.: Smart cities at the forefront of the future internet. In: Domingue, J., et al. (eds.) The Future Internet Assembly, pp. 447–462. Springer, Heidelberg (2011). https://doi.org/10.1007/978-3-642-20898-0_32
20. Schaffers, H., Komninos, N., Pallot, M., Trousse, B., Nilsson, M., Oliveira, A.: Smart cities and the future internet: towards cooperation frameworks for open innovation. In: Domingue, J., et al. (eds.) The Future Internet Assembly, pp. 431–446. Springer, Heidelberg (2011). https://doi.org/10.1007/978-3-642-20898-0_31
21. Xylomenos, G., et al.: A survey of information-centric networking research. IEEE Commun. Surv. Tutor. **16**(2), 1024–1049 (2014)

Shinobi: A Novel Approach for Context-Driven Testing (CDT) Using Heuristics and Machine Learning for Web Applications

Duc-Man Nguyen[1]([⊠]) [iD], Hoang-Nhat Do[2], Quyet-Thang Huynh[3] [iD],
Dinh-Thien Vo[2], and Nhu-Hang Ha[1]

[1] Duy Tan University, Da Nang 550000, Vietnam
{mannd,hatnhuhang}@duytan.edu.vn
[2] MeU Solutions, Ho Chi Minh City, Vietnam
{nhatdo,thienvo}@meu-solutions.com
[3] Ha Noi University of Science and Technology, Ha Noi, Vietnam
thanghq@soict.hust.edu.vn

Abstract. Context-Driven Testing is widely used in the Agile World. It optimizes the testing value and provides an effective way to detect unexpected bugs. Context-driven testing requires the testing team to leverage the full knowledge and skills to solve the problem or to make a decision. In this paper, we propose an approach for Context-Driven Testing using Heuristics and Machine Learning for web applications with a framework called Shinobi. The framework can detect web controls, suggest a set of heuristic values, recognize the meaningful input data, and detect changes of application to recommend test ideas. In the context of improvising the testing performance, Shinobi is considered as Test Assistant for context-driven testers. Shinobi is a PoC to prove the idea of using Machine Learning to develop a Virtual Tester to improve the test quality and train junior testers as responsible testers. The framework is well integrated into all eCommerce projects at MeU Solutions which is a value-added advantage for testing.

Keywords: Shinobi · MeU-Solutions · Context-driven testing
Machine learning · Exploratory testing · Software testing · Web testing

1 Introduction

In the digital age, the speed is uppermost to a successful operation. For software development, agile is the way to reach market quickly, forcing businesses to become as fleet-footed as possible to retain their market leadership position [1].

Supported by MeU Solutions.

However, testing in an agile environment is one of the most common and significant challenge that impacts most enterprises [2,3,36]. The bigger they are, the more time they take to test with agility, therefore, automation is the most feasible option to meet the entire business objectives.

According to the World Quality Report 2017–2018, only 16% are automated testing activities while 60% of companies that are practicing Agile in the North American and European markets [4]. Other studies designate that lack of automation is obstructing uptake of Agile practices, as well as digital transformation initiatives. The report points out that automation testing is needed to succeed in today's digital-oriented era. Automated testing is still relatively low. New ways to speed up automated testing are required like the use of BOTs, artificial intelligence, and machine learning to optimize batch testing without compromising quality [4]. While companies are appealing for methods that will quickly and efficiently optimize the testing cycle, the study indicates that Artificial Intelligence (AI), and Machine Learning(ML) processes can save at least 35% in testing efforts. The result is the quite unusual combination of development that is at once (1) faster (agility), (2) better (quality), and (3) less expensive (efficiency) [5,6]. Testing is often not practical in agile environments in large enterprises. Many companies have backlogs of thousands of test-cases, and so they need automation to address the issue. AI and ML can eliminate human errors, reduce duplication and other problems, and improve traceability. AI-powered automated testing enhances test case quality, reduces the time, cost and scalability shortages of traditional testing approaches [37,38].

As the complexity of web applications continues to increase and as release delivery cycles shorten, testers need to provide quality feedback to developers instantaneously. Moreover, there are now too many tools, frameworks, and technologies for web application development to meet the fast-changing needs of the stakeholders as well as the integrality and flexibility of developing applications. According to Gerd Weishaar-Chief Product Officer at Tricentis [6–8], the key to making software testing smarter and more efficient is AI. By incorporating machines that can accurately emulate human behavior, testers can move beyond traditional, manual testing models and toward automated ones. Customers will be able to automate test cases fully. By adding AI to test creation, execution, and data analysis, testers can more quickly identify controls, spot links between defects and other components, and eliminate the need to update test cases by manual means continually [38,39].

Context-Driven Testing (CDT) is a combination of skills, technique, documentation that depends on the specific situation. Testing is a solution to the given problems. It must be suitable to the context of the project, and therefore testing is a human activity that requires a bundle of skill to do well. CDT suggests a process of critical and creative thinking by using heuristics to solve a problem or to make a decision [9].

Heuristic refers to experience-based techniques for problem-solving, discovery, and learning. As an exhaustive search is unrealistic, heuristic methods are used to speed up the process of finding a satisfactory solution. Examples of this include using a rule of thumb, an educated guess, an intuitive judgment, or common sense [10,11]. In some cases, heuristics testing is more efficient in solving problems as compared to blind testing. Test heuristics are vital when it comes to exploratory testing in an agile environment. When there is not enough time to construct test cases and the product is continuously evolving. There is no need to just rely on pre-designed test scripts, but to use domain knowledge (Test Oracles) and previous testing experience (Test Heuristics) to quickly design and execute tests simultaneously along with learning about the product [12,13,40]. Heuristics help testers reduce the personal bias and direct the analysis and decision making. Thus, heuristics make testers more productive and reliable.

A tester performs CDT by building test models, using heuristics and skills to explore, reason and evaluate the product. Therefore, a heuristic is essential in context-driven testing. But in reality, not all the testers can use heuristics well or recognize what heuristics should be used for in a given context. Moreover, not all of the time, we have a mature context-driven tester who is very experienced with this kind of testing to coach and train other junior testers. Even a knowledgeable tester is sometimes prone to intentional blindness to recognize obvious issues or suffers from confirmation bias that continually validates their beliefs without using a heuristic to challenge them [11,13,14].

In this paper, we propose a novel approach for CDT using Heuristics and Machine Learning with a framework called Shinobi for web applications testing. Shinobi employs ML to overcome challenges from technology changes and makes smarter from learning experience from testers over the time on the same application. Shinobi provides a set of many features that can be categorized into three groups: (i) using ML to detect controls and their types, and use the heuristic library to recommend test ideas to testers. (ii) Using ML to capture and learn all test executions, recognize bias and recommend test ideas for current test execution. (iii) Reporting test execution results combined with heuristics learned from all test executions in the form of mind-map. Currently, Shinobi stays at Proof of Concepts (PoC) with some user stories which are selected from the group (i) and group (ii) to prove the idea's capability. These user stories include: (1) to detect web controls by using Tensorflow and Faster R-CNN and suggests a set of heuristic values for numeric and DateTime controls, (2) to recognize if data entered into them meaningful, (3) to detect changes between application versions through by ML to recommend test ideas. Regarding improving the testing performance, Shinobi can be considered as a Test Assistant for context-driven testers, a Virtual Tester who improves the test quality and coaches junior testers to be the responsible testers.

Organization of the Paper. The rest of paper is structured as follows. Section 2 presents related terminologies and literature review. Section 3 outlines a solution by applying heuristics and Machine learning for testing. Section 4 presents the results and analysis a case study at MeU Solution. Section 5 concludes the paper with a future scope.

2   Related Terminologies and Literature Review

Machine Learning is organized as taxonomy based on expected outcome of the algorithm [12,15–17,41].

Supervised Learning: In supervised learning, user gives a set of input variables (X) and a set of output variables (Y) and use the algorithm to learn the mapping function from the input to the output. $Y = f(X)$. The goal is to approximate the mapping function so well that a new input set is created for which the application can predict outcomes. Supervised learning can be divided into two categories: regression and classification [41]

- Classification is when output set is a category, such as: recognize handwriting, because outputs is a category of A–Z and 0–9.
- Regression is when the output set is a real value, for example: predict the stock's value of a company in next year.

Unsupervised Learning: In Unsupervised learning, user gives a set of input variables (X) and there is no corresponding outcome. For example: Classification of customers based on their respective behaviors.

TensorFlow is an open source deep learning software library for defining, training and deploying machine learning models using data flow graphs. Nodes in the graph represent mathematical operations, while graph edges represent multi-dimensional data arrays communicated between them. The flexible architecture allows you to deploy computation to one or more CPUs or GPUs in a desktop, server, or mobile device with a single API [28,29,43].

Faster R-CNN was originally published in NIPS 2015. Faster R-CNN is the third iteration of the R-CNN papers by Ross Girshick. Everything started with "Rich feature hierarchies for accurate object detection and semantic segmentation" (R-CNN) in 2014, which used an algorithm called Selective Search to propose possible regions of interest and a standard Convolutional Neural Network (CNN) to classify and adjust them. It quickly evolved into Fast R-CNN, published in early 2015, where a technique called Region of Interest Pooling allowed for sharing expensive computations and made the model much faster [30–32,42]. There were number of approaches to combine the tasks of finding the object location and identifying the object to increase speed and accuracy. We implemented Faster R-CNN in Shinobi, a virtual tester toolkit based on TenorFlow which makes it easy to train, monitor and use these types of models [33,34]. Applying Faster R-CNN model is illustrated as below [46,47]:

(1) *Pre-Train VGG16. The feature map from conv53 is used for region proposals and fed into region-of-interest (RoI) pooling.*

(2) *Change the ROI pooling layer, use crop_and_resize operator, which crops and resizes feature maps to 14 × 14, and then max-pools them to 7 × 7 to match the input size of fc6.*
Sample R = 256 regions from N = 1 instead of R = 128 regions and N = 2 as in the original implementation.
(3) *Train a Fast R-CNN object detection model using the proposals generated by the current RPN.*
(4) *Then use the Fast R-CNN network to initialize RPN training. While keeping the shared convolutional layers, we only fine-tune the RPN-specific layers. At this stage, RPN and the detection network have shared convolutional layers.*
(5) *Finally fine-tune the unique layers of Fast R-CNN.*
(6) *Step 4–5 can be repeated to train RPN and Fast R-CNN, alternatively, if needed.*

Literature Review. Several engaging endeavors have already been made for automating the process of software testing [17]. ML as a subdomain of AI is widely used in various phases of the software development lifecycle, especially for automating software testing processes [12,18–21]. According to Wegener and Ahmed [22,23], evolutionary algorithms have been employed for automating test case generation. von Mayrhauser et al. [24]. Artificial Neural Networks have been used to build a model for estimating the effectiveness of the generated test cases. Briand et al. in [25] have proposed a method based on the C4.5 decision tree algorithm for predicting potential bugs in a software system and localizing the bugs to reduce the debugging process time. All related research work demonstrates the employment of machine learning techniques in automation testing processes is highly effective [26,27].

However, most of the studies focus solely on testing techniques based on "scripted testing" such as test cases generation, verification, model checking [19, 27]. The application of ML in the CDT approach is limited. Meanwhile, with the development of AI, CDT, Agile, the application of ML, Heuristics in the CDT and Agile should be of great interest and should include support frameworks for testers or QA, as well as developers. The trend is ML that is being investigated by the researchers for the field of software testing, test automation [44,45].

3   The Proposed Method

The prerequisite of all user story implementations is to recognize the web controls and their types in a web application. It becomes impossible to analyze HTML, JavaScripts and CSS. As it is highly complex to deal with changes with respect to technology and structure especially in custom controls. There is no standard to create a custom control so that a developer can create his control with any CSS structure. A new technology (React JS) requires the user to re-implement or add more new controls (that are the same with a human in term of the shape) for the application. Therefore, using ML approach can solve this problem because of its way of recognizing the objects like the human way.

As a fact, all web controls can be categorized into TextBox, Dropdown list box, CheckBox, Radio button, Button, Image, Label, Panel, ListBox Control, DateTime Control, and Hyperlink. It is observed that these control shapes are not changed even when the technology changes or even when a developer creates a custom control, it should be of similar shape as standard ones.

By this fact, Supervised Learning has been chosen as a solution to detect web objects from an application. Figure 1 shows the framework of the proposed approach for web applications testing and illustrates how Shinobi works. It is integrated into MeU's projects to experiment the design and efficiency. Shinobi is now a proposed integral component of all eCommerce projects at the company which is a value-added advantage for testing. The diagram also illustrates the approach implemented in Shinobi to recognize web objects.

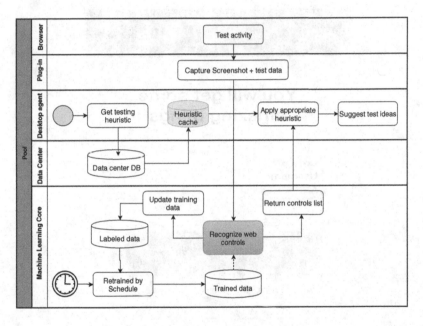

Fig. 1. Framework of Shinobi

3.1   Building and Training Object Detection Classifier

Data Collection: Tensorflow is used to train an Object Detection Classifier. We gathered many images containing web controls that are used to train the proposed detection classifier. The images selected with web objects are outlined differently. For example textbox type images with different text fonts, text sizes, or different borders that were captured from Blogs, Product Management Web page. These are used to generate training data and test data. The training data required depends both on the complexity of the problem as well as algorithm. For

Shinobi, the training set that used to train the classifier in this work, generated from about 500 manually-selected images captured from different web pages.

Label Data and Generate Training Data: The next step is concerned with labeling all the desired web objects in every image. Labeling creates the outcome set. The training data requires high level of labeling quality, so supervised learning does not fall apart and ensures that models can predict, classify or analyze the phenomenon of interest with accuracy. When all images are labeled, the last thing before training execution is to create a Label Map that identifies what kind of control is, by defining a mapping of class names to class ID numbers (Fig. 2).

Fig. 2. Labeling textbox controls and button that contain text inside

Selecting the model is important to train to reduce loss drops rate to less than 0.05. As there are lots of models, some of them are very accurate but have slow speed. On the other hand, there are models which are fast but have low accuracy. In Shinobi, Fast R-CNN has been employed as the model for learning because the input images (control shapes) are not very complicated and it is slightly fast in detecting the objects. Figure 3 highlights the image of Fast R-CNN model.

Training the Classifier and Monitoring the Loss. When everything is established, it is ready to start training. During this progress, the loss at each

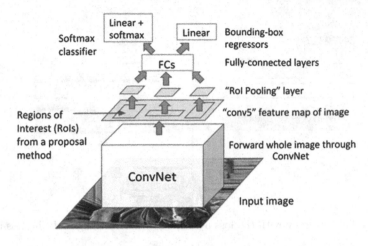

Fig. 3. Fast R-CNN model [35]

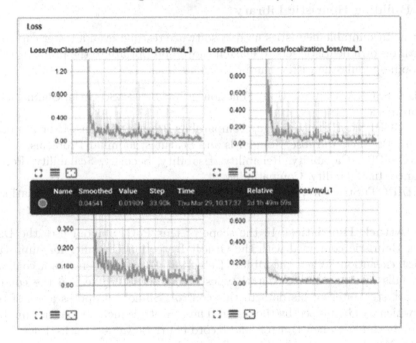

Fig. 4. The loss is dropping at a given number

step of training is viewed. The loss function tells Tensor-Flow how good or bad the predictions are, as compared to the desired result. The loss must consistently drop a lower given number (here is 0.05 as the target). Figure 4 highlights the loss observed from training execution.

The Fig. 5(a) and (b) illustrate that the accuracy depends on the loss. When the loss hits 0.02, all controls are recognized at 99% of its accuracy.

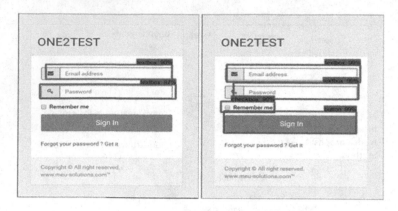

Fig. 5. (a) The accuracy with the loss at 0.8, (b) The accuracy with the loss at 0.02

3.2   Building Heuristic Library

There is uncountable heuristics used in software testing; even a tester can come up with his own heuristic sets to make it more appropriate to the product under test. Some popular heuristics can be listed such as:

- DUFFSSCRA: Domain, User, Function, Flow, Stress, Scenario, Claims, Risk, Automatic.
- HICCUPPSF: History, Image, Comparable Product, Claims, User Expectations, Product, Purpose, Standards and Statutes, Familiar Problems.
- CRUSSPIC: Capability, Reliability, Usability, Security, Scalability, Performance, Installability, Compatibility.
- SFDIPOT: Structure, Operation, Function, Time, Data, Interface, Platform.

Data Attack Heuristics. In the scope of this PoC, Shinobi uses the Data Attack Heuristic combined with the Object Detection Classifier for improving the test execution. When the Object Detection Classifier identifies a control - which falls into one of the control types listed below - based on the context when a tester entering his data to this control. Shinobi proposes a set of test ideas which can be missed by the tester. These controls include, but not limited, as Shinobi can learn new controls and record heuristic values: Paths/Files, Time & Date, Number (Textbox), Strings (Textbox).

Detect Meaningless Data. Bad behavior tester enters meaningless data to just fill the layout. Although it is to save time for data entry, it prevents the tester to generate test ideas and to think more about context when the end user uses the application.

   When controls' types are recognized, Shinobi, then, analyzes data filled in the active page to see, if it is meaningful. The analysis is heuristic, based on some factors such as number of meaningful words with some meaningless words, number of controls, number of controls with its data filled, how fast the data is entered.

All Web Objects Must Be Tested. Many testers ignore some fields during the testing. There are numerous reasons. The first, they believe that these fields are not very important (to test) or not being "dangerous" (no bug) to the application. The second, they pay too much attention to one thing that can result in not seeing objects, even if they are quite salient -this is known as intentional blindness. It is always a good idea to have any mechanics to detect this missing. Shinobi uses this heuristic all the time when a tester performs his test. It reports a list of objects which are not interacted by the tester. These objects can be visible or invisible.

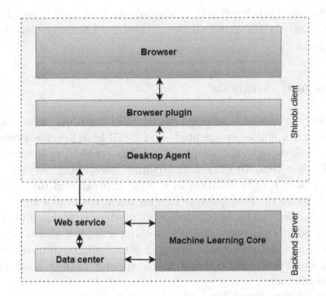

Fig. 6. Shinobi high level architecture

3.3   Shinobi High Level Design

Figure 6 demonstrates Shinobi high level Architecture.

Shinobi is organized into seven elements of testing covering heuristics. These elements are enough to reflect all angles of the product. Any item found by Shinobi is categorized, stored, analyzed and reported in one of these seven elements. They include:

- Structure: Test what the product is made of.
- Function: Test what the product does.

- Data: Test what input/output for the product.
- Interface: Test how the product looks.
- Platforms: Test what the product depends upon.
- Operation: Test how the product is used.
- Time: Test how the product is affected by time.

Shinobi includes three main components of Browser Plug-in (client side), Desktop Agent (client side) and Back-end Server (server) as shown in Fig. 6.

- Browser Plug-in: A chrome plug-in built to analyze and retrieve data from the web page.
- Desktop Plug-in: Component to analyze tester's activities and test data prior to passing them to the server for Machine Learning Core to proceed or to store.
- Back-end Server: Web Service to communicate with the Agent; Data Center to proceed, analyze and store all data; Machine Learning Core to recognize all web objects.

4   Results and Analysis

The methodology proposed was tested out at MeU Solution - a leading ITO vendor of cost-effective, efficiency-oriented, and reliable delivery services. At MeU Solutions, we believe that the proposed solution could bring most success to a company. Rather than to offer a cheap and low-quality service, we employ right people to offer a right solution in a right context and invest much in our people in order to be the best because human resources play a key driver in proposed business model. All the solutions are value-driven, stable and efficient [48]. MeU Solutions adopts the context-driven testing school in almost agile projects. By their expertise, MeU Solutions invested into building innovative tools like One2Explore, One2Test, and Shinobi to sharp and perfect their testing in a context-driven manner.

After three weeks of training with about 500 images and more than 100K steps executed, the loss consistently dropped at 0.02. We have used a test set of 300 images from different web pages so that diverse genres were represented in the test set. These genres include Personal Blogs, Product Management Web pages, Travel System web pages, News. All controls could be detected.

Also, MeU Solutions has used the system PHPTravels (www.phptravels.net/) as a product under test for their experiment with Shinobi. The experiment indicates that all controls in this website have been detected and recognized with high accuracy. Figure 7 illustrates the recognition of the tool.

With the power of this object detection, Shinobi also applied successfully to determine heuristics of Data Attack and of detected all meaningless data to generate more test ideas to assist testers to conduct context-driven testing more effectively as well as increasing the testing coverage. By results from the experiment with PHPTravel, testers have got the following benefits:

Fig. 7. A result of recognizing web controls from the test data

- Recommend more test data which falls into various partitions, that increases the opportunity to catch unexpected bugs.
- Make the testing conducted in manner with meaningful data which provides more insight of its real context.
- Increase UI Test coverage, decrease risks of missing tests from controls that users can interact with.

Fig. 8. Shinobi has detected controls were not interacted with on the left-hand side

Figure 8 shows results from Shinobi running on PHP Travel System. There are 12 controls that were not interacted. All controls are highlighted in the screenshot. These findings are merged into the final report in the form of mindmap.

In Figs. 9 and 10, Shinobi is not just based on analyzing each control; this analysis is using many factors to make sure that data is entered unmeaning fully with the intention.

Fig. 9. Shinobi has detected all control types and used the heuristic of data attack to recommend appropriate values

Fig. 10. Detecting test data has been entered without context (unmeaningful)

The testing to mark the efficiency of Shinobi was carried out by two teams at MeU Solutions within one week. One Blue team uses Shinobi and One Red team performs exploratory testing without any support from the tool. These two teams have been chosen with criteria:

- Similar knowledge background.
- Similar experiences in exploratory testing (about three years) and software testing (about five years).
- Having more than one year working together.

Both of two teams performed the same five test charters. Throughout the testing with results collected from reviewing their test artifacts and interviewing testers, Shinobi brings up more benefits directly to the Exploratory Testing that is acknowledged as follows:

- More than 30% context has been generated from Shinobi recommendations. These contexts mostly fall into UI Testing.
- Much more test data type has been generated for each individual control that was in various equivalences.
- Red Team spent more time on discovering edge cases and less useful data. While Blue team performed testing more efficiently with diverse contexts.

On learning perspective, Shinobi motivated testers in discovering and applying heuristics to testing web controls. It triggered testers to think by recommending more ideas and other different contexts.

5 Conclusion and Future Scope

In this paper, we have introduced how Shinobi is built, works and what its benefits are. At present, Shinobi is a PoC to prove the idea "Using AI-Machine Learning to develop a Human-Machine Tester (Virtual Tester) who improves the test quality and coach junior testers to be the responsible testers". Shinobi has just been completed its implementation with three heuristics which are just emphasizing on UI elements of web applications. Many other heuristics must be leveraged into Shinobi to cover other aspects of SFDIPOT (Structure - Function - Data - Interface - Platform - Operation - Time). Also, the performance is a challenge. Even when using Fast R-CNN, Shinobi sometimes raises issues with its performance. Some other models will be employed to gear up its speed. In addition, distributing business and user transaction to clients and server to improve the performance is the next step of our implementation.

In the context of improving the testing performance and based on results of the experiment at MeU Solutions, Shinobi with combining Machine Learning algorithms and Heuristics can be considered as Test Assistant for all context-driven testers.

In the near future, we explore more possibilities for implementing Shinobi in other types of testings and in several other testing and software engineering firms and observe the results by comparing the traditional approach as compared to machine learning approach in terms of accuracy of bug detection, error-free software.

References

1. Zang, J.: Financial organization transformation strategy. In: Concas, G., Damiani, E., Scotto, M., Succi, G. (eds.) XP 2007. LNCS, vol. 4536, pp. 188–192. Springer, Heidelberg (2007). https://doi.org/10.1007/978-3-540-73101-6_34
2. Rajasekhar, P., Shafi, R.M.: Agile software development and testing: approach and challenges in advanced distributed systems. Glob. J. Comput. Sci. Technol. B Cloud Distrib. **14**(1), 6–10 (2014)

3. Vijay Anand, R., Dinakaran, M.: Issues in scrum agile development principles and practices in software development. Indian J. Sci. Technol. **8**(35) (2015). https://doi.org/10.17485/ijst/2015/v8i35/79037
4. CapGemini, Sogeti and Micro Focus: World Quality Report 2017-18, 9th edn (2017)
5. Papadopoulos, P., Walkinshaw, N.: Black-box test generation from inferred models. In: 4th IEEE/ACM International Workshop on Realizing AI Synergies in Software Engineering, pp. 19–24 (2015)
6. Shaukat, H., Marselis, R.: Testing of Artificial Intelligence-AI Quality Engineering Skills - An Introduction. SOGETI (2017)
7. Eguide, T.: The Impact of Software Failure - And How Automated Testing Reduces Risks. Tricentis (2017). https://www.stickyminds.com/tricentis-eguide-impact-software-failure-and-how-automated-testing-reduces-risks
8. Tricentis: Exploratory Testing: The Heart of All Things Testing (2016). https://www.tricentis.com/resource-assets/exploratory-testing-whitepaper/
9. Bach, B.J., Bolton, M.: A Context - Driven Approach to Automation in Testing, vol. 2016. Satisfice Inc. (2016). http://www.satisfice.com/articles/cdt-automation.pdf
10. Johnson, K.N.: Software Testing - Heuristics and Mnemonics (2012). http://karennicolejohnson.com/wp-content/uploads/2012/11/KNJohnson-2012-heuristics-mnemonics.pdf
11. Baller, H., Lity, S. Lochau, M., Schaefer, I.: Multi-objective test suite optimization for incremental product family testing. In: Proceedings - IEEE 7th International Conference on Software Testing, Verification and Validation, ICST 2014, pp. 303–312 (2014)
12. Ding, J., Zhang, D.: A machine learning approach for developing test oracles for testing scientific software. In: The 28th International Conference on Software Engineering and Knowledge Engineering, SEKE 2016, pp. 390–395 (2016)
13. Bach, J.: Heuristic Test Strategy Model. Satisfice, Inc. (2002). http://www.satisfice.com/tools/htsm.pdf
14. Ghazi, A.N., Garigapati, R.P., Petersen, K.: Checklists to support test charter design in exploratory testing. In: Baumeister, H., Lichter, H., Riebisch, M. (eds.) XP 2017. LNBIP, vol. 283, pp. 251–258. Springer, Cham (2017). https://doi.org/10.1007/978-3-319-57633-6_17
15. Zhang, D.: Applying machine learning algorithm in software development. The Effects of Brief Mindfulness Intervention on Acute Pain Experience: An Examination of Individual Difference, vol. 1, pp. 1689–1699 (2003). https://doi.org/10.1017/CBO9781107415324.004
16. Hormozi, H., Hormozi, E., Nohooji, H.R.: The classification of the applicable machine learning methods in robot manipulators. Int. J. Mach. Learn. Comput. **2**(5), 560–563 (2012)
17. Noorian, M., Bagheri, E.B., Du, W.: Machine learning-based software testing: towards a classification framework. In: SEKE 2011 - Proceedings of the 23rd International Conference on Software Engineering and Knowledge Engineering, pp. 225–229 (2011)
18. Briand, L.C.: Novel applications of machine learning in software testing. In: The Eighth International Conference on Quality Software, pp. 3–10 (2008)
19. Joshi, N.: Survey of rapid software testing using machine learning. Int. J. Trend Res. Dev. **3**(5), 91–93 (2016)
20. Raghuwanshi, V.: AI and software testing. In: 17th Annual International Software Testing Conference (2017)

21. Bhateja, N., Sikka, S.: Achieving quality in automation of software testing using Ai based techniques. Int. J. Comput. Sci. Mob. Comput. **6**(5), 50–54 (2017)
22. Sahoo, R.K., Ojha, D., Mohapatra, D.P., Patra, M.R.: Automated test case generation and optimization: a comparative review. Int. J. Comput. Sci. Inf. Technol. **8**(5), 19–32 (2016)
23. Wegener, J., Baresel, A., Sthamer, H.: Evolutionary test environment for automatic structural testing. Inf. Softw. Technol. **43**(14), 841–854 (2001)
24. von Mayrhauser, A., Anderson, C., Mraz, R.: Using a neural network to predict test case effectiveness. In: The IEEE Aerospace Applications Conference Proceedings, no. level 1, pp. 77–91 (1995)
25. Briand, L.C., Labiche, Y., Liu, X.: Using machine learning to support debugging with tarantula. In: The 18th IEEE International Symposium on Software Reliability, (ISSRE 2007), pp. 137–146 (2007)
26. Sathyavathy, V.: Evaluation of software testing techniques using artificial neural network. Int. J. Electr. Comput. Sci. **6**(3), 20617–20620 (2017)
27. Akmel, F., Birihanu, E., Siraj, B.: A literature review study of software defect prediction using machine learning techniques. Int. J. Emerg. Res. Manag. Technol. **6**(6), 300–306 (2017)
28. Ramsundar, B.: TensorFlow Tutorial (2016)
29. Goldsborough, P.: A Tour of TensorFlow (2016). https://arxiv.org/abs/1610.01178
30. Girshick, R.: Fast R-CNN. In: Proceedings of the IEEE International Conference on Computer Vision, pp. 1440–1448 (2015)
31. Huang, J., et al.: Speed/accuracy trade-offs for modern convolutional object detectors (2016). https://arxiv.org/abs/1611.10012
32. Chen, X., Gupta, A.: An Implementation of Faster RCNN with Study for Region Sampling (2017). https://arxiv.org/abs/1702.02138
33. Wongsuphasawatl, K., et al.: Visualizing dataflow graphs of deep learning models in TensorFlow. IEEE Trans. Vis. Comput. Graph. **24**(1), 1–12 (2018)
34. Yaman, F., Oates, T., Burstein, M.: A context driven approach for workflow mining. In: Proceedings of the 21st International Joint Conference on Artifical Intelligence, IJCAI 2009, pp. 1798–1803 (2009)
35. Ren, S., He, K., Girshick, R., Sun, J.: Faster R-CNN: towards real-time object detec-tion with region proposal networks. IEEE Trans. Pattern Anal. Mach. Intell. **39**(6), 1137–1149 (2017)
36. Ghahrai, A.: How to Overcome Agile Testing Challenges. https://www.testingexcellence.com/agile-testing-challenges-qa-agile-projects/. Accessed 28 Apr 2018
37. Mathuria, M.: AI and Machine Learning to Optimize Software Testing. https://www.readitquik.com/articles/ai/ai-and-machine-learning-to-optimize-software-testing/. Accessed 28 Apr 2018
38. Sypolt, G.: AI Test Automation: The AI Test Bots Are Coming. https://saucelabs.com/blog/ai-test-automation-the-ai-test-bots-are-coming. Accessed 28 Apr 2018
39. Wandile, P.: Overcoming Testing Challenges In Agile. https://dzone.com/articles/overcoming-testing-challenges-in-agile. Accessed 28 Apr 2018
40. Ghahrai, A.: What Are Test Oracles and Test Heuristics? https://www.testingexcellence.com/test-oracles-test-heuristics/. Accessed 28 Apr 2018
41. Brownlee, J.: A Tour of Machine Learning Algorithms. https://machinelearningmastery.com/a-tour-of-machine-learning-algorithms/. Accessed 28 Apr 2018

42. Faster R-CNN: Down The Rabbit Hole of Modern Object. https://tryolabs.com/blog/2018/01/18/faster-r-cnn-down-the-rabbit-hole-of-modern. Accessed 28 Apr 2018

43. Tensorflow. https://www.tensorflow.org/. Accessed 28 Apr 2018

44. From the Experts: Top 5 Trends Shaping the Future of Software Testing. https://www.qasymphony.com/blog/5-trends-future-software-testing/. Accessed 28 Apr 2018

45. RapidValue: New age Software Testing with Artificial Intelligence and Machine Learning. https://www.rapidvaluesolutions.com/new-age-software-testing-artificial-intelligence-machine-learning/. Accessed 28 Apr 2018

46. Object Detection with Faster R-CNN in Chainer. https://github.com/mitmul/chainer-faster-rcnn. Accessed 28 Apr 2018

47. Weng, L.: Object Recognition for Dummies Part 3: R-CNN and Fast/Faster/Mask R-CNN and YOLO. https://lilianweng.github.io/lil-log/2017/12/31/object-recognition-for-dummies-part-3.html#faster-r-cnn. Accessed 28 Apr 2018

48. MeU-Home - Meu Solutions. http://meu-solutions.com/. Accessed 28 Apr 2018

DOA Estimation of Underwater Acoustic Signals Using Non-uniform Linear Arrays

Sang Van Doan[1](✉), Trang Cong Tran[1], and Van Duc Nguyen[2]

[1] Vietnam Naval Academy, Nha Trang, Vietnam
doansang.gl@gmail.com
[2] School of Electronics and Telecommunication,
Hanoi University of Science and Technology, Hanoi, Vietnam

Abstract. This paper proposes to use the multiple signal classification (MUSIC) algorithm applied to non-uniform linear array (NLA) to estimate the direction of arrival (DOA) for passive sonar systems. The performance of the DOA estimation obtained by the MUSIC algorithm is investigated by applying different configurations of the antenna array parameters. For the purpose of comparison, uniform linear arrays (ULAs) are also considered. With same number of elements and antenna parameter configuration, the simulation results show, that the NLA significantly provides a better performance in terms of DOA estimation than that obtained by the ULA. In addition, the accuracy and resolution of the DOA, as well as the number of detectable signal sources can be increased, if we enlarge the antenna spacing as well as the number of the antenna elements.

Keywords: Ambiguity · DOA estimation · MUSIC method
Non-uniform linear antenna array · Uniform linear array

1 Introduction

The direction of arrival (DOA) is an important parameter of passive sonar systems [1]. Most of traditional sonar systems usually employ antenna array or passive beamforming to estimate the DOA of an acoustic signal [2, 3]. In addition, the DOA can be estimated by using the phase difference or the propagation delay difference between the received signals from two different acoustic signal sensors [4–7]. The conventional methods using amplitude, phase or propagation delay of the received signal are suitable for detecting the single source of narrow band acoustic signals. However, these methods are not suitable for the DOA estimation of acoustic signals from multiple sources.

In literature, the multiple signal classification (MUSIC) and estimation of signal parameters via rotational invariance technique (ESPRIT) are two well-known high-resolution DOA estimation algorithms. These techniques use the antenna array signal processing for detecting multi-wideband signals [8–10]. In addition, with ULA method, the angular ambiguity can be avoided, if the inter-element spacing d is designed not to be larger than a half of wavelength ($\lambda/2$) of the received signal. For high frequency signals, it is very difficult to set an acoustic sensor array with the condition of $d \leq \lambda/2$.

© ICST Institute for Computer Sciences, Social Informatics and Telecommunications Engineering 2019
Published by Springer Nature Switzerland AG 2019. All Rights Reserved
T. Q. Duong and N.-S. Vo (Eds.): INISCOM 2018, LNICST 257, pp. 103–110, 2019.
https://doi.org/10.1007/978-3-030-05873-9_9

For understanding this difficulty, we consider the following example. If the frequency of the received acoustic signal is 200 kHz, then the corresponding wavelength in underwater propagation is 0.75 cm. Therefore, the ULA configuration with element spacing fulfilled condition of $d \leq \lambda/2$ is impractical for detecting the DOA of a high frequency acoustic signal. In addition, a small array leads to a lower accuracy and resolution of the DOA estimation [11–13]. For resolving this problem, several NLAs with suitable configurations will be investigated in this study.

In this paper, the MUSIC algorithm is simulated in Matlab environment to evaluate the DOA estimation accuracy and resolution of the NLAs, which can be applied in passive sonar systems. Based on the MUSIC algorithm, the performance of various array configurations for DOA estimation is evaluated. Firstly, the performance of the ULAs using different number of elements is studied. Secondly, by using the same antenna configuration, we compare the performance of the ULA with the NLA one. Finally, the various configurations of the NLA are investigated to find out the optimal antenna configuration for the DOA estimation in terms of accuracy and resolution, as well as, the number of signal sources could be detected. The simulation results show that the NLA provided more precise and better resolution than those obtained by the ULA. The higher SNR and wider NLA aperture, the higher accuracy and resolution can be obtained. These simulation results show that the NLA is promising for application in designing passive sonar systems.

The rest of this paper is organized as follows: Sect. 2 describes briefly the signal model and the MUSIC algorithm. In Sect. 3, the NLAs are compared with each other and with the ULAs in terms of the DOA estimation performance. Finally, conclusion is given in Sect. 4.

2 Signal Model and MUSIC Algorithm

2.1 Signal Model Description

For modelling the signals, a linear array is considered to investigate the MUSIC method. If the signals of far-field sources are mixed at each linear array sensor, then the outputs are modeled by a sum of steering vectors together with the additive noise:

$$\mathbf{x}(t) = \mathbf{A}(\theta) \cdot \mathbf{s}(t) + \mathbf{N}(t) \tag{1}$$

where $\mathbf{A}(\theta) = [\mathbf{a}(\theta_1)\ldots\mathbf{a}(\theta_P)]$ and $\mathbf{a}(\theta_i)$ are the $M \times P$ steering matrix, and the $M \times 1$ steering vector corresponding the ith source, respectively. The signal outputs of the antenna array, the corresponding signal amplitudes, and the additive noise are represented by the $M \times 1$ vector $\mathbf{x}$, the $P \times 1$ vector $\mathbf{s}$, and the $M \times 1$ vector $\mathbf{N}$, respectively. The number of sensor elements, M, is usually chosen to be larger than the number of sources, P. In the expression of the steering matrix, θ_i is the DOA of the ith source in the azimuth. In our research work, the elevation angle is not considered.

If the array is an ULA, then the steering vector can be written as [13]:

$$\mathbf{a}(\theta_i) = \begin{bmatrix} 1 & e^{-j\pi v \sin \theta_i} & \cdots & e^{-j\pi v(M-1)\sin \theta_i} \end{bmatrix}^T \tag{2}$$

where, $v = 2\,d/\lambda$, and d is the spacing between sensor elements.

If the array is an NLA, the steering vector can de described as [13]:

$$\mathbf{a}(\theta_i) = \begin{bmatrix} 1 & e^{-j2\pi \frac{d_1}{\lambda} \sin \theta_i} & \cdots & e^{-j2\pi \frac{d_{M-1}}{\lambda} \sin \theta_i} \end{bmatrix}^T \tag{3}$$

where, d_m is the distance between the mth element and the first element.

2.2 A Brief Introduction of the MUSIC Algorithm

According to the signal model, the correlation matrix is represented as following [13]:

$$\mathbf{R}_{xx} = E[\mathbf{x} \cdot \mathbf{x}^H] = \mathbf{A} \cdot E[\mathbf{s} \cdot \mathbf{s}^H] \cdot \mathbf{A}^H + E[\mathbf{N} \cdot \mathbf{N}^H] \tag{4}$$

$$\mathbf{R}_{xx} = \mathbf{A} \cdot \mathbf{R}_{ss} \cdot \mathbf{A}^H + \mathbf{R}_{NN} \tag{5}$$

$$\mathbf{R}_{ss} = E[\mathbf{s} \cdot \mathbf{s}^H] \tag{6}$$

whereby, H denotes the Hermitian conjugate transpose. E[.] represents statistical averaging operation. $\mathbf{R}_{ss}$ is the signal correlation matrix with the dimension of $P \times P$. $\mathbf{R}_{NN}$ corresponds to noise correlation matrix with the size of $M \times M$.

The matrix $\mathbf{R}_{xx}$ has P dominant eigenvalues whose corresponding eigenvectors $\mathbf{E}_P$ represent a signal subspace. The steering vectors $\mathbf{a}(\theta_i)$ span the signal subspace. The remaining $M - P$ eigenvectors $\mathbf{E}_N$ span the noise subspace. The noise subspace eigenvectors are orthogonal to the array steering vectors at the angles of arrival $\theta_1, \theta_2, \ldots, \theta_P$. The MUSIC spatial spectrum can be given as:

$$\mathbf{P}_{MUSIC}(\theta) = \frac{1}{\mathbf{a}(\theta)^H \mathbf{E}_N \mathbf{E}_N^H \mathbf{a}(\theta)} \tag{7}$$

3 A Comparison of NLA with ULA in Terms of DOA Estimation Performance

To evaluate the performance of the NLA versus ULA, we investigate the following cases: case (1) the ULA is considered for different number of elements (4, 5, ..., 8 elements); case (2) the NLA is simulated with the constant number of elements (e.g. 4 elements), however, the element distance is varied.

3.1   Simulation Results and Discussion of the ULA Performance

The Fig. 1a shows the simulation results in terms of the spectrum power of the received signal obtained by the MUSIC algorithm for the case 4-element and 7-element array, where the inter-element spacing d is $\lambda/2$. The system is simulated at the condition of SNR = 10 dB. It can be seen, that the ULA with 4 elements can detect up to 3 signal sources (full curve line), whereas the ULA with 7 elements can detect up to 6 signal sources (dash curve line).

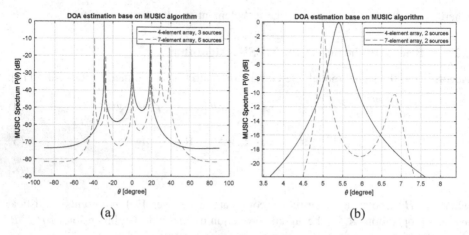

(a)   (b)

Fig. 1. The MUSIC spectrum function obtained using four-element array (full line), and 7-element array (dash line) with inter-element spacing $d = \lambda/2$; SNR = 10 dB; The simulation results are depicted for (a) maximum number of signal sources, which can be detected; (b) the DOA resolution possibility obtained by the ULAs with different number of elements

Figure 1b validated the DOA resolution possibility of array. The 4-element ULA (full curve line) cannot distinguish two signal sources in the conditions of DOA = 5° and 7°, while the 7-element ULA (dash curve line) can estimates the DOAs of these two sources with relatively high accuracy. It can be concluded that a larger number of sensor elements provides better a DOA resolution.

Regarding to the DOA estimation accuracy of the ULA array, the ULAs with different number of elements were considered. Specifically, the ULAs with 4, 5, 6 and 8 elements were chosen for simulation and comparison. Figure 2 illustrated the dependence of DOA accuracy on the SNR values. It is obvious that the ULA with 4 elements gives the lower accuracy than other array, and the ULA with 8 elements provides the higher accuracy than other array. It also demonstrated, that the larger number of sensor elements as well as the wider aperture of array, the higher accuracy of estimated DOA can be obtained.

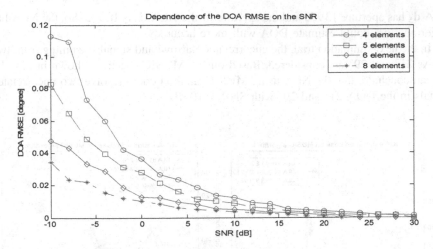

Fig. 2. The obtained DOA RMSE versus the SNR

3.2 Simulation Results and Discussion of the 4-Element NLA Performance

This section investigates the performance of 4-element array configured by different inter-element spacing. We conducted several trials. In the first trial, three array configurations of $d_1 = [0\ 1\ 2\ 3]\ \lambda/2$, $d_2 = [0\ 2\ 4\ 6]\ \lambda/2$ and $d_3 = [0\ 3\ 6\ 9]\ \lambda/2$) are compared to each other. The first trial simulation result plotted in Fig. 3a shows, that the ULAs generates ambiguity of the estimated DOA, if the inter-element spacing is greater than a half of wavelength. Particularly, the first array d_1 is estimated appropriately unambiguous DOA = 19°; the second array d_2 with $d = 2\ \lambda/2$ is estimated appropriately ambiguous DOA = 19° and 42°; and the third array d_3 with $d = 3\ \lambda/2$ provides appropriately three DOA = −20°, 19° and 83°.

The second simulation trial shows a comparison in performance between NLA and ULA. Following that, three arrays are chosen to simulate and compare with array d_1. The three arrays were $d_4 = [0\ 1\ 3\ 8]\ \lambda/2$, $d_5 = [0\ 3\ 8\ 13]\ \lambda/2$ and $d_6 = [0\ 9\ 12\ 16]\ \lambda/2$. These chosen configurations are satisfied the condition of the remainder theorem [14]. The array d_6 is chosen following the principle of $[0\ m^2\ m \cdot n\ n^2]\ \lambda/2$, where m and n are integer numbers whose greatest common divisor is 1 and $m < n$ [4, 5]. A signal is emitted from one source, which is placed in the direction 9° with SNR = 10 dB. The MUSIC spectrum for this case is shown in Fig. 3b.

Figure 3b shows that NLAs achieved higher accuracy than the ULA in case of same number of sensor elements. To evaluate the accuracy of different array configurations Monte-Carlo simulation is performed with 10000 trials for each array. The RMSEs of the DOA estimation for each array were shown in Fig. 4a.

Figure 4a indicates that three NLAs d_4, d_5 and d_6 have better RMSEs than the ULA d_1. The NLA d_6 has highest accuracy, the second is NLA d_5 and the third is d_4. This is due to the difference of array aperture. The NLA d_4 has aperture equal to 8 $\lambda/2$, the

NLA $\mathbf{d}_5$ has aperture 13 $\lambda/2$ and the aperture of the NLA $\mathbf{d}_6$ is 16 $\lambda/2$. So, the NLA has larger aperture could estimate DOA with more accuracy.

In the next simulation trial, the uncorrelated narrowband signals incoming from two DOAs 21° and 29° are considered. Based on the MUSIC spectrum shown in Fig. 4b, we can conclude, that the NLA using MUSIC method could resolve two uncorrelated signals in the DOA 21° and 29° with SNR = 10 dB.

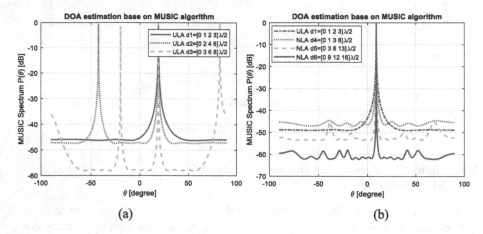

(a)   (b)

Fig. 3. (a) Ambiguity of DOA estimation using MUSIC algorithm for ULA with $d > \lambda/2$; (b) The MUSIC spectrum for various NLAs (SNR = 10 dB)

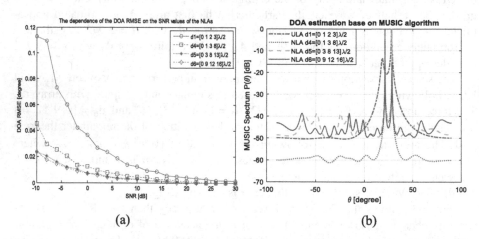

(a)   (b)

Fig. 4. (a) The dependence of the DOA RMSE on the SNR values of the NLA; (b) The DOA resolution possibility of the NLAs (SNR = 10 dB)

In the following, we compare the probability of DOA resolution for different configurations. The probability of DOA resolution represents the possibility of resolution of two close signal sources in dependence on the SNR values. In the simulations, the DOAs of $-2°$ and $2°$ are considered. The Monto-Carlo simulations conducted for 10000 trials. This comparison results are illustrated in Fig. 5. It reveals that the NLAs has a higher probability of DOA resolution than that obtained by the ULA. Moreover, the array with wider aperture will provide a higher probability as well as higher accuracy in terms of the estimated DOA.

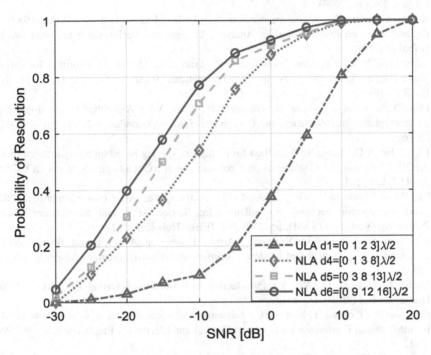

Fig. 5. The probability of DOA resolution of different NLAs

4 Conclusion

The paper analyzes the performance of the MUSIC algorithm, which is applied to estimate the DOA of acoustic signal. The DOA estimation possibility of the ULAs and NLAs are evaluated from points of view of the accuracy and resolution. The MUSIC algorithm is simulated to compare the performance of the various array configurations with the purpose of the DOA estimation. The simulation results show that the NLA provides more precise and better resolution than the ULA. The higher signal to noise ratio (SNR) and/or the wider aperture of NLA, the higher accuracy and resolution can be obtained. These simulation results are promising for developing and designing a NLA in passive sonar systems, in which the DOA of the acoustic signal can accurately be estimated.

Acknowledgment. This work has been partially supported by the project of Faculty of Communication and Radar, Naval Academy, Nha Trang City, Khanh Hoa Province, Vietnam.

References

1. Richard, G.: ELINT: The Interception and Analysis of Radar Signals. Artech House Radar Library, Boston (2006)
2. Robert, G.: Principles of Underwater Sound, 3rd edn. McGraw-Hill, New York (1983)
3. Richard, P.: Underwater Acoustic: Analysis, Design, and Performance of Sonar. Wiley, Hoboken (2010)
4. Doan V.S., Vesely, J., Janu P., Hubacek, P., Tran X.L.: Optimized algorithm for solving phase interferometer ambiguity. In: 17th International Radar Symposium (IRS), pp. 1–6. IEEE (2016)
5. Doan V.S., Vesely, J., Janu, P., Hubacek, P., Tran, X.L.: Algorithm for obtaining high accurate phase interferometer. In: Conference Radioelektronika, vol. 26, pp. 433–437. (2016)
6. Li, Y., He, X.D., Tang, B.: A method for ambiguity solving based on rotary interferometer. In: 2013 International Conference on Communications, Circuits and Systems (ICCCAS), pp. 152–155 (2013)
7. Ly, P.Q.C., Elton, S.D., Gray, D.A., Li, J.: Unambiguous AOA estimation using SODA interferometry for electronic surveillance. In: Sensor Array and Multichannel Signal Processing Workshop (SAM), pp. 277–280. IEEE, Hoboken (2012)
8. Chandran, S.: Advances in Direction-of-Arrival Estimation. Artech House, Boston (2006)
9. Friedlander, B.: Classical and Modern Direction-of-Arrival Estimation. Academic, Boston (2009)
10. Chan, Z., Gokeda, G., Yu, Y.: Introduction to Direction-of-Arrival Estimation. Artech House, Boston (2010)
11. El Kassic, C., Picheral, J., Mokbel, C.: Advantages of nonuniform arrays using root-MUSIC. In: International Conference on Industrial Control and Electronics Engineering, pp. 904–907 (2012)
12. Zhang, H., Zhang, F., Chen, M., Fu, H.: Non-uniform linear sonar array based DOA estimation. In: Proceedings of the 33rd Chinese Control Conference, pp. 7240–7243 (2014)
13. Tran, X.L., Vesely, J., Hubacek, P., Doan, S.V.: DOA estimation with different NLA configurations. In: The 18th International Radar Symposium IRS 2017, Prague, Czech Republic (2017)
14. Ding, C., P'ei, T., Salomaa, A.: Chinese Remainder Theorem: Applications in Computing, Coding, Cryptography. World Scientific, River Edge (1996)

Performance Analysis of the Access Link of Drone Base Station Networks with LoS/NLoS Transmissions

Huazhou Li[1(✉)], Ming Ding[2], David López-Pérez[3], Azade Fotouhi[4],
Zihuai Lin[1], and Mahbub Hassan[4]

[1] University of Sydney, Sydney, Australia
{huazhou.li,zihuai.lin}@sydney.edu.au
[2] Data61, CSIRO, Sydney, Australia
Ming.Ding@data61.csiro.au
[3] Nokia Bell Labs, Dublin, Ireland
david.lopez-perez@nokia.com
[4] University of New South Wales, Sydney, Australia
azade.fotouhi@gmail.com, mahbub.hassan@unsw.edu.au

Abstract. In this paper, we provide performance analysis for drone base station (DBS)-enabled wireless communication networks. The lower bound performance of such networks has been previously obtained in the literature, assuming DBSs are statically hovering and randomly distributed according to a homogeneous Poisson point process (HPPP). We derive the upper bound performance of such networks assuming a teleportation mode, i.e., DBSs can instantaneously move to the positions directly overhead ground users (UEs). By considering both line-of-sight (LoS) and non-line-of-sight (NLoS) transmissions in the access links between DBSs and ground UEs, coverage probability and area spectral efficiency (ASE) are derived in closed-form expressions based on stochastic geometry analysis. The characterization of both the lower and upper bound performances of DBS networks indicates the performance region of practical DBS network operations. Moreover, our analytical and simulation results in this paper provide guidelines for performance optimization of further DBS networks.

Keywords: DBS networks · Performance analysis
LoS/NLoS transmissions

1 Introduction

All but unheard of until just recently, drones – also known as unmanned aerial vehicles (UAVs) – are now envisioned to shape the future of technology [1]. They are among the best candidates to automate emergency search-and-rescue missions, ease crowd management, and act as relays to provide a cellular coverage

T. Q. Duong and N.-S. Vo (Eds.): INISCOM 2018, LNICST 257, pp. 111–121, 2019.
https://doi.org/10.1007/978-3-030-05873-9_10

extension and an ad-hoc capacity boost. A vast growth in the UAV business is also likely to open attractive vertical markets and new revenue opportunities for both mobile network vendors and operators. However, for these technological and commercial visions to turn into a reality, UAVs will require a reliable control and fast wireless connectivity.

Terrestrial cellular networks are well positioned to provide a communication link towards UAVs flying up to an altitude of few hundred meters [2,3]. However, although connecting UAVs through cellular technologies has key potential advantages – such as reusing existing spectrum resources and network infrastructure – it also involves important challenges [4]. Indeed, UAVs may undergo radio propagation characteristics that are profoundly different from those encountered by conventional ground user equipment (UE). UAVs could be placed in locations considerably above ground, experiencing favourable LoS propagation conditions with a vast number of cells. As a result, a UAV transmitting uplink information could create significant interference to a plurality of neighbouring cells receiving ground transmissions. Conversely, cells communicating with their ground UEs could severely disrupt the downlink of a UAV associated to a neighbouring cell [5]. Similar problems arise in operating UAVs as relays, with the addition of energy consumption and autonomy concerns.

With the aim of integrating UAV communications in cellular networks and address those issues, the third generation partnership project (3GPP) has been gathering key industrial players to collaborate on a work item on enhanced cellular support for aerial vehicles [6]. Such ongoing joint effort has already produced systematic measurements and accurate modelling of UAV-to-cell-tower channels, also defining the various UAV link types along with their respective requirements. Simultaneously, the academic community is providing a large number of analytical studies to assess the potential of using UAVs as mobile base stations (BSs). In this context, remarkable progress has been made in optimising the position and trajectory of these flying relays.

Among those, it is worth highlighting the works in [7] and [8]. In the former, the positions of drone base stations (DBSs) were modelled as a three-dimensional Poisson point process (3D-PPP), and a performance analysis was carried for different DBS heights. In the latter, a dynamic re-positioning DBS algorithm was proposed to increase networks spectral efficiency.

In this paper, to characterise the potential gains of DBS networks, we analyse theoretically – using a stochastic geometry analysis (SGA) – the performance upper bound of the DBS to ground UE access link in sparse, dense and ultra-dense networks[1]. In more detail, this performance upper bound is derived, assuming that DBSs can instantaneously move to the positions directly overhead of the ground UEs (teleportation). This study complements that in [9], where a performance lower bound was derived, considering that DBSs were hovering randomly following a homogeneous Poisson point process (HPPP). Importantly, our stochastic geometry model accounts for *(i)* a realistic 3GPP path loss model

---

[1] Note that the ground BS to DBS backhaul links are considered to be ideal in this paper, and further extensions of this work will study its non-negligible impact.

with both LoS and NLoS transmissions between the DBSs and ground UEs, where a probabilistic function governs the switch between them, and *(ii)* idle modes at the DBSs to switch them off and save energy/mitigate interference when they serve no ground UEs. Using this framework, we derive coverage probability and area spectral efficiency (ASE) expressions, as a function of the DBS and ground UE density. From our analytical and simulation results, the maximum coverage probability and ASE are provided for further investigations on DBS optimization problems.

2 System Model

In this paper, we consider the access link of DBS networks, as shown in Fig. 1. The DBSs are all located at the same height, which is denoted as h_{DBS}. The ground UEs are also all located at the same height, which is denoted as h_{UE}. The absolute antenna height difference between a DBS and a ground UE, i.e., $h_{\mathrm{DBS}} - h_{\mathrm{UE}}$, is denoted by L. In practice, DBSs should not fly too low because of the obvious safety reasons or too high due to the potential performance loss in the backhaul link [9].

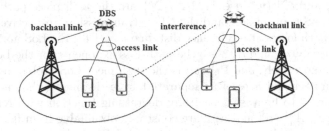

Fig. 1. DBS networks.

The two-dimensional (2D) distance between a DBS and a ground UE is denoted by r. Thus, the 3D distance between a DBS and a ground UE can be expressed as $w = \sqrt{r^2 + L^2}$.

The DBS deployment follows a HPPP distribution with a density λ in an infinite 2D space, while ground UEs are Poisson distributed with a density of λ_{UE} in an infinite 2D space. Note that λ_{UE} may or may not be sufficiently larger than λ, thus there may be DBSs without associated ground UEs in its coverage area.

In practice, a BS will enter an idle mode if there is no UE connected to it, which reduces the interference to UEs in neighbouring cells as well as the energy consumption of the network. As a result, the UE distribution and UE association strategy (UAS) determines the set of active BSs. In this paper, we assume that each ground UE associates with the DBS having the smallest path loss.

Based on the previous considerations and assumptions, the set of active DBSs also follows a HPPP distribution [10], the density of which is denoted by $\tilde{\lambda}$

DBSs/km^2, where $\tilde{\lambda} \leq \lambda$ and $\tilde{\lambda} \leq \lambda_{\text{UE}}$, since one UE is served by at most one DBS. From [10,11], $\tilde{\lambda}$ can be calculated as

$$\tilde{\lambda} = \lambda \left[1 - \frac{1}{(1 + \frac{\lambda_{\text{UE}}}{q\lambda})^q} \right],$$ (1)

where according to [11], q also depends on the path loss model.

To derive a performance upper bound of the access link, we can assume that there is always a typical DBS above the head of the typical UE. In this case, we have $r = 0$ and $w = L$. Thus, the DBS and ground UE point processes are identical in the 2D domain, except for height difference of L. This concept is further developed in the next section.

As for the path loss model, in this paper, we use a very general and practical one, in which the path loss $\zeta(w)$ associated with distance w can be segmented into N pieces, and each piece $\zeta_n(w), n \in \{1, 2, \ldots, N\}$ is modelled as

$$\zeta_n(w) = \begin{cases} \zeta_n^{\text{L}}(w) = A_n^{\text{L}} w^{-\alpha_n^{\text{L}}}, & \text{LoS: } \text{Pr}_n^{\text{L}}(w) \\ \zeta_n^{\text{NL}}(w) = A_n^{\text{NL}} w^{-\alpha_n^{\text{NL}}}, & \text{NLoS: } 1 - \text{Pr}_n^{\text{L}}(w) \end{cases},$$ (2)

where $\zeta_n^{\text{L}}(w)$ and $\zeta_n^{\text{NL}}(w), n \in \{1, 2, \ldots, N\}$ are the n-th piece path loss functions for the LoS transmission and the NLoS transmission, respectively, A_n^{L} and A_n^{NL} are the path losses at a reference distance $w = 1$ for the LoS and the NLoS cases, respectively, α_n^{L} and α_n^{NL} are the path loss exponents for the LoS and the NLoS cases, respectively, and $\text{Pr}_n^{\text{L}}(w)$ is the n-th piece LoS probability function that a transmitter and a receiver separated by a distance w has a LoS path, which is assumed to be a monotonically decreasing function with regard to w. In practice, A_n^{L}, A_n^{NL}, α_n^{L} and α_n^{NL} are constants obtainable from field tests.

Moreover, the multi-path fading between a DBS and a UE is modelled as independently identical distributed (i.i.d.) Rayleigh fading in this paper. Results with Rician fading are left for the journal version of this work.

3   Main Results

Using a 3D SGA based on the HPPP theory, we study the performance of a DBS network by considering the performance of a typical UE located at the origin o.

We first investigate the coverage probability, which is defined as the probability that this UE's signal-to-interference-plus-noise ratio (SINR) is above a per-designated threshold γ:

$$p^{\text{cov}}(\lambda, \gamma) = \text{Pr}[\text{SINR} > \gamma],$$ (3)

where the SINR is calculated as

$$\text{SINR} = \frac{P\zeta(w)h}{I_{\text{agg}} + N_0},$$ (4)

where h is the channel gain, modelled as an exponential random variable (RV) with the mean of one (Rayleigh fading), P and N_0 are the transmission power of each DBS and the additive white Gaussian noise (AWGN) power at each UE, respectively, and I_{agg} is the cumulative interference given by

$$I_{\mathrm{agg}} = \sum_{i:\, b_i \in \Phi \backslash b_o} P\beta_i g_i, \qquad (5)$$

where b_o is the BS serving the typical UE located at distance w from the typical UE, and b_i, β_i and g_i are the i-th interfering BS, the path loss associated with b_i and the multi-path fading channel gain associated with b_i, respectively.

When DBSs are deterministically hovering right on top of the UEs, the access link performance reaches an upper bound, as DBSs are as close to the UEs as they could be. The realisation of such mobile DBSs hovering just above the UEs is difficult, as further considerations on DBSs mobility control management are needed. In this paper, for simplicity, we assume a DBS teleportation model, where DBSs can instantaneously move to the positions just above the UEs, allowing us to derive the upper bound performance of the access links.

Based on the existing expression of $p_{\mathrm{low}}^{\mathrm{cov}}(\lambda, \gamma)$ [12,13], i.e., the lower bound coverage probability, we present our main result on $p_{\mathrm{up}}^{\mathrm{cov}}(\lambda, \gamma)$, i.e., the upper bound coverage probability, in Theorem 1.

Theorem 1.

$$p_{\mathrm{up}}^{\mathrm{cov}}(\lambda, \gamma) = \Pr\left[\frac{P\zeta_n^{\mathrm{L}}(L) h}{I_{\mathrm{agg}} + N_0} > \gamma \right], \qquad (6)$$

where

$$\Pr\left[\frac{P\zeta_n^{\mathrm{L}}(L) h}{I_{\mathrm{agg}} + N_0} > \gamma \right] = \exp\left(-\frac{\gamma N_0}{P\zeta_n^{\mathrm{L}}(L)} \right) \mathscr{L}_{I_{\mathrm{agg}}}^{\mathrm{L}}(s), \qquad (7)$$

where $s = \frac{\gamma}{P\zeta_n^{\mathrm{L}}(L)}$, *and* $\mathscr{L}_{I_{\mathrm{agg}}}^{\mathrm{L}}(s)$ *is the Laplace transform of* I_{agg} *for LoS signal transmission evaluated at* s, *which can be further written as*

$$\mathscr{L}_{I_{\mathrm{agg}}}^{\mathrm{L}}(s) = \exp\left(-2\pi\tilde{\lambda} \int_0^{+\infty} \frac{\Pr^{\mathrm{L}}\left(\sqrt{u^2 + L^2}\right) u}{1 + \left(sP\zeta^{\mathrm{L}}\left(\sqrt{u^2 + L^2}\right)\right)^{-1}} du \right)$$

$$\times \exp\left(-2\pi\tilde{\lambda} \int_0^{+\infty} \frac{\left[1 - \Pr^{\mathrm{L}}\left(\sqrt{u^2 + L^2}\right)\right] u}{1 + \left(sP\zeta^{\mathrm{NL}}\left(\sqrt{u^2 + L^2}\right)\right)^{-1}} du \right), \qquad (8)$$

Proof. As we can see in Theorem 1 of [12], the calculation of the coverage probability is accumulated by components of the coverage probability for the case when the signal comes from the n-th piece LoS path and the n-th piece NLoS path between the typical DBS and the typical UE, respectively.

To derive a performance upper bound of the access links, we can assume that there will always be a DBS located right on top of the typical UE. Under such condition of $r = 0$, i.e., $w = L$, $r_1 = 0$, and $\Pr_1^{\mathrm{L}}(w) = 1$, we can obtain Theorem 1 from Theorem 1 in [12].

According to [12,13], we also investigate the ASE in bps/Hz/km^2, which can be computed as

$$A^{\text{ASE}}(\lambda, \gamma_0) = \tilde{\lambda} \int_{\gamma_0}^{+\infty} \log_2(1+\gamma) f_\Gamma(\lambda, \gamma) \, d\gamma, \tag{9}$$

where γ_0 is the minimum working SINR for the considered DBS, and $f_\Gamma(\lambda, \gamma)$ is the probability density function (PDF) of the SINR observed at the typical UE for a particular value of λ.

In the following and to finish this section, we present three concepts that are key to understand the performance behaviour of the studied network, i.e., coverage probability and ASE, which are functions of the DBS density [12–14]. Although originally named after the ASE, these concepts are highly related and apply to the coverage probability explanations too.

■ The ASE Crawl

A much shorter distance between a UE and its serving DBS in ultra-dense networks implies high probabilities of strong LoS transmissions. Generally speaking, LoS transmissions are helpful to improve the signal power, but they aggravate the interference too. Thus, the ASE will suffer from a slow growth or even a decrease when the DBS density is sufficiently large, and the stronger interference paths transition from NLoS to LoS. This performance behaviour is referred to as the ASE Crawl [12].

■ The ASE Crach

The existence of a non-zero antenna height difference between UEs and DBSs leads to a non-zero cap on the minimum distance between them, and thus a cap on the signal power strength. Although each inter-cell interference power strength is subject to the same cap, the aggregated inter-cell interference power will overwhelm the signal power in an ultra-dense network due to the sheer number of strong interferers. Thus, the ASE will suffer from a significant loss when the DBS density is sufficiently large. This performance behaviour is referred to as the ASE Crash [14].

■ The ASE Take-off

When the number of DBSs is larger than that of UEs, the surplus of DBSs encourages idle mode operations to mitigate unnecessary inter-cell interference and reduce energy consumption. Consequently, the SINR performance benefits from *(i)* a DBS diversity gain in UEs selecting a good serving DBS, and *(ii)* a decreased inter-cell interference, which is bounded by the active UE density. As a result, the signal power continues increasing with the network densification, while the interference power reduces or remains at a constant level due to the idle mode capability. This performance behaviour is referred to as the ASE Take-off [13].

4   Simulation Results

In this section, we use simulation results to verify the accuracy of our analysis. It is important to note that there are no specific system model recommendations available for a DBS network. Fortunately, however, models for the channel between the terrestrial BSs and the ground UEs are provided for different scenarios [15], which can be reused in our simulation, treating terrestrial BSs as DBSs. Specifically, and according to 3GPP [15], we reuse the urban macro (UMa) and the urban micro (UMi) terrestrial BS channel models, adopting the parameters shown in Table 1.

Table 1. Parameters.

	UMa model	UMi model
h_{DBS}	25 m	10 m
α_n^L	2.2	2.1
α_n^{NL}	3.908	3.53
A_n^L	$10^{-2.8-2\log_{10}f_c}$	$10^{-3.24-2\log_{10}f_c}$
A_n^{NL}	$10^{-1.354-2\log_{10}f_c}$	$10^{-2.24-2.13\log_{10}f_c}$
P	46 dBm	41 dBm
h_{UE}	1.5 m	1.5 m

Moreover, we adopt the following parameter values: $N_0 = -95\,\mathrm{dBm}$, $q = 3.5$, $\lambda_{\mathrm{UE}} = 300\,\mathrm{UEs/km}^2$, $\gamma = 0\,\mathrm{dB}$ and $f_c = 2\,\mathrm{GHz}$. The LoS probability function for both models is given by [15] and shown as below:

$$\mathrm{Pr}^L\left(\sqrt{r^2+L^2}\right) = \begin{cases} 100\%, & r \leq 18\,\mathrm{m} \\ (18/r + \exp(-r/63)*(1-18/r)), & r > 18\,\mathrm{m} \end{cases}, \quad (10)$$

where r is the 2D distance, and $L = h_{\mathrm{DBS}} - h_{\mathrm{UE}}$. Note that $\mathrm{Pr}^L\left(\sqrt{r^2+L^2}\right) = 1$ in the teleportation mode, since $r = 0$.

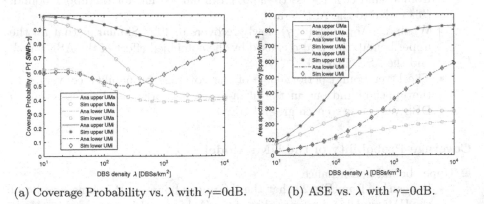

(a) Coverage Probability vs. λ with γ=0dB.   (b) ASE vs. λ with γ=0dB.

Fig. 2. Networks performance

4.1   Coverage Probability Analysis

As can be seen from Fig. 2a, we analyse $p_{\text{up}}^{\text{cov}}(\lambda, \gamma)$ and $p_{\text{low}}^{\text{cov}}(\lambda, \gamma)$ for the UMa and UMi models, respectively. It is important to note that the result of $p_{\text{up}}^{\text{cov}}(\lambda, \gamma)$, given by Theorem 1, perfectly match the simulation results, which validates the accuracy of our analysis.

Coverage Probability of the UMa Model

■ Upper bound performance
  - At low DBS densities, the probability of coverage, $p_{\text{up}}^{\text{cov}}(\lambda, \gamma)$, is high, around 97%. This is due to the high signal power, provided by the DBS overhead the typical UE, and the low interference power, common in sparse networks. The majority of interference links are NLoS ones.
  - As λ increases, the distances between DBSs become smaller and thus the inter-cell interference increases, and as a result, $p_{\text{up}}^{\text{cov}}(\lambda, \gamma)$ monotonically decreases. The major decrease happens when $\lambda \approx 100\,\text{DBSs/km}^2$. This is due to the ASE Crawl [12], i.e., a large number of interference transit from NLoS to LoS.
  - When $\lambda > \lambda_{\text{UE}}$, $p_{\text{up}}^{\text{cov}}(\lambda, \gamma)$ decreases at a much slower rate due to the combined effect of the ASE Crash and the ASE Take-off [13,14], with the former effect being stronger. Although the antenna height difference will lead to the severe loss of coverage probability in the ultra-dense network, the idle mode at BSs effectively mitigates the inter-cell interference.
■ Lower bound performance
  - The result of $p_{\text{low}}^{\text{cov}}(\lambda, \gamma)$ is obtained from [14], where it was assumed that the DBSs are randomly deployed.
  - When λ is around $30\,\text{DBSs/km}^2$, $p_{\text{low}}^{\text{cov}}(\lambda, \gamma)$ moderately increases due to the enhancement of LoS signal power.
  - When $\lambda \in [30, 300]\,\text{DBSs/km}^2$, $p_{\text{low}}^{\text{cov}}(\lambda, \gamma)$ suffers from a significant loss, from around 60% to less than 40%. Similar as that for the upper bound, this is due to the ASE Crawl.
  - When $\lambda > \lambda_{\text{UE}}$, $p_{\text{low}}^{\text{cov}}(\lambda, \gamma)$ slowly recovers to 40%. Similar as that for the upper bound, this is also caused by the combined effect of the ASE Crash and the ASE Take-off.
  - The lower bound performance of the coverage probability converges to the upper bound one in the ultra-dense networks, as there is always a DBS very close to each ground UE.

Coverage Probability of the UMi Model

■ Upper bound performance
  - We can see from Fig. 2a that the the upper bound performances trend of the UMi model is similar to that of the UMa one. However, it is important to note that $p_{\text{up}}^{\text{cov}}(\lambda, \gamma)$ of the UMi model is always better than that of the UMa one, since a lower antenna height provides a stronger signal power and postpones the ASE Crash [14].

■ Lower bound performance
  • The lower bound performance trend of the UMi model is also similar to the UMa one. Due to the enhancement of LoS signal link, $p_{\text{low}}^{\text{cov}}(\lambda, \gamma)$ also slowly increases when $\lambda \in [10, 30]$DBSs/km^2. Then, $p_{\text{low}}^{\text{cov}}(\lambda, \gamma)$ first decreases due to the ASE Crawl and ASE Crash, and then increases due to the ASE Take-off. The last increase is more significant as the ASE Crash effect is weaker due to the lower antenna height in the UMi model. Moreover, note that $p_{\text{low}}^{\text{cov}}(\lambda, \gamma)$ of the UMi model is less than $p_{\text{low}}^{\text{cov}}(\lambda, \gamma)$ of the UMa one until $\lambda > 40$ DBSs/km^2. After that, it is the other way around. This is due to lower transmission power and different LoS probability functions, as shown in Table 1.

4.2 Area Spectral Efficiency Analysis

In Fig. 2b, we present the results of the ASE performance for both the UMa and UMi model, respectively. As we can see from Fig. 2b, the analytical results and simulation results are perfectly matched.

ASE of the UMa Model

■ Upper bound performance
  • It can be seen from Fig. 2b that the ASE is about 80 bps/Hz/km^2 when the network is sparse, and the ASE keeps increasing to the maximum value near 300 bps/Hz/km^2 when λ is about 300 DBSs/km^2 due to the larger spectrum spatial reuse.
  • The ASE upper bound decreases its growth rate when $\lambda > 300$ DBSs/km^2 due to the degraded coverage probability as a result of the combined effect of the ASE Crash and the ASE Take-off.
■ Lower bound performance
  • The ASE is around 10 bps/Hz/km^2 when the network is sparse, and then it keeps monotonically increasing with the DBS density, reaching around 200 bps/Hz/km^2 when λ is around 10^4 DBSs/km^2. There is no decrease as in the upper bound, because the coverage probability remains almost constant with the BS density in this case, and the spatial reuse dominates the ASE performance. The gap to the upper bound performance in this ultra-dense network is about 90 bps/Hz/km^2.

ASE of the UMi Model

■ Upper bound performance
  • As shown in Fig. 2b, the upper bound performance of ASE for the UMi model is always better than that of the UMa model due to the superior performance of coverage probability, as shown in Fig. 2a.
  • Note that compared with the results for the UMa model, the upper bound performance of ASE for the UMi model grows faster for all DBS densities. This is due to the smaller antenna height difference, which leads to the stronger signal power and postpones the ASE Crash.

■ Lower bound performance
  • The lower bound performance of ASE for the UMi model shows a similar trend compared with that for the UMa model when $\lambda < 100\,\mathrm{DBSs/km^2}$. But after that, it exceeds the UMa lower bound, which is also caused by different coverage probability performance of the two models.

5 Conclusion

In this paper, we derive an upper bound for the performance of DBS networks, by assuming DBSs can move instantaneously over the serving UEs' head (teleportation mode). Both coverage probability and ASE are derived using a practical channel model adopted by the 3GPP. Numerical results characterize the theoretical performance limit that can be achieved by future DBS networks, with various DBS densities, heights and DBS trajectory optimizations.

References

1. New America: Drones and aerial observation: new technologies for property rights, human rights, and global development - a primer, July 2015
2. Mozaffari, M., Saad, W., Bennis, M., Debbah, M.: Communications and control for wireless drone-based antenna array. arXiv:1712.10291, December 2017
3. Lyu, J., Rui Zhang, Y.Z.: UAV-aided offloading for cellular hotspot. IEEE Trans. Wirel. Commun. **17**, 3988–4001 (2018)
4. Geraci, G., Garcia Rodriguez, A., Giordano, L.G., López-Pérez, D., Björnson, E.: Understanding UAV cellular communications: from existing networks to massive MIMO. arXiv:1804.08489, April 2018
5. Geraci, G., Garcia Rodriguez, A., Giordano, L.G., López-Pérez, D., Björnson, E.: Supporting UAV cellular communications through massive MIMO. In: Proceedings of the IEEE ICC Workshops, May 2018. arXiv:1802.01527 (2018, to appear)
6. 3GPP Technical Report 36.777: Technical specification group radio access network; Study on enhanced LTE support for aerial vehicles (Release 15), December 2017
7. Zhang, C., Zhang, W.: Spectrum sharing for drone networks. IEEE J. Sel. Areas Commun. **35**(1), 136–144 (2017)
8. Fotouhi, A., Ding, M., Hassan, M.: Dynamic base station repositioning to improve spectral efficiency of drone small cells. In: IEEE 18th International Symposium on A World of Wireless. Mobile and Multimedia Networks (WoWMoM), pp. 1–9 (2017)
9. Ding, M., López-Pérez., D.: Please lower small cell antenna heights in 5G. In: IEEE Global Communications Conference (GLOBECOM), pp. 1–6 (2016)
10. Mozaffari, M., Saad, W., Bennis, M., Debbah, M.: Drone small cells in the clouds: design, deployment and performance analysis. In: 2015 IEEE Global Communications Conference (GLOBECOM), pp. 1–6 (2015)
11. Madhusudhanan, P., Restrepo, J.G., Liu, Y., Brown, T.X., Baker, K.R.: Downlink performance analysis for a generalized Shotgun cellular system. IEEE Trans. Wireless Commun. **13**(12), 6684–6696 (2014)
12. Ding, M., Wang, P., López-Pérez, D., Mao, G., Lin, Z.: Performance impact of LoS and NLoS transmissions in dense cellular networks. IEEE Trans. Wirel. Commun. **15**(3), 2365–2380 (2016)

13. Ding, M., López-Pérez, D., Mao, G., Lin, Z.: Performance impact of idle mode capability on dense small cell networks. IEEE Trans. Veh. Technol. **66**(11), 10446–10460 (2017)
14. Ding, M., López-Pérez, D.: Performance impact of base station antenna heights in dense cellular networks. IEEE Trans. Wirel. Commun. **16**(12), 8147–8161 (2017)
15. 3GPP: TR 38.901(v14.0.0) Study on channel model for frequencies from 0.5 to 100 GHz (2017)

Development of the Rules for Transformation of UML Sequence Diagrams into Queueing Petri Nets

Vu Van Doc, Huynh Quyet Thang[(⊠)] [ID], and Nguyen Trong Bach

Hanoi University of Science and Technology,
No. 1, Dai Co Viet, Hanoi, Vietnam
vvdoc@uneti.edu.vn, thanghq@soict.hust.edu.vn,
bachnt9x@gmail.com

Abstract. Sequence diagrams (SDs) are an abstraction of communication modeling between different entities, objects or classes. SDs are used to describe an execution trace of a particular system at a particular point in time. Queueing Petri Nets (QPNs) are graphical formalisms, at a lower level of abstraction, for which efficient and mature simulation-based solution techniques are available. This paper defines and explains the relationship between sequence diagrams and QPNs. Our approach can be used to transform sequence diagrams into QPNs. We presented the development of the model transformation solution to translate UML Sequence Diagrams (SDs) into equivalent QPNs. A case study of a new account opening for banking management system is used to illustrate the transformation rules.

Keywords: Model transformation · Sequence Diagrams · Queueing Petri Nets

1 Introduction

An ordinary Petri Net is a mathematical modeling language for systems description. Petri Net consists of a set of places and a set of transitions. The places contain a certain number of tokens. At the start of the analysis, the number of tokens is determined by an initial marking function. The forward incidence function defines how many tokens a transition requires in each connected place to be ready to fire. When a transition fires, it deducts that number of tokens from each incoming place and deposits new tokens in other places if a backward incidence function is defined [8].

Colored Petri Net uses token colors to distinguish different token classes. The initial marking and incidence functions are now defined with respect to multiple different colors. The different possibilities of firing a transition are referred to as modes. Modes also have firing weights that influence which mode is chosen when multiple modes are ready to fire [8]. Generalized Stochastic Petri Net introduces timed and immediate transitions. Queueing Petri Nets (QPNs) are built on Colored, a combination of Colored Petri Nets and Generalized Stochastic Petri Nets [8]. Queueing Networks and Generalized Stochastic Petri Nets are among the most popular modelling formalisms that have been used in the past decade, but both have some serious short-comings [8].

T. Q. Duong and N.-S. Vo (Eds.): INISCOM 2018, LNICST 257, pp. 122–144, 2019.
https://doi.org/10.1007/978-3-030-05873-9_11

Queueing Networks provide a powerful mechanism for modelling resource contention and scheduling strategies. However, they are not as suitable for representing the blocking and synchronization of processes. Similarly, Generalized Stochastic Petri Nets are efficient for modelling blocking and synchronization aspects but they have drawbacks in representing scheduling strategies [8]. QPNs introduce a new type of place: the queueing place. A queueing place includes a queue, a server, and a depository. The server processes the tokens in the queue according to a certain scheduling strategy. The time a token occupies the server is defined through a statistical distribution. When the time of a token finished, the token is put into the depository, which then behaves like an immediate place for connected transitions. Only tokens in the depository are considered as available for the incidence function [8]. QPNs combines Queueing Networks and Petri Nets into a single formalism and eliminates the above disadvantages. QPNs allow queues to be integrated into places of Petri Nets. This enables easily representing of scheduling strategies and brings the benefits of Queueing Networks into the world of Petri Nets.

Sequence Diagram is a type of UML Interaction Diagram [1]. SDs are two-dimensional charts where the vertical axis represents the time and the horizontal axis represents interaction. Sequence Diagrams are commonly used to depict the flow of information in a system through communication between objects.

Model transformations from SDs into Queueing Petri Nets support mediate the discrepancies between different models by making them consistent with each other while supporting model simulation Queueing Petri Nets.

This paper makes the following contributions: (i) A formal mapping from SDs to QPNs in the form of transformation rules; (ii) Implementation of transformation from SDs to QPNs software tool and analyzing a typical case study.

The remainder of the paper is organized as follows: Sect. 2 introduces related works; Sect. 3 explains the transformation approach method with a detailed explanation for each SDs to QPNs; In Sect. 4, we present a typical case study scenario illustrating the trans-formation; Sect. 5 concludes the work and discuss our directions of future works.

2 Related Work

Some recent researches discuss the transforming SDs (scenario-based model) into Petri Nets (state-based models). In Ref. [2] the authors analyze the advantage and shortage of UML and Petri net in complex software modeling and propose translation rules from UML to Petri net. In Ref. [7] the author proposes model transformation from Sequence Diagrams to Petri Nets and an approach using Labeled Event Structures (LES), as well as methods to translate both Sequence Diagrams and Petri Nets into LES.

The author also develops an application of model transformation from UML Sequence Diagrams to Petri Nets, named SD2PN. In Ref. [1] the author proposes an effective method for transforming SDs into Petri nets is presented. Other works [3, 6, 11, 12] the authors introduce a solution for transformation most relevant features of UML Sequence Diagrams for modelling distributed systems into equivalent Coloured Petri Nets that accepts the same execution traces (event sequences) as the original models. UML Sequence Diagrams created with the Papyrus visual modelling tool are converted to Coloured Petri Nets executable with CPN Tools. The transformation rules are implemented in ETL (Epsilon Transformation Language). The authors define and describe the transformation from UML2 sequence diagrams to Coloured Petri nets and prove the syntactic and semantic correctness of the transformation [4, 6]. They also present SD2CPN, a scenario-based model transformation tool with analysis capabilities. Paper [8] contributes a formal mapping from PCM to QPN models, implemented by means of an automated model-to-model transformation as part of a new PCM solution method based on simulation of QPNs. Other authors proposed IMPACT [10], a plug-in tool that makes use of the modeling capabilities of the Papyrus tool and the relational QVT model transformation implementation of mediniQVT to produce an LQN performance model of a MARTE annotated UML design. Following the approach for model transforming from SDs into Petri Nets and CPNs, we develop the rules to transform each of SDs to QPNs. Detailed information is presented in Sect. 3. Based on [14], we develop transformation rules for 7 groups of SDs: Interaction and Interaction, Lifeline, Occurrence and Execution, Workload, Message, General Ordering, and Loops.

3   Transformation Approach and Developed Rules

In this section, we show how to transform all of the components available in the UML 2.5 SDs, into a behaviorally equivalent QPN. To accomplish this, we explain the semantics of the component. We describe how the transformation is achieved and, show the result of applying these ideas to illustrative components. The nodes and edges are typically drawn in a UML sequence diagram include lifeline, execution specification, message, combined fragment, interaction use, general ordering [12]. We added the Workload component, which is not a part of SDs. So, we can build the transformation rules from SDs to QPN for these 7 different groups. The description of each rule is presented in the following sections, which includes a graphical representation of the rule and a textual description as well as additional conditions.

3.1   Interaction and Interaction Use

Each interaction in a SD is converted into a QPN space. An interaction use used in a SD is mapped to a subnet place (s-place) in the QPN space. Figures 1 and 2 shows the interaction and interaction use transformation rules respectively.

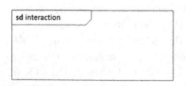 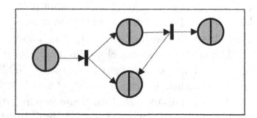

Fig. 1. Interaction transformation rule

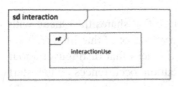

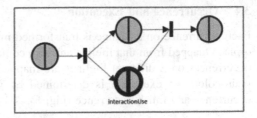

Fig. 2. Interaction use transformation rule

3.2 Lifeline

Each lifeline in an interaction is converted into a queueing place (q-place) in the QPN space. This q-place contains a main-color (characteristic color) represents for a lifeline. Figure 3 denotes the lifeline transformation rule.

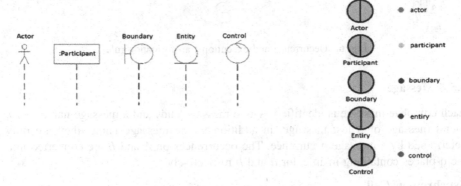

Fig. 3. Lifeline transformation rule

3.3 Workload

The workload isn't a part of SD, however, it is necessary for our method to ensure that the transformed QPN is live. In this section, we discuss two types of workload: Open Workload and Closed Workload.

Open Workload specified by a distribution of the arrival speed of the request, which shows the duration between the two continuous requests (Figs. 4 and 5).

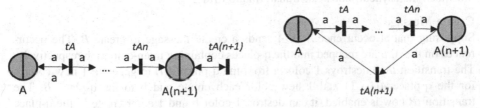

Fig. 4. Open workload rule   **Fig. 5.** Closed workload rule

3.4   Occurrence and Execution

Each occurrence on a lifeline is transformed into a q-place that shares the Queue with q-place mapped from that lifeline and also contains the main-color of that lifeline. Two occurrences on 2 different lifelines are mapped into 2 q-places that distinguish about main-color. An execution is determined by two execution occurrences – the start occurrence and finish occurrence (Fig. 6).

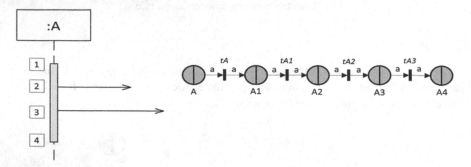

Fig. 6.   Occurrence and execution transformation rule

3.5   Message

Each complete message is identified by two message ends and a message name. For a found message or a lost message, in addition to the message name, they are only determined by a message occurrence. The occurrences on A and B are converted into the q-places containing main-color a and b respectively.

Synchronous Call

Supposing that at occurrence $A(i)$, A sends a synchronous call m to B. The message end of m on B is $B(j)$. Figure 7 illustrates the synchronous call and reply message transformation rules. After a period of time greater than 0, B moves from occurrence $B(j)$ to occurrence $B(k)$ where B sends a reply message r to A. The message end of r on A is $A(i+1)$.

Asynchronous Call

Supposing that at occurrence $A(i)$, A sends an asynchronous call $m1$ to B. The message end of $m1$ on B is $B(j)$. A does not wait for any reply messages, at occurrence $A(i+1)$, A can send a synchronous call $m2$ to C. The message end of $m2$ on C is $C(k)$. Figure 8 illustrates the asynchronous call transformation rule.

Create Message

Supposing that at occurrence $A(i)$, A sends a create message to create B. The occurrences on the A, B are mapped into the q-places containing main-color a, b respectively. The transition $tA(i)$ destroys 1 color a from the q-place $A(i)$ to generate 1 new color a for the q-place $A(i+1)$ and 1 new color each create and b to the q-place B. The transition tB now is enabled, it can destroy 1 color b and 1 color create in the q-place

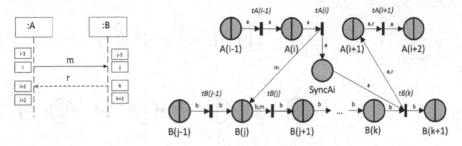

Fig. 7. Synchronous call transformation rule

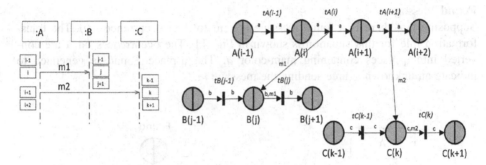

Fig. 8. Asynchronous call transformation rule

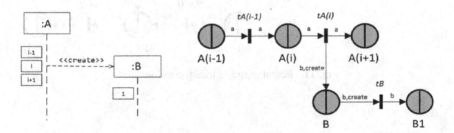

Fig. 9. Create message transformation rule

B to generate 1 new color b for the q-place *B1*. Figure 9 shows the create message transformation rule.

Delete Message

Supposing that at occurrence *A(i)*, *A* sends a delete message to destroy *B*. One of the message end of this delete message is in occurrence *B(k)* on *B*. Figure 10 shows the delete message transformation rule.

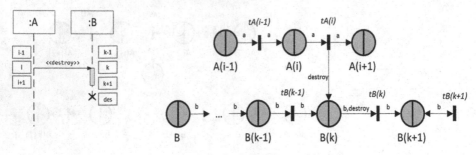

Fig. 10. Delete message transformation rule

Found Message

Supposing that there is a found message *m* sent to *A* at occurrence *A(i)*. The transformation rule for this situation is shown in Fig. 11. The occurrences on *A* are converted into q-places containing main-color *a*. The q-place Found is generated to indicate an unknown source sending the message *m*.

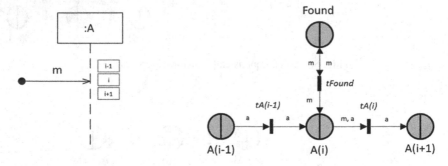

Fig. 11. Found message transformation rule

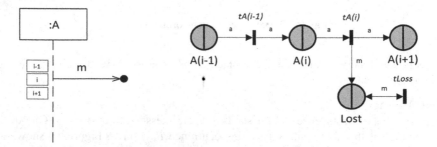

Fig. 12. Lost message transformation rule

Lost Message

Supposing that at occurrence $A(i)$, A sends a lost message m. The transformation rule for this case is shown in Fig. 12.

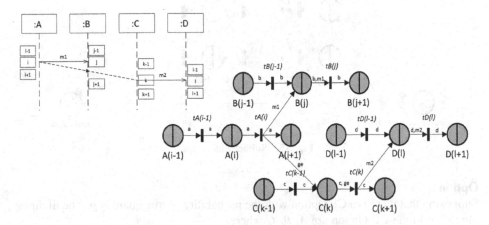

Fig. 13. General ordering transformation rule

3.6 General Ordering

Supposing that at occurrence $A(i)$, A sends a message $m1$ to B with the message end at occurrence $B(j)$ on B. At the occurrence $C(k)$, C sends a message $m2$ to D with the message end at occurrence $D(l)$ on D. The occurrence $A(i)$ connects under general ordering with the occurrence $C(k)$ to indicates that $A(i)$ occurs before $C(k)$.

Figure 13 shows the general ordering transformation rule. The occurrences on the A, B, C, and D are converted into the q-places containing the main-color a, b, c, and d respectively.

3.7 Combined Fragment

In the Combined Fragment (CF) of SD, there are one or more guards that are an expression that returns the Boolean value. We only focus on the probability of the event where "the guard value is true" (true guard) (Fig. 14).

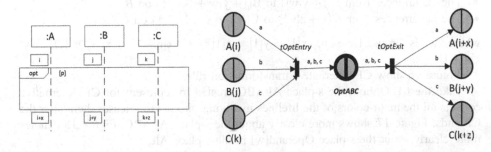

Fig. 14. CF Option transformation rule

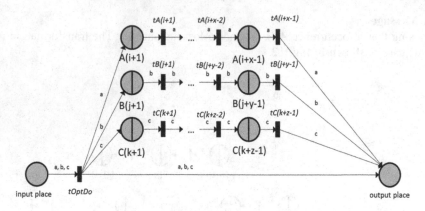

Fig. 15. Subnet place Opt

Option
Supposing that there is a CF Option with the probability of true guard is p. The lifelines involving in the CF Option are A, B, C where:

– A joins in the CF Option from the occurrence $A(i+1)$ to $A(i+x-1)$.
– B joins in the CF Option from the occurrence $B(j+1)$ to $B(j+y-1)$.
– C joins in the CF Option from the occurrence $C(k+1)$ to $C(k+z-1)$.

The s-place OptABC generated to represent for the CF Option will collect all of the main-color of the lifelines involving into it and release them after the CF ends. Figure 15 shows more clearly about the s-place OptABC.

Alternative
Supposing that there is a CF Alternative including n operands with the probabilities of true guards (from top to bottom) are $p1, p2, \ldots, pn$, where: $p1 + p2 + \ldots + pn = 1$.
The lifelines joining in this CF Alternative are A, B, C.

– A joins in the CF Alternative from the occurrence $A(i+1)$ to $A(i+x-1)$.
– B joins in the CF Alternative from the occurrence $B(j+1)$ to $B(j+y-1)$.
– C joins in the CF Alternative from the occurrence $C(k+1)$ to $C(k+z-1)$.
– Operand[w] in the CF Alternative has:
  • The occurrences from $A(i+x[w])$ to $A(i+x[w+1]-1)$ on A
  • The occurrences from $B(j+y[w])$ to $B(j+y[w+1]-1)$ on B
  • The occurrences from $C(k+z[w])$ to $C(k+z[w+1]-1)$ on C

where w is from 1 to n, $x[1] = y[1] = z[1] = 1$, $x[n+1] = x$, $y[n+1] = y$, $z[n+1] = z$.
Figure 16 show CF Alternative transformation rule.

Like the CF Option, the s-place AltABC created to represent for CF Alternative collects all the main-colors of the lifelines involving into it and releases them after the CF ends. Figure 17 shows more clearly about the s-place AltABC. Figure 18 denotes more clearly about the s-place Operand[w] in the s-place Alt.

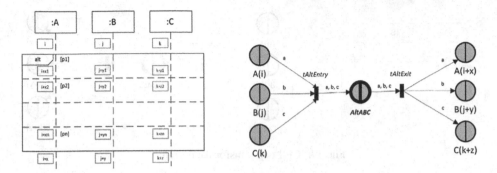

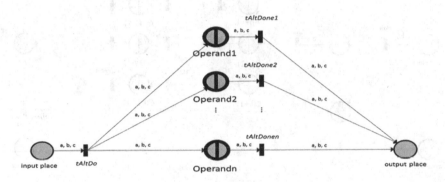

Fig. 16. CF Alternative transformation rule

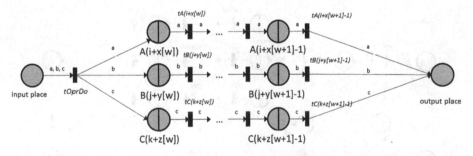

Fig. 17. Subnet place Alt

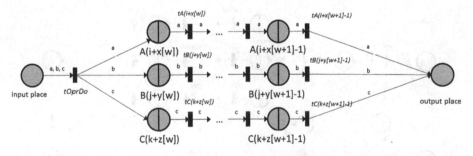

Fig. 18. Subnet place Operand[w]

Loop

There are many options for CF Loop related to bounds and guard, however we restrict that the number of iteration in CF Loop is fixed. Suppose that there is a CF Loop (n) with the probability of true guard is p. Lifelines involving in this CF Loop are A, B, C. The s-place LoopABC generated to represent the CF loop will collect all the main-colors of the

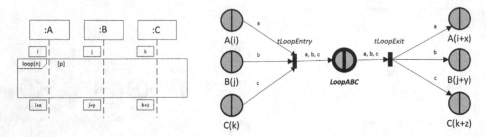

Fig. 19. CF Loop transformation rule

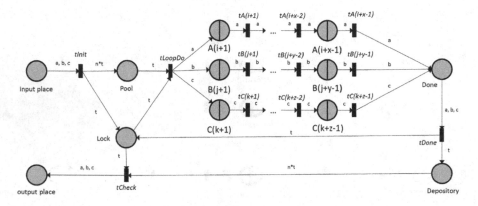

Fig. 20. Subnet place loop with p = 1

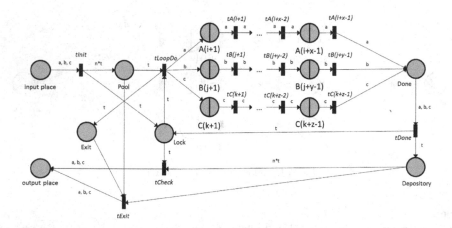

Fig. 21. Subnet place loop with p < 1

lifelines involving into it and releases them after the CF ends. Figure 20 shows the s-place LoopABC when p = 1 and Fig. 21 shows the s-place LoopABC when p < 1 (Fig. 19).

Break

Supposing that there is a CF Loop (n) with the probability of true guard is p. Lifelines involving in this CF Loop are A, B, C.

- A joins in the CF Loop from the occurrence $A(i+1)$ to $A(i+x-1)$.
- B joins in the CF Loop from the occurrence $B(j+1)$ to $B(j+y-1)$.
- C joins in the CF Loop from the occurrence $C(k+1)$ to $C(k+z-1)$.

This CF loop has 1 CF Break with the probability of true guard is q, and may have:

- The occurrences from $A(i+x[2])$ to $A(i+x[3]-1)$ on A.
- The occurrences from $B(j+y[2])$ to $B(j+y[3]-1)$ on B.
- The occurrences from $C(k+z[2])$ to $C(k+z[3]-1)$ on C.

Figure 22 shows the mapping of CF Loop in this case. The o-place Break is generated to collect all main-color of the lifelines involving in the CF Loop, marking the state before deciding whether to go to the CF Break or not (Fig. 23).

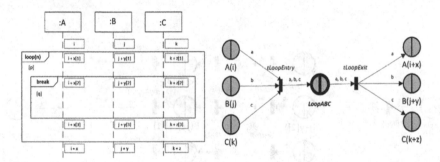

Fig. 22. CF Loop contains CF Break transformation rule

Figure 24 illustrates more detail about the s-place *Break*.

Parallel

Supposing that there is a CF Parallel including n operands. The lifelines involving in this CF Parallel are A, B, C in which:

- A joins in the CF Parallel from the occurrence $A(i+1)$ to $A(i+x-1)$.
- B joins in the CF Parallel from the occurrence $B(j+1)$ to $B(j+y-1)$.
- C joins in the CF Parallel from the occurrence $C(k+1)$ to $C(k+z-1)$.
- Operand[w] in the CF Parallel has:
- The occurrences from $A(i+x[w])$ to $A(i+x[w+1]-1)$ on A
- The occurrences from $B(j+y[w])$ to $B(j+y[w+1]-1)$ on B
- The occurrences from $C(k+z[w])$ to $C(k+z[w+1]-1)$ on C where w is from 1 to n, $x[1]=y[1]=z[1]=1$, $x[n+1]=x$, $y[n+1]=y$, $z[n+1]=z$.

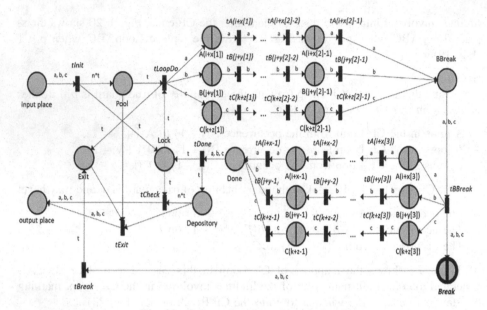

Fig. 23. Subnet place loop (contain break)

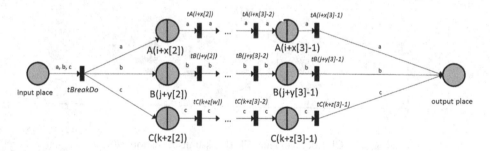

Fig. 24. Subnet place break

Figure 25 denotes the CF Parallel transformation rule.

The s-place ParABC created to represent for the CF Parallel will collect all main-color of the lifelines involving in it and releases them after the CF ends. Figure 26 shows more clearly about the s-place ParABC.

Strict Sequencing

Supposing that there is a CF Strict Sequencing consisting of n operands. The lifelines involving in CF Strict Sequencing are A, B, C. Figure 27 show the CF Strict Sequencing transformation rule.

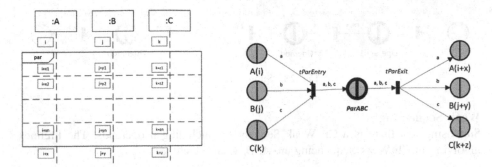

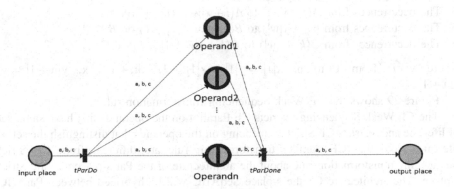

Fig. 25. CF Parallel transformation rule

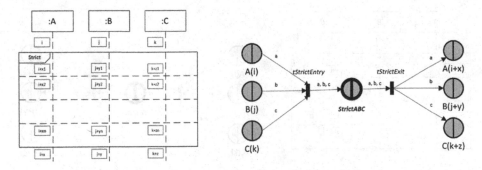

Fig. 26. Subnet place ParABC

Fig. 27. CF Strict sequencing transformation rule

The s-place *StrictABC* created to represent for the CF Strict Sequencing will collect all main-color of the lifeline involving in it and releases them after the CF ends. Figure 28 shows more clearly about the s-place *StrictABC*.

Fig. 28. Subnet place strict

Weak Sequencing

Supposing that there is a CF Weak Sequencing including n operands. The lifelines involving in CF Weak Sequencing are *A, B, C* in which:

- A joins in the CF Weak Sequencing from the occurrence $A(i+1)$ to $A(i+x-1)$.
- B joins in the CF Weak Sequencing from the occurrence $B(j+1)$ to $B(j+y-1)$.
- C joins in the CF Weak Sequencing from the occurrence $C(k+1)$ to $C(k+z-1)$.
- Operand[w] in CF weak sequencing has:
- The occurrences from $A(i+x[w])$ to $A(i+x[w+1]-1)$ on A
- The occurrences from $B(j+y[w])$ to $B(j+y[w+1]-1)$ on B
- The occurrences from $C(k+z[w])$ to $C(k+z[w+1]-1)$ on C

where w is from 1 to n, $x[1] = y[1] = z[1] = 1$, $x[n+1] = x$, $y[n+1] = y$, $z[n+1] = z$.

Figure 29 shows the CF Weak Sequencing transformation rule.

The CF Weak Sequencing restricts CF Parallel on the operands that have same set of life-line and restricts CF Strict Sequencing on the operands that distinguish the set of life-line. Note that CF Parallel transformation rule are different from CF Strict Sequencing transformation rule about the architecture of the Par s-place and the Strict s-place. The architecture for the s-place SeqABC will be hybrided between ParABC and StrictABC (Fig. 30). The different sequence of the operands in the CF Weak Sequencing lead to the different s-places SeqABC.

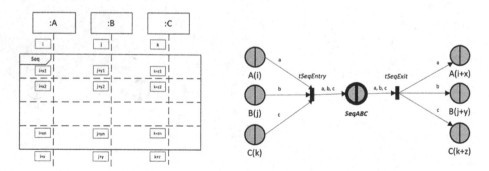

Fig. 29. CF Weak Sequencing transformation rule

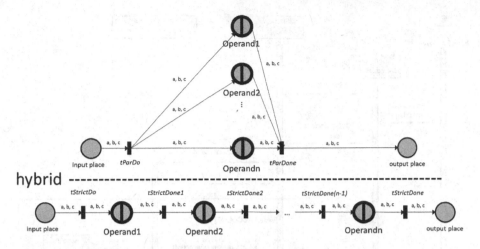

Fig. 30. Hybrid between subnet place Par and subnet place strict

In this section, we introduce 20 rules to transform SDs components into QPN. These rules are divided into 7 different groups. With this we successfully create an equivalent QPN to the initial SD that is interpretable and executable by QPME Tool.

4   Tool Development and Experiment

We implemented a tool named SD2QPN based on Eclipse Java Platform for our transformation method. This tool converts an XML file exported from a UML Sequence Diagram which designed by using Papyrus [15] to another XML file denotes a QPN. We use a tool named QPME [16] to see the graphical view of this obtained QPN.

4.1   Case Study

In fact, it is complicated to find a UML sequence diagram which has all the elements we mentioned in this paper. A UML sequence diagram for a typical bank opening a new account is used to illustrate our transformation method [1]. This case study on Fig. 31 has 6 participants: Bank Employee, Open New Account UI, Open New Account, ExistingAccountList, Account, and Error Log.

First, the participant Open New Account sends a create message to create a business interface named Open New Account UI. Once the interface is created, a reply message is sent to the participant Open New Account. Then the main actor Bank Employee can send a synchronous call Enterdetails (details) to the participant Open New Account UI. The participant Open New Account UI sends a similar message to the participant Open New Account which afterward sends a synchronous message Check (details) to the participant Existing Account List. If the account is valid, a reply will be sent from the participant Existing Account List to the participant Open New Account. Next, a create message will be sent to create the participant Account by the

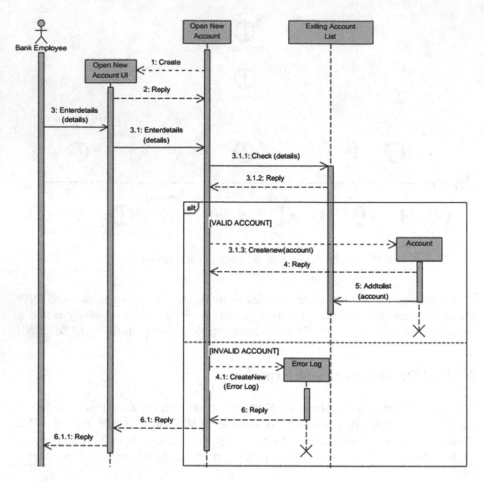

Fig. 31. New account open sequence diagram (created by using Papyrus [13])

participant Open New Account and then the participant Account will send a message Addtolist (account) to the participant Existing Account List. If the account is invalid, a participant Error Log will be created.

In this case study, we use rules to transformation SD into QPN as follow: 3.5 Synchronous Call, Asynchronous Call, Create Message, Delete Message and 3.7 Combined Fragment: Alternative to transform this SD into QPN. This SD has 3 Synchronous Calls: The Synchronous Call Enterdetails from the *Bank Employee Actor* to the *Open New Account UI, the* Enterdetails from *Open New Account UI* to the *Open New Account*, The *Check (details)* from the *Open New Account* to the *Existing Account List*.

The Fig. 32 shows details about the Synchronous Call Enterdetails from the *Bank Employee Actor* to the *Open New Account UI*. The SD has 3 create messages: The participant Open New Account sends a create message to the Open New Account UI, the participant Open New Account sends a create message to create the participant

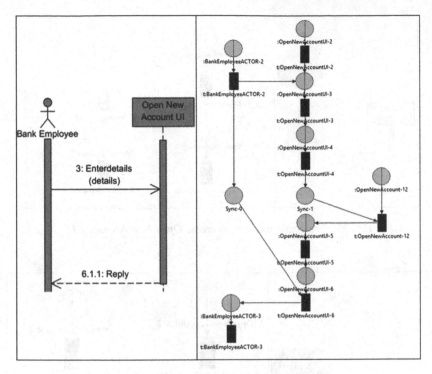

Fig. 32. The synchronous call Enterdetails

Account, The participant Open New Account sends a create message to create the participant Error Log. Applying the 3.5 Synchronous Call for transformation, we recieved the QPN as shown in Fig. 32.

Figure 33 show details about the participant Open New Account sends a create message to create a business interface named Open New Account UI. Having 4 Asynchronous Calls in the SD: *the* Asynchronous call *Reply* from the *Open New Account UI* to the *Open New Account, the Addtolist (account)* from the *Account* to the *Existing Account List, the Reply* from the *Account* to the *Open New Account, the Reply* from *Error Log* to the *Open New Account.* The result of transformation is shown in Fig. 33.

The Fig. 34 shows details about Asynchronous Call Enterdetails *Open New Account UI* to *Open New Account.*

More details about s-place ALT-0 is showed in Fig. 35. This s-place has 2 s-place: VALID ACCOUNT and INVALID ACCOUNT corresponding with 2 operands of CF Alt in the case study. To transform this s-place ALT-0 we firstly apply the rules to the s-place VALID ACCOUNT and INVALID ACCOUNT. After that we compose the obtained QPN into a Subnet Place ALT-0. Figure 36 represents the s-place VALID ACCOUNT and Fig. 37 represents the s-place INVALID ACCOUNT.

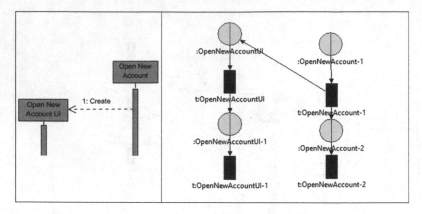

Fig. 33. The create message to create *Open New Account UI*

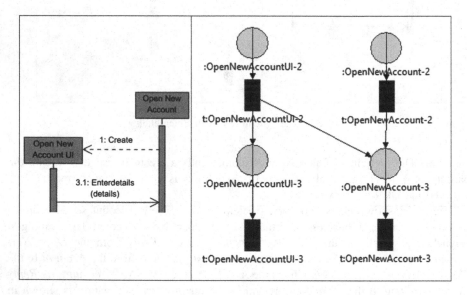

Fig. 34. The asynchronous call enter details

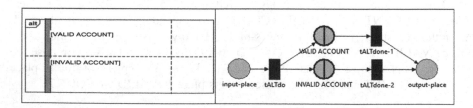

Fig. 35. Subnet place ALT-0

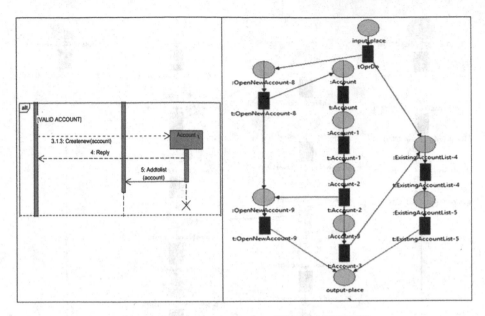

Fig. 36. Represent the s-place VALID ACCOUNT

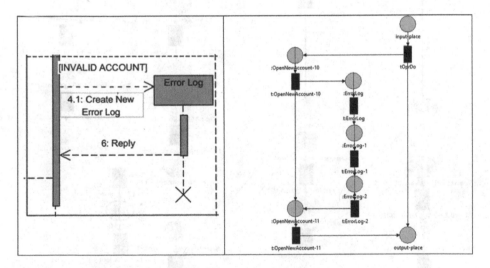

Fig. 37. Represent the s-place INVALID ACCOUNT

4.2 Results

We have used the base rulers in the Sect. 3 to transform all of the components in the sequence diagram in Fig. 29 into a behaviorally equivalent QPN. Figure 38 shows the QPN corresponding to this sequence diagram. It has 25 q-places, 3 o-places, 1 s-places (*ALT-0*), 25 transitions, 13 colors.

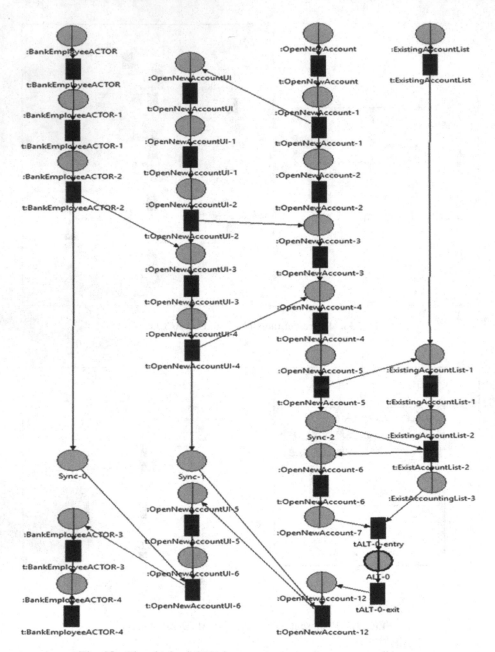

Fig. 38. The obtained QPN for new account open sequence diagram

5   Conclusions

In this paper, we present a transformation from UML Sequence Diagram to Queueing Petri nets. We believe it is feasible to use Labelled Event Structure (LES) [7] to prove the correctness of the transformation rules. A typical case study for banking applications is used to illustrate our approach. We also implemented a tool named SD2QPN that allows to convert from the XML file which denotes a sequence diagram to another XML file which denotes QPN and used this tool to experiment the case study.

By using our method, the obtained QPN can be used for software performance prediction. In the future, we will research more about other UML diagrams to find new approaches to convert them into QPN, combine these methods with the mapping in this paper to gain a set of transformations from UML to QPN that supports to evaluate the software performance.

References

1. Tony Spiteri Staines: Transforming UML sequence diagrams into petri nets. J. Commun. Comput. **10**, 72–81 (2013)
2. Wang, C.J., Fan, H.J., Pan, S.: Research on mapping UML to petri-net in system modeling. In: MATEC Web of Conferences, vol. 44, p. 02038 (2016)
3. Soares, J.A.C., Lima, B., Faria, J.P.: Automatic model transformation from UML sequence diagrams to coloured petri nets. In: 6th International Conference on Model-Driven Engineering and Software Development MODELSWARD (2018)
4. Meedeniya, D.A.: Correct model-to-model transformation for formal verification. A Thesis Submitted for the Degree of Ph.D. at the University of St. Andrews (2013). https://research-repository.st-andrews.ac.uk/handle/10023/3691
5. Meedeniya, D., Perera, I., Bowles, J.: Tool support for transforming unified modelling language sequence diagram to coloured petri nets. Maejo Int. J. Sci. Technol. **10**(03), 272–283 (2016)
6. Lian-Zhang, Z., Fan-Sheng, K.: Automatic conversion from UML to CPN for software performance evaluation. Int. Work. Inf. Electron. Eng. (IWIEE) **29**, 2682–2686 (2012)
7. Ameedeen, M.A.: A model driven approach to analysis and synthesis of sequence diagrams. A thesis submitted to The University of Birmingham for the degree of Doctor of Philosophy (2011). http://etheses.bham.ac.uk/3282/
8. Meier, P., Kounev, S., Koziolek, H.: Automated transformation of component-based software architecture models to queueing petri nets. In: IEEE 19th Annual International Symposium, pp. 339–348 (2011)
9. Pasqua, R., Foures, D., Albert, V., Nketsa, A.: From sequence diagrams UML 2.x to FD-DEVS by model transformation. In: European Simulation and Modelling Conference, vol. 12481, pp. 37–43 (2012)
10. El-kaedy, R.A., Sameh, A.: Performance analysis and characterization tool for distributed software development. Int. J. Res. Rev. Comput. Sci. **2**(3), 906–915 (2011)
11. Ribeiro, O.R., Fernandes, J.M.: Some rules to transform sequence diagrams into coloured petri nets. In: Jensen, K., Aarhus, D., (eds.), 7th Workshop and Tutorial on Practical Use of Coloured Petri Nets and the CPN Tools (CPN 2006), pp. 237–256 (2006)

12. Fernandes, J.M., Tjell, S., Jorgensen, J.B., Ribeiro, O.: Designing tool support for translating use cases and UML 2.0 sequence diagrams into a coloured petri net. In: Sixth International Workshop on Scenarios and State Machines (2007). https://doi.org/10.1109/scesm.2007.1
13. https://www.omg.org/spec/MARTE/About-MARTE. Accessed 21 Apr 2018
14. OMG Unified Modeling Language TM (OMG UML) Version 2.5.1, December 2017. https://www.omg.org/spec/UML/About-UML/. Accessed 22 Apr 2018
15. Papyrus. https://www.eclipse.org/papyrus/. Accessed 22 Feb 2018
16. QPME. https://se.informatik.uni-wuerzburg.de/tools/qpme/. Accessed 18 Jan 2018

HHUSI: An Efficient Algorithm for Hiding Sensitive High Utility Itemsets

Vy Huynh Trieu[1(✉)], Chau Truong Ngoc[2], Hai Le Quoc[3],
and Long Nguyen Thanh[3]

[1] Pham Van Dong University, Quang Ngai City, Quang Ngai Province, Vietnam
htvy@pdu.edu.vn
[2] Da Nang University, Da Nang City, Vietnam
truongngocchau@yahoo.com
[3] Quang Tri Teacher Training College, Dong Ha City,
Quang Tri Province, Vietnam
{hai_lq,long_nt}@qtttc.edu.vn

Abstract. Itemset hiding is a technique that modifies data in order to remove sensitive itemsets from a database. The traditional frequent itemset hiding algorithms cannot be applied directly into high utility itemset hiding problem. In order to solve this problem, Tseng et al. [8] proposed HHUIF and MSICF algorithms. The important target of high utility itemset hiding process is to minimize the side effects caused by data distortion, including missing itemsets, ghost itemsets, remaining sensitive itemsets, and database accuracy. In this paper, we propose an algorithm, named HHUSI, for hiding high utility sensitive itemsets. The method consists of two steps: (1) identify victim transaction and victim item and (2) modify internal utility of the victim item in the victim transaction. Experiment shows that the performance of this method is better than HHUIF and MSICF.

Keywords: High utility itemset · High utility itemset mining
High utility sensitive itemset · Privacy preserving in data mining

1 Introduction

In the traditional itemset mining, the role of frequent itemsets is the same in database [1]. However, in reality, the role of itemsets are different if their profits are taken into account. Mining itemsets according with their profits give knowledge that is more meaningful. For example, each transaction in a supermarket database contains not only quantity of products but also their profit. Yao et al. [2] proposed a model for mining itemsets based on their utility called high utility itemset mining. Basing on this idea, many efficient algorithms were then proposed [3–10].

Recently, in the global business cooperation, data sharing is an important condition to leverage collaborative relationship between parties. However, sharing data may reveal the sensitive information, which is inferred from sensitive high utility itemsets. This could put the data owner party at a disadvantage.

© ICST Institute for Computer Sciences, Social Informatics and Telecommunications Engineering 2019
Published by Springer Nature Switzerland AG 2019. All Rights Reserved
T. Q. Duong and N.-S. Vo (Eds.): INISCOM 2018, LNICST 257, pp. 145–154, 2019.
https://doi.org/10.1007/978-3-030-05873-9_12

In order to preserve the sensitive knowledge from revealing to competitors, the database must be modified in such a way that all sensitive high utility itemsets are hidden in the sanitized database. The hiding process aims to remove sensitive itemsets from database while minimizing the size effects, which are defined as missing itemsets, ghost itemsets, remaining sensitive itemsets, and database accuracy. In 2010, Yeh et al. [11] were the first researchers who proposed the high utility itemset hiding. Their method is to reduce quantity value of an item or remove item in some transaction containing the sensitive itemset. In the two proposed algorithms, named HHUIF (Hiding High Utility Item First) and MSICF (Maximum Sensitive Itemsets Conflict First), the item having highest utility is selected as victim item for the modification process.

Basing on HHUIF and MSICF algorithms [11], Selvaraj et al. [12] proposed an improvement for selecting victim items in case of their utility are the same, namely MHIS (Modified HHUIF Algorithm with Item Selector). Lin et al. [13] proposed a method for hiding sensitive itemsets by inserting fake transactions. In this research, the authors based on genetic algorithm to identify number of fake items to insert into database in order to hide high utility sensitive itemsets. In 2016, Lin et al. [14] proposed two algorithms named MSU-MAU (Maximum Sensitive Utility-MAximum item Utility) and MSU-MIU (Maximum Sensitive

Utility-MInimum item Utility). These algorithms apply Max-Min feature of utility to minimize side effects and speed-up data modification process comparing to HHUIF and MSICF proposed in [11]. Relying on a genetic algorithm, in 2017, Lin et al. [15] proposed PPUMGAT (Privacy-Preserving Utility Mining by adopting a GA-based approach to find appropriate transactions to be deleted) algorithm. This algorithm hides high utility sensitive itemsets by deleting victim items or modifying their internal utility. The genetic method is applied into the victim item identification.

Generally, a high utility sensitive itemset hiding includes two stages: (1) Victim item selection and transaction selection, and (2) Modifying victim item from victim transaction. The side effects of an algorithm for hiding high utility itemset depend on the first stage.

Suppose that S_i is a sensitive itemset, the algorithms proposed in [11, 12, 14] perform repeat the data modification until the utility of S_i is smaller than minimum utility threshold. They hide S_i after four steps as follows:

In the first step, they compute the utility difference (*diff*) of S_i compare to minimum utility threshold ε ($diff = u(S_i) - \varepsilon$). The second step specifies the victim transaction (T_{vic}) and the victim item (i_{vic}) for the data modification. The third step modifies the internal utility of i_{vic} in T_{vic}:

- If $u(i_{vic}, T_{vic}) < diff$, reduce the internal utility of i_{vic} in T_{vic} to zero (this means i_{vic} is removed from T_{vic}).
- Otherwise, reduce an enough utility of S_i in such a way that it can make the utility of S_i less than minimum utility threshold but still keep i_{vic} appear in T_{vic}.

The fourth step aims at updating $diff = diff - u(i_{vic}, T_{vic})$ when the internal utility of i_{vic} in T_{vic} is reduced to zero, otherwise $diff = 0$.

In the fourth step, the value to update *diff* is not exact because if i_{vic} is removed from T_{vic} then T_{vic} will not contain S_i and the difference of utility of S_i must be $diff - u$ (S_i, T_{vic}). Reducing a smaller value as $u(i_{vic}, T_{vic})$ causes higher side effects. We can examine this by investigating an example as follows.

Example 1. Given a database as in Tables 1 and 2, the minimum utility $\varepsilon = 25$. Suppose that itemset *{B, D}* is a high utility sensitive itemset. Applying the algorithms in [11, 12, 14] to hide *{B, D}*, itemset hiding process modifies data two times. The first time, they modify the internal utility of item *B* at transaction T_9 from 5 by 0 and update a new value for *diff* as *diff = 3*. Because *diff = 3 > 0*, so they repeat the data modification in the second time by modifying the internal utility value of item *B* at transaction T_3 from 3 by 2. However, in the first modification, the itemset *{B, D}* is hidden because the utility of *{B, D}* after the first modification is 20 < ε. The second modification is redundant. Therefore, it increases the side effects.

In this paper, we propose a novel algorithm to specify exactly the victim item, the victim transaction and compute correctly the difference value (*diff*) for the data modification. The next section presents some related works corresponding to high utility itemset hiding, then the novel algorithm is proposed in Sect. 3, and Sect. 4 is a conclusion.

2 Related Works

2.1 High Utility Itemset Mining

This section represent some definitions and concepts about the high utility itemset mining proposed by Yao et al. [2].

Definition 1. Let $I = \{i_1, i_2, \cdots, i_m\}$ be a finite set of items, where each item $i_l \in I$ has the external utility $p(i_l)$. An itemset $X = \{i_1, i_2, \cdots, i_k\}$ is a set of k distinct items, where $i_j \in I$, $1 \le j \le k$ and k is the length of X. A transaction database is a set of transactions $D = \{T_1, T_2, ..., T_n\}$, where each transaction $T_c \subseteq I$, $1 \le c \le n$ has a unique identifier *id*, called *Tid*. Each item i_p in the transaction T_c is associated with a weight indicator called quantity $q(i_p, T_c)$, which is the number of item i_p appearing in the transaction T_c.

Table 1. Transaction data set D

Tid	Transaction	Tid	Transaction
T1	A(4), C(1), E(6), F(2)	T6	B(1), F(2), H(1)
T2	D(1), E(4), F(5)	T7	D(1), E(1), F(4), G(1), H(1)
T3	B(3), D(1), E(5), F(1)	T8	B(1), D(1), E(1)
T4	D(1), E(2), F(6)	T9	B(5), D(4), G(10)
T5	A(3), C(1), E(1)		

Table 2. External utility of transaction data set D

Item	A	B	C	D	E	F	G	H
Utility	3	4	5	2	1	1	1	2

The data set given in Tables 1 and 2 will be used for all examples in this paper.

Example 2. $q(A, T_1) = 4$ and $p(A) = 3$; $q(C, T_1) = 1$ and $p(C, T_1) = 5$.

Definition 2. The utility of an item i in a transaction T_c, denoted as $u(i, T_c)$, is defined as: $u(i, T_c) = q(i, T_c) * p(i)$.

Example 3. $u(A, T_1) = q(A, T_1) * p(A) = 3 \times 4 = 12$; $u(C, T_1) = q(C, T_1) * p(C) = 1 * 5 = 5$.

Definition 3. The utility of an itemset X in a transaction T_c, denoted as $u(X, T_c)$, is defined as $u(X, T_c) = \sum_{i \in X} u(i, T_c)$.

Example 4. $u(\{A, C\}, T_1) = u(A, T_1) + u(C, T_1) = 17$; $u(\{A, C\}, T_5) = u(A, T_5) + u(C, T_5) = 14$.

Definition 4. The utility of an itemset in transaction database D, denoted as $u(X)$, is defined as: $u(X) = \sum_{X \subseteq T_c \wedge T_c \in D} u(X, T_c)$.

Example 5. $u(\{A, C\}) = u(\{A, C\}, T_1) + u(\{A, C\}, T_5) = 17 + 14 = 31$.

Definition 5. An itemset X is said to be a high utility itemset if the utility of X is not less than a minimum utility threshold ε given by the user. Let $SetHUI^D$ is a set of high utility itemsets then $SetHUI^D = \{X \mid X \in I, u(X) \geq \varepsilon\}$.

Example 6. Table 3 is a set of high utility mined from data set given in Tables 1 and 2 when setting the minimum utility threshold $\varepsilon = 25$.

Table 3. The set $SetHUI^D$ mined from D with $\varepsilon = 25$

Itemset	Utility	Itemset	Utility	Itemset	Utility
AC	31	B	40	BDG	38
AE	28	BD	48	DEF	36
ACE	38	BG	30	EF	36
ACEF	25	BDE	26		

2.2   High Utility Sensitive Itemset Hiding

High utility sensitive itemset hiding is a process that sanitizes database in order to reduce the utility of sensitive itemset under minimal utility threshold [11].

Definition 6. Given that $S \in SetHUI^D$ as a sensitive itemset in the database D. The set of high utility sensitive itemsets, denoted as $SetHUSI^D$, is defined as: $SetHUSI^D = \{S | S \in SetHUI^D\}$.

We have $SetHUSI^D \subseteq SetHUI^D$.

In order to evaluate the performance of a high utility sensitive itemset hiding algorithm, Yeh [11] proposed three measurements, including: Hiding Failure (*HF*); Miss Cost (*MC*) and Difference between the Original and Sanitized (*DIF*).

Definition 7. Given that HF is the ratio of high utility sensitive itemsets in original database and sanitized database and is defined as: $HF = \frac{|HS^D|}{|HS^{D'}|}$, where HS^D and $HS^{D'}$ are the set of sensitive itemsets in original database D and sanitized database D', respectively.

Definition 8. Given that *MC* is a ratio of high utility non-sensitive itemsets that are missedly hidden:

$$MC = \frac{|\sim HUI^D - \sim HUI^{D'}|}{|\sim HUI^D|},$$

where $\sim HUI^D$ and $\sim HUI^{D'}$ are the set of high utility non-sensitive itemsets mined from the original database D and sanitized D', respectively.

Definition 9. Given that *DIF* is the ratio of the difference between original database D and the sanitized database D' for hiding the sensitive itemsets, and is defined as:

$$DIF = \frac{1}{\sum\limits_{i=1}^{n} f_D(i)} \left(\sum_{i=1}^{n} [f_D(i) - f_{D'}(i)] \right)$$

Where $f_D(i)$ and $f_{D'}(i)$ are the frequency of item i in the database D and D', respectively, n is a number of total items existing in D.

3 Algorithm Proposal

In order to hide high utility sensitive itemset, we reduce the utility of sensitive itemset by reducing the internal utility of some items that belong to sensitive itemset from a set of transactions containing them until its utility is less than minimum utility threshold.

Definition 10. The difference between the utility values of sensitive itemset S_i compare to the minimum utility threshold ε, denoted as *diff*. A sensitive itemset S_i is hidden if and only if *diff* < 0.

To make *diff* < 0, we have to modify repeatedly the original data by reducing the internal utility of the victim item (i_{vic}) at victim transaction (T_{vic}):

If $u(i_{vic}, T_{vic}) > diff$, the internal utility value of the item i_{vic} at transaction T_{vic} needed to be reduced as $k = \left\lceil \frac{diff}{p(i_{vic})} \right\rceil$, that is $q(i_{vic}, T_{vic}) = q(i_{vic}, T_{vic}) - \left\lceil \frac{diff}{p(i_{vic})} \right\rceil$

Otherwise, the internal utility value of item i_{vic} at transaction T_{vic} needed to be reduced to zero (i.e. removing item i_{vic} from transaction T_{vic}). Thus, $diff$ will be reduced a value $u(S_i, T_{vic})$ after removing item i_{vic}, that is $diff = diff - u(S_i, T_{vic})$.

3.1 HHUSI Algorithm

Input: Original database D; Minimum utility threshold ε; a set of high utility sensitive itemset $SetHUSI^D = \{S_1, S_2, ..., S_l\}$
Output: Sanitized database D' (S_i is hidden in D').

```
1  For each  S_i ∈ SetHUSI^D {
2      diff = u(S_i) − ε ;
3      While (diff >=0) {
4          T_vic = T_c | max{u(S_i,T_c), S_i ⊆ T_c ∧ T_c ∈ D} ;
5          i_vic = i_k | max{u(i_k,T_vic), 1 ≤ k ≤ |S_i|, i_k ∈ S_i} ;
6          if ( u(i_vic, T_vic) > diff ) {

7              if (diff == 0)
8                  q(i_vic, T_vic) = q(i_vic, T_vic) − 1 ;
9              else

10                 q(i_vic, T_vic) = q(i_vic, T_vic) − ⌈ diff / p(i_vic) ⌉ ;

11             diff = −1 ;
12         }
13         else {
14             diff = diff − u(S_i, T_vic) ;
15             q(i_vic, T_vic) = 0 ;
16         }
17     }
18 }
```

Running HHUSI for Example 1, we have:

Input: $SetHUSI^D = \{\{B, D\}\}$; Line 1: $S_i = \{B, D\}$; Line 2: $diff = 23$; Line 3: $diff = 23 > 0$; Line 4: $T_{vic} = T_9$; Line 5: $i_{vic} = B$; Line 6: $u(i_{vic}, T_{vic}) = 20 < diff$ jump to line 14; Line 14: $diff = diff - u(\{BD\}, T_9) = 23 - 28 = -5$; Line 15: Remove item B at transaction T_9; Back to line 3: $diff < 0$, end the while loop.

Therefore, itemset *{B, D}* is hidden after the first time of data modification at transaction T_9 while algorithms in [11, 12, 14] executes the second time of data modification.

3.2 Experiment Results

Experiment Data Description: The HHUSI algorithm was performed with four datasets, including: retail, foodmart, and two randomly generated datasets named T1k100I20MT and T1k50I15MT. The internal utility value of items in datasets were generated randomly in a range from 1 to 10, the external utility of items were generated randomly in the range from 1 to 30. The detail of data sets are described in the Table 4. Sensitive itemsets were selected randomly with 5, 10, 15, 20, 25, 30 itemsets from high utility itemsets mined by the FHM algorithm [4]

Experiment System Description: CPU Core I3 2.4 GHz, RAM 2 GB, Windows 10.

Experiment results and the comparison between proposed algorithm when running with different data sets.

The results show that high utility sensitive itemset hiding by modifying the internal utility value of sensitive items in order to reduce the utility of sensitive itemset does not cause the following side effects:

- The ghost itemsets: The data modification does not make utility of itemsets in database to be increased.
- The hiding failure: This method will repeat the data modification process in order to hide sensitive itemsets one by one until their utility is lower than the minimal utility threshold. Moreover, this method does not make utility of any itemset to be increase. Therefore, since the sensitive itemset is hidden, it will not appear again as high utility after hiding process.

Table 4. The data sets description

Databases	#$|D|$	#$|I|$	#AvgLen	#MaxLen
Retail	88,162	16,470	10.3	76
Foodmart	4,141	1,559	4	11
T1k100I20MT	1,000	100	8	20
T1k50I15MT	1,000	50	7.6	15

The experiment results are described in the following Figs. 1, 2 and 3:

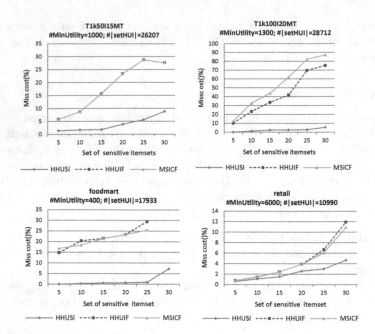

Fig. 1. Missing cost under various sensitive itemsets

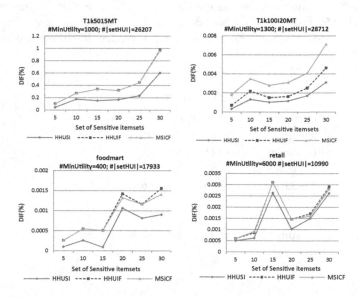

Fig. 2. Difference rate between original database and the sanitized database under various sensitive itemsets

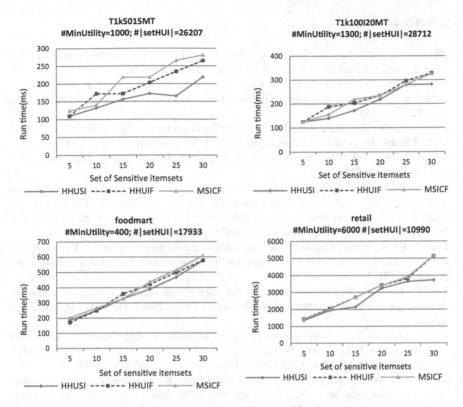

Fig. 3. Runtime under various sensitive itemsets

The figures show that the performance of HHUSI algorithm is better than previous proposed algorithms named HHUIF and MSICF. The side effects caused by HHUSI when running with four datasets are lower than HHUIF and MSICF. The results also show that HHUSI runs faster than HHUIF and MSICF when running with every data sets.

4 Conclusion

This paper proposed a new algorithm for hiding high utility sensitive itemset, named HHUSI, based on the data modification method. It aims to modify internal utility of victim item at victim transaction in order to reduce the utility of sensitive itemset below the minimal utility threshold. In HHUSI, the victim transaction is specified as the transaction that the sensitive itemset in this transaction have highest utility. The victim item is identified as the item having highest utility among items belong to the sensitive itemset in victim transaction. In HHUSI algorithm, the variable for specifying number of data modification, *diff*, is computed exactly compare to the *diff* variable in HHUIF and MSICF. The proposed algorithm was executed with HHUIF and MSICF with four datasets. The experiment results show that the performance of HHUSI is better than HHUIF and MSICF algorithms in minimizing side effects and running time.

References

1. Agrawal, R., Imieliński, T., Swami, A.: Mining association rules between sets of items in large databases. In: ACM SIGMOD Record, pp. 207–216 (1993)
2. Yao, H., Hamilton, H.J., Butz, C.J.: A foundational approach to mining itemset utilities from databases. In: Proceedings of the 2004 SIAM International Conference on Data Mining, pp. 482–486 (2004)
3. Erwin, A., Gopalan, R.P., Achuthan, N.R.: Efficient mining of high utility itemsets from large datasets. In: Washio, T., Suzuki, E., Ting, K.M., Inokuchi, A. (eds.) PAKDD 2008. LNCS (LNAI), vol. 5012, pp. 554–561. Springer, Heidelberg (2008). https://doi.org/10.1007/978-3-540-68125-0_50
4. Fournier-Viger, P., Wu, C.-W., Zida, S., Tseng, Vincent S.: FHM: faster high-utility itemset mining using estimated utility co-occurrence pruning. In: Andreasen, T., Christiansen, H., Cubero, J.-C., Raś, Zbigniew W. (eds.) ISMIS 2014. LNCS (LNAI), vol. 8502, pp. 83–92. Springer, Cham (2014). https://doi.org/10.1007/978-3-319-08326-1_9
5. Liu, M., Qu, J.: Mining high utility itemsets without candidate generation. In: Proceedings of the 21st ACM International Conference on Information and Knowledge Management, pp. 55–64 (2012)
6. Liu, Y., Liao, W.-K., Choudhary, A.: A fast high utility itemsets mining algorithm. In: Proceedings of the 1st international Workshop on Utility-Based Data Mining, pp. 90–99 (2005)
7. Liu, Y., Liao, W.-k., Choudhary, A.: A two-phase algorithm for fast discovery of high utility itemsets. In: Ho, T.B., Cheung, D., Liu, H. (eds.) PAKDD 2005. LNCS (LNAI), vol. 3518, pp. 689–695. Springer, Heidelberg (2005). https://doi.org/10.1007/11430919_79
8. Tseng, V.S., Wu, C.-W., Shie, B.-E., Yu, P.S.: UP-growth: an efficient algorithm for high utility itemset mining. In: Proceedings of the 16th ACM SIGKDD International Conference on Knowledge Discovery and Data Mining, pp. 253–262 (2010)
9. Wu, C.W., Shie, B.-E., Tseng, V.S., Yu, P.S.: Mining top-k high utility itemsets. In: Proceedings of the 18th ACM SIGKDD International Conference on Knowledge Discovery and Data Mining, pp. 78–86 (2012)
10. Zida, S., Fournier-Viger, P., Lin, J.C.-W., Wu, C.-W., Tseng, V.S.: EFIM: a highly efficient algorithm for high-utility itemset mining. In: Sidorov, G., Galicia-Haro, S.N. (eds.) MICAI 2015. LNCS (LNAI), vol. 9413, pp. 530–546. Springer, Cham (2015). https://doi.org/10.1007/978-3-319-27060-9_44
11. Yeh, J.-S., Hsu, P.-C.: HHUIF and MSICF: Novel algorithms for privacy preserving utility mining. Expert Syst. Appl. 37, 4779–4786 (2010)
12. Selvaraj, R., Kuthadi, V.M.: A modified hiding high utility item first algorithm (HHUIF) with item selector (MHIS) for hiding sensitive itemsets (2013)
13. Lin, C.-W., Hong, T.-P., Wong, J.-W., Lan, G.-C., Lin, W.-Y.: A GA-based approach to hide sensitive high utility itemsets. Sci. World J. **2014** (2014)
14. Lin, J.C.-W., Wu, T.-Y., Fournier-Viger, P., Lin, G., Zhan, J., Voznak, M.: Fast algorithms for hiding sensitive high-utility itemsets in privacy-preserving utility mining. In: Engineering Applications of Artificial Intelligence, vol. 55, pp. 269–284 (2016)
15. Lin, J.C.-W., Hong, T.-P., Fournier-Viger, P., Liu, Q., Wong, J.-W., Zhan, J.: Efficient hiding of confidential high-utility itemsets with minimal side effects. J. Exp. Theor. Artif. Intell. **29**, 1225–1245 (2017)

Converting the Vietnamese Television News into 3D Sign Language Animations for the Deaf

Quach Luyl Da[1(✉)], Nguyen Hua Duy Khang[2],
and Nguyen Chi Ngon[2]

[1] Department of Engineering Technology, Tay Do University,
Can Tho, Vietnam
luyldaquach@gmail.com
[2] College of Engineering Technology, Can Tho University, Can Tho, Vietnam
ncngon@ctu.edu.vn

Abstract. Sign language (SL) is based on the movement of hands, body gestures and motions to express information instead of speech or written forms. SL has been used since 384-322 BC and developed in many countries, including Vietnam. In some countries, there are several applications of using technology to develop SL and communicate within the deaf community. However, there is lack of studies in these areas in Vietnam. This study is approaching of using 3D avatar of the HamNoSys and building a dictionary for Vietnamese SL animation. The study builds a data set and applies the machine learning decision tree (ID3) to convert the Vietnamese normal sentences into the short sentences of the deaf. The conversion process generates SiGML language to express the SL by a virtual signer leading to present the television news. The assessment results by experts in SL institutions indicate high precision of proposed solution and show the needs of the study.

Keywords: Converted syntax · ID3 · Machine learning · Decision tree
Vietnamese sign language

1 Introduction

Sign language (SL) or body language is used in the deaf community to convey information through gestures, postures and facial expressions. Normally, text expresses information for ordinary people, however, it is difficulty for SL. The word orders of the deaf are different from the normal grammar, as shown in Table 1. Therefore, researchers have used ID3 algorithm [1] to solve the problem of grammar conversion and expression of SL by virtual signer.

The conversion from text to SL is interested by many researchers, such as the project to build ViSiCast tools [2], SiSi software – Say it Sign it System [3], Vcom3D software [4], Sign-to-me [5], etc. In addition, some research groups have successfully developed tools for building graphical SL for German [6], French [7] and Chinese [8]. Practices have proved the usefulness of these tools for the society in general and the deaf community in particular. Each tool converting from text to SL has its advantages and disadvantages.

© ICST Institute for Computer Sciences, Social Informatics and Telecommunications Engineering 2019
Published by Springer Nature Switzerland AG 2019. All Rights Reserved
T. Q. Duong and N.-S. Vo (Eds.): INISCOM 2018, LNICST 257, pp. 155–163, 2019.
https://doi.org/10.1007/978-3-030-05873-9_13

Table 1. Comparison between normal text grammar and SL grammar.

Vietnamese grammar	Vietnamese SL grammar
Nhân dịp đầu năm mới 2013, xin chúc các vị đại biểu Quốc hội và gia đình sức khỏe, hạnh phúc	Năm mới 2013, Quốc hội đại biểu, gia đình, sức khỏe, hạnh phúc, chúc
Translated: On the occasion of New Year 2013, wishing the congressmen and family health and happiness	Translated: New Year 2013, congressman, family, healthy, happiness, wishing

Based on previous results, application of computer character animation or virtual signer is a good access, flexibility and convenience in building SL [4] that is also approached in this study. Concerning the issue of Vietnamese language study, many authors were interested in the past years, such as Le *et al.* [9], Vinh *et al.* [10], etc. The most brilliant study could be the project named "Research on developing some essential products for Vietnamese speech and text processing" under a national research project KC01/06-10 which has been very successful in the "Vietnamese text processing" [11].

Fig. 1. Pictures from VTV2 news.

According to the Vietnamese General Statistics Office, in 2009, Vietnam has about 6.7 million disabled people, included more than 1 million the deaf approximating of 6.3% of the population. The government has issued guidelines and policies to help them for better integrating into society. For example, Vietnam Television is actively developing a program of "Teaching sign language on television" and the Vietnam Television News VTV2 provides daily news in SL (Fig. 1). However, using real people will take time and high expenses. So, the urgent need is to research and use virtual characters for this purpose instead of using SL experts.

Automatic conversion of Vietnamese text into SL for the deaf is still a new approaching in Vietnam that requires the movements expressing SL must be smoothly animated by a 3D virtual signer. This study aims to develop a tool that can support producing SL news based on texts used in television. And the most difficulty of this study is to convert the Vietnamese normal sentences into the short sentences of the deaf.

2   Methods

2.1   Vietnamese Sign Language

SL is developed with a little difference between regions, cultures and habits leading to the differences between the vocabulary system and SL grammar. The Vietnamese SL is also different from local areas such as Ha Noi, Hai Phong, Thai Binh, Da Nang, Binh Duong, Ho Chi Minh city, etc. Among these areas, there are three main dialects used in Ha Noi, Hai Phong and Ho Chi Minh city. In SL, the signs are generally considered as the words of speech. However, information is received by the eyes, so it also has distinctive features. In the structure, information in SL is made up of five elements: the symbol positions, shapes, motions, directions, and expressions not by hand [12]. However, it is lack of research to develop a tool for conversion of the Vietnamese text into SL using 3D virtual signers.

2.2   VnTokenizer Tool

VnTokenizer [10] is a tool to segment the Vietnamese. It parses input text into lexical phrases and other patterns using pre-defined regular expressions. It has developed based on the principles of word segmentation [13], including: compounds, derivation, multiword expressions, proper names and regular patterns. It is a technique based on the linear interpolation adjusts well bigram and unigram probabilities that improves the estimation and accuracy of the model. The precision is computed as the count of common tokens over tokens of the automatically segmented files, recall as the count of common tokens over tokens of the manually segmented files, and F-measure is computed as usual from these two values. The word segmentation tool has an accuracy rate of about 96% [10].

2.3   JVntagger Tool

JVntagger is a tool used for tagging Vietnamese words using CRFs (Conditional Random Fields) [14]. It has given correct results in the word separation, phrase assignment, etc. Give $s = (s_1, s_2, \ldots, s_T)$ is a sequence of states, based on that when knowing the sequence of observations $o = (o_1, o_2, \ldots, o_T)$, CRFs determine the conditional probabilities as follows:

$$p_\theta(s|o) = \frac{1}{Z(o)} \exp\left[\sum_{t=1}^{T} \sum_{k} \lambda_k f_k(s_{t-1}, s_t, o, t)\right] \tag{1}$$

Therein, $Z(o)$ is the normalization factor on all possible label strings (2)

$$Z(o) = \sum_{s'} \exp\left(\sum_{t=1}^{T} \sum_{k} \lambda_k f_k(s'_{t-1}, s'_t, o, t)\right) \tag{2}$$

In (1) and (2), each feature f_k has a weight λ_k. Two types of features are considered in f_k included pre-state features (3) and transition features (4).

$$f_k^{(per-state)}(s_t, o, t) = \delta(s_t, l)x_k(o, t) \tag{3}$$

$$f_k^{(transition)}(s_{t-1}, s_t, t) = \delta(s_{t-1}, l')\delta(s_t, l) \tag{4}$$

where, δ is Kronecker-δ [15]. Each feature (3) incorporates with l of current state s_t and a context predicate – a binary function $x_k(o, t)$ determines the critical context of observation o at position t. A transition features (4) represents chain dependencies by combining labels l' in previous state s_{t-1} and label l in state s_t.

CRFs are typically used by maximizing the likelihood function by training data using the L-BFGS optimization technique [16]. Based on the model learned to find the corresponding label chain of a series of observations. For CRFs, the Viterbi dynamic planning algorithm [17] is used to argue with new data. The precision when tagging type in Vietnamese is 91.98% [11].

2.4   HamNoSys and 3D Avatar

There exist some SL recording systems, especially for HamNoSys [18–20]. It was developed in the 1980s at the German SL Institute of Hamburg University to express symbols in SL (Fig. 2), that is inherited for the Vietnamese SL.

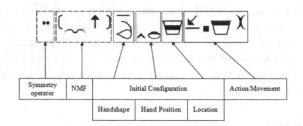

Fig. 2.   Format of the HamNoSys.

2.5   Decision-Tree Learning Machine

Decision-tree learning machine is among the top-10 key algorithms of data mining. The model of decision-tree has a shape of a tree, in which the leaf nodes are sticked on the corresponding data and the layer is integrated into the testing conditions to fork as presenting. A decision in a form of IF-THEN is created from implementing AND together with the conditions of the path from the root node to the leaf node. Decision-tree construction algorithms consist of 2 major steps: building the tree (top-down), cutting branches (bottom-up) to avoid learning by rote. It classifies data based on best attribute p in partition data [1], and gain of the Shannon's entropy function [21].

$$Info(D) = -\sum_{i=1}^{k} p_i log_2(p_i) \tag{5}$$

Chaos measure after using the attribute A to partition the data D into V share is calculated by using the formula:

$$Info_A(D) = \sum_{j=1}^{v} \frac{|D_j|}{D} \times Info(D_j) \qquad (6)$$

The helpful information rate when being selected by the attributes A will partition the data D into the component V. It is calculated as (7):

$$Gain(A) = Info(D) - Info_A(D) \qquad (7)$$

Due to the converted data file has more value than other attributes, the useful information gaining set increases on the attributes with a lot of partition value. To reduce this bias, the study uses the helpful ratio to perform the process of the partition during the transition process in (8) and (9).

$$IntrinsicInfo(A) = -\sum_{j=1}^{v} \frac{|D_j|}{D} log_2 \frac{|D_j|}{D} \qquad (8)$$

$$GainRation(A) = \frac{Gain(A)}{IntrinsicInfo(A)} \qquad (9)$$

3 Conversion System

3.1 Constructing Training Database

The study conducted a text process for 740 input sentences getting from the popular words on Vietnamese daily news. That process is presented in Fig. 3, where all complex sentences were analyzed and separated words. These words are analyzed and tagged to label word items (Table 2) by using label-algorithms with the enclosed position. After this process, a training data set is generated corresponding to 740 sentences which were also checked again by some SL experts.

Table 2. Label word in JvnTagger.

Symbol	Explanation	Symbol	Explanation	Symbol	Explanation
N	Noun	P	Pronoun	A	Adjective
Np	Person Noun	L	Attribute	I	Interjection
Nc	Classification Noun	M	Numeral	Y	Abbreviation
Nu	Unit Noun	R	Adjunct	X	Un-known
V	Verb	E	Preposition	C	Conjunction

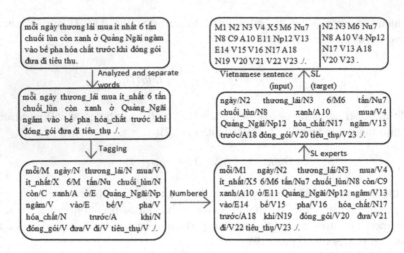

Fig. 3. Illustrating the process of building the training data.

3.2 Training Process

The training process for converting Vietnamese grammar structure to the shortened the grammar structure of SL is implemented as shown in Fig. 4. In that process, the Data set is generated by the process showed in Fig. 4; the preprocessed data splits words and removes punctuation characters; the Training extracts grammar structures and sentence structures; and the Rule set performs the converting rules.

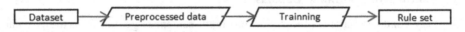

Fig. 4. Training scheme converter.

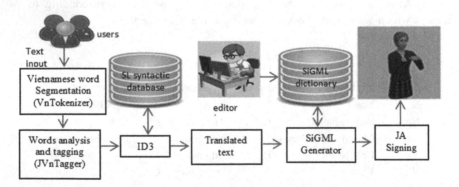

Fig. 5. Converting the Vietnamese text from television news into SL animations.

3.3   Process of Implementing Program

The implementing process of the conversion can be summarized in Fig. 5, where, users type text into the program; separating words to analyze and separate Vietnamese word items; the Words analysis and tagging to label word items (Table 2) by using label-algorithms with the enclosed position; interacting with the converted data set aims to analyze and match to order to select the corresponding sentences; ID3 makes up a constructed dictionary, and a synonym-set which has been built upon [22] with 12,675 words. The study has built a data set of 2,558 SL words based on [23, 24] and combined with the expression not by hands from HamNoSys. If the words do not exist, it will separate words by spelling characters, on the opposite action, it will take the SiGML results. Finally, this process creates a SL file in SiGML form, that will be used for JASigning to generate HamNoSys codes to animate a 3D signer.

4   Results and Discussion

The actual results were made on a sample of 390 sentences, of which 8 were in each of five different sets of data, each of which was randomly assigned to 5 sets of data. 1/3 to do the test, each set was taken 2/3 to do the training and 1/3 to do the testing. The accuracy of the three algorithms is carried out by the use of the three TER tools [25] to evaluate the results of the converted data is presented in Tables 3 and 4.

Table 3. Result of the testing with 390 sentences (%).

Dataset	Set 1	Set2	Set 3	Set 4	Set 5	Average
Result	86.99	86.99	86.63	58.61	58.61	**75.57**

Table 4. Processing time with 740 sentences.

Num.	Process steps	Preprocessor	Execution time(s)
01	Text processing	Separation of words	4.446
		POS tagging	3.271
		Assign sequential numbers from the sentence	0.097
		Selection results process	0.035
02	Transition process	ID3	0.124

The study was tested on 3 TV news of the Can Tho City Television that the compilation was processed, tested by the SL experts. The SL animation video outputs were integrated on the TV news as presented in Table 5. Using above clips, the study conducted a survey of 98 deaf people, including SL Club of Can Tho City, CED and SL Club of Ho Chi Minh City with 13 multiple choice questions about content. The survey results show that the level of understanding the content of 3 clips in Table 5 meets average of 97.06% and the deaf are expecting to use this software.

Table 5. Video results.

Video 1	Video 2	Video 3

5 Conclusion

This study applied an application of machine translation in the conversion of the Vietnamese normal grammar into the SL structures. With a desire to create a prerequisite step for developing and unifying Vietnamese SL, the study proposed a solution to develop a system that builds a data set and applies ID3 to convert structure of sentences into reduced forms of SL. The conversion process generates SiGML code to express the SL by a virtual signer leading to present the television news for the deaf. Experimental results indicate the feasibility of the proposed solution. In the future, the study is continued to develop and integrate in a website, so that the community can join in edition, complement more for Vietnamese SL dictionaries. When the database is large enough, the application of the machine translation will become easier, higher accuracy and faster processing speed.

References

1. Quinlan, J.: C4.5, Programs for Machine Learning. Morgan Kaufmann Publishers, Burlington (1993)
2. Hanke, T. (ed.): Interface Definitions, Virtual Signing: Capture, Animation, Storage and Transmission, Deliverable D5-1 (2001)
3. Al-Ohali, Y.: Identification of most desirable parameters in SIGN language tools: a comparative study. Glob. J. Comput. Sci. Technol. **10**(6), 23–29 (2010)
4. Stewart, J.: VCom3D - Sign Smith studio (2012)
5. Waterfall Rainbows, Sign to Me (BSL) - User Guide (2005)
6. Kipp, M., Heloir, A., Nguyen, Q.: Signing avatars: a feasibility study. In: DFKI – Embodied Agents Research Group, Saarbruecken, Germany (2011)
7. Courty, N., Gibet, S.: Why is the creation of a virtual signer challenging computer animation? In: Boulic, R., Chrysanthou, Y., Komura, T. (eds.) MIG 2010. LNCS, vol. 6459, pp. 290–300. Springer, Heidelberg (2010). https://doi.org/10.1007/978-3-642-16958-8_27

8. Chai, X., Li, G., et al.: Sign language recognition and translation with kinect. In: The FiDiPro Program of Tekes and Natural Science Foundation of China under contracts, Nos. 61001193 and 60973067. Microsoft Research Asia (2013)
9. Hong Thanh, L., Rathany Chan, S., Hoan Cong, N., Thuy Thanh, N.: Named entity recognition in Vietnamese text using label propagation. In: Proceedings of the 5th International Conference of Soft Computing and Pattern Recognition, SoCPaR (2013)
10. Hong Phuong, L., Thi Minh Huyen, N., Roussanaly, A., Vinh, H.T.: A hybrid approach to word segmentation of Vietnamese texts. In: Martín-Vide, C., Otto, F., Fernau, H. (eds.) LATA 2008. LNCS, vol. 5196, pp. 240–249. Springer, Heidelberg (2008). https://doi.org/10.1007/978-3-540-88282-4_23
11. Bao, H.T.: About Vietnamese processing in ICT, JAIST, VLSP-KC01/06-10 (2012). (in Vietnamese)
12. Phuong, N.T., Ton, N.D.: Some problems of Vietnamese sign language syntaxes. J. Ngon Ngu (4), 17–31 (2012). ISSN 0866-7519. (in Vietnamese)
13. ISO/TC 37/SC 4 AWI N309: Language Resource Management - Word Segmentation of Written Texts for Mono-lingual and Multi-lingual Information Processing. Technical report. ISO (2006)
14. Lafferty, J., McCallum, A., Pereira, F.: Conditional random fields: probabilistic models for segmenting and labeling sequence data. In: Proceedings of International Conference on Machine Learning, ICML, pp. 282–289 (2001)
15. Agacy, R.L.: Generalized Kronecker delta and permanent deltas, their spinor and tensor equivalents and applications. J. Math. Phys. 40(4), 2055–2063 (1999)
16. Byrd, R.H., Lu, P., Nocedal, J.: A limited memory algorithm for bound constrained optimization. SIAM J. Sci. Stat. Comput. 16(5), 1190–1208 (1995)
17. Forney Jr., G.D.: The Viterbi algorithm. Proc. IEEE 61, 268–278 (1973)
18. Hanke, T.: Interface Definitions, Virtual Signing: Capture, Animation, Storage and Transmission, Deliverable D5-1 (2001)
19. Hanke, T., Popescu, H.: Intelligent Sign Editor. ESIGN D2.3 Report, Institute of GermanSign Language and Deaf Communication, University of Hamburg (2003)
20. Hanke, T., Marshall, I., Safar, E., Schmaling, C., Langer, G., Metzger, C.: Interface definitions. ViSiCAST Report D5.1 (2001)
21. Shannon, C.-E.: A mathematical theory of communication. Bell Syst. Technol. J. 27, 379–423 (1948). pp. 623–656
22. Bau, N.T.: Vietnamese synonym an antonym dictionary. Ha Noi Social Sciences Publishing House (2016). (in Vietnamese)
23. VNIES - Trung tam nghien cuu giao duc tre khuyet tat, Ky hieu cu chi dieu bo cua nguoi diec Viet Nam, Ha Noi (2002). (in Vietnamese)
24. Ngon, N.C., Da, Q.L.: Application of HamNoSys and avatar 3D JASigning to construction of Vietnamese sign language animations. J. Sci. Technol. 1(11), 61–65 (2017). ISSN 1859 – 1531. (in Vietnamese)
25. Snover, M., Dorr, B., Schwartz, R., Micciulla, L., Makhoul, J.: A study of translation edit rate with targeted human annotation. In: Proceedings of Association for Machine Translation in the Americas (2006)

Outage Performance of the Downlink NOMA Relay Networks with RF Energy Harvesting and Buffer Aided Relay

Xuan Nam Tran[1(✉)], Tran Manh Hoang[1], Nguyen Ba Cao[1], and Le The Dung[2]

[1] Le Quy Don Technical University, Hanoi, Vietnam
namtx@mta.edu.vn, tranmanhhoang@tcu.edu.vn, bacao.sqtt@gmail.com
[2] Chungbuk National University, Cheongju, Korea
dung.t.le@ieee.org

Abstract. In this paper, we investigate performance of a decode-and-forward Non-Orthogonal Multiple Access (NOMA) relay system using Simultaneous Information and Power Transfer (SWIPT). In the considered system a source node transmits data in the downlink simultaneously to two destination nodes via an energy-harvesting relay node. In order to cope with fading channels the relay is assumed to have an infinite capacity buffer to store data if the transmission link is in outage. We analyze outage performance of the system and obtain the closed-form expressions for the system outage probability for two cases, i.e. the relay is equipped and not equipped with the buffer. Numerical results are provided to demonstrate the merit of using the buffer-aided relay.

Keywords: NOMA · Energy harvesting
Successive interference cancellation · Power allocation
Buffer aided relay

1 Introduction

Recently, Non-Orthogonal-Multiple-Access (NOMA) technique has been realized as a promising multiuser communication technique for the fifth-generation (5G) mobile network due to its superior spectral efficiency [1]. The key idea of NOMA is to utilize power domain to differentiate users based on their power levels [2]. Recent researches have shown that NOMA can be applied not only to point-to-point but also relay networks [3,4]. The work [5] considered a conventional cooperative NOMA system with buffer-aided relaying and adaptive transmission which can operate in different modes for individual time slots. The authors of [4] proposed a dual-hop cooperative relaying scheme using NOMA, where the two terminal nodes communicate with each other simultaneously via a relay over the same frequency band. In this scheme, the received symbols with different power levels from two source nodes are encoded into a superposition symbol based on

---

Invited paper.

© ICST Institute for Computer Sciences, Social Informatics and Telecommunications Engineering 2019
Published by Springer Nature Switzerland AG 2019. All Rights Reserved
T. Q. Duong and N.-S. Vo (Eds.): INISCOM 2018, LNICST 257, pp. 164–177, 2019.
https://doi.org/10.1007/978-3-030-05873-9_14

the NOMA principle and then forwarded back to the destination nodes. All users are assumed to be allocated with an equal power level.

On the other hand, harvesting energy from the ambient radio frequency (RF) environment has become a promising solution for energy-constrained electronic devices which are normally supported by a limited power source such as battery [6]. In many cases, especially ad-hoc wireless sensor networks or body area networks, charging battery is too expensive or even impossible. Meanwhile, natural energy sources such as solar, wind and radio wave can be effectively exploited for energy harvesting (EH). Among these sources, RF energy harvesting [7] has more advantages such as reliability and green energy. Recently, RF energy harvesting has also been considered for the NOMA relay networks [8,9] to prolong their lifetime. In [8] a simultaneous wireless information and power transfer (SWIPT) scheme was proposed for NOMA networks in which a base station serves both the relay and far users. The work [10] considered a NOMA system in which users near the source can act as the EH relays to assist far users in forwarding data to the destination.

In this paper, we apply SWIPT to NOMA relay systems which are supported by RF energy harvesting. In our proposed system, the relay is assumed to be able to harvest RF energy from the source to support its operation. Moreover, it is provided with a buffer for data processing and uses a cooperative decode-and-forward scheme to transmit superposition symbols from the source to end users in the context of a downlink NOMA relay system. We then derive the closed-form expression for the system outage probability and propose an optimal power allocation scheme to optimize performance of end users.

The main contributions of the paper can be summarized as follows:

- First, in order to support the sustainable operation of relay, we propose to use a time-switching SWIPT scheme in a downlink NOMA relay system.
- Using the Markov chain model to describe the operation of the buffer-aided relay we can derive the closed-form expression for the system outage probability and compare it with the case without using the buffer.
- Performance of the system in terms of outage probability over the Rayleigh fading channel is analyzed and verified using computer simulations.

The remainder of the paper is organized as follows. Section 2 presents the system model of the considered system. Outage performance of the system is analyzed in Sect. 3. Numerical results are shown in Sect. 4 and finally conclusions are drawn in Sect. 5.

2   System Model

2.1   System Configuration

The system model of a downlink NOMA relay network considered in this paper is shown in Fig. 1. In this model, a source node (basestation) S wishes to send its messages to two end users (destination nodes) D_1 and D_2 simultaneously via a

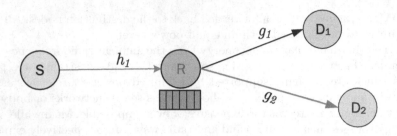

Fig. 1. Wirelessly powered NOMA downlink relaying network.

decode-and-forward (DF) relay node R. It is assumed that all three nodes S, D_1, and D_2 are powered by mains electricity while the relay is self-powered by the energy harvested from the RF transmitted signal of the source. In addition, the relay is assumed to have an unlimited-size buffer to store the received messages. We assume that the direct link between the source and the two destination nodes is not available and that the communication between them is always via the relay node. The channels between two arbitrary nodes are subject to block and flat Rayleigh fading. The assumption of block fading means that the channel coefficients are constant during each data block transmission interval τ but vary from one to another. All nodes are equipped with a single antenna and operate in a half-duplex mode.

In the case of buffer-aided relay, it is assumed that the relay has perfect channel state information (CSI) of the links $S \rightarrow R$ and $D_1, D_2 \rightarrow R$ at the start of each time slot by using some pilot-assisted channel method. Based on assumption, the relay node R can decide to operate in the transmit or receive mode [11]. The complex channel coefficient of the link $S \rightarrow R$ is denoted by $h_1 \sim \mathcal{CN}(0, \Omega_0)$ while that of the link $R \rightarrow D_i$ is $g_i \sim \mathcal{CN}(0, \Omega_i)$ where $i = \{1, 2\}$, $\Omega_1 = \mathbb{E}\{|g_1|^2\}$, $\Omega_2 = \mathbb{E}\{|g_2|^2\}$. The additive white Gaussian noise (AWGN) at R, D_1 and D_2 is respectively denoted by $w_\Lambda \sim \mathcal{CN}(0, N_0)$, where $\Lambda \in \{R, D_1, D_2\}$ and N_0 is the single-sided AWGN power spectral density. Without loss of generality, we assume that the channel gains g_i are sorted in the descending order as follows: $|g_1|^2 > |g_2|^2$.

2.2   Data Buffer at Relay

Since the relay operates in the half-duplex mode, the received messages from the source node S need to be stored in the buffer and process before forwarding to the destination nodes D. Using RF energy harvesting, the relay node can be powered by the harvested energy directly or via a storage device such as battery or high volume capacitor [12]. The system operates in the time-switching mode, in which each operation period τ is divided into two portions depending on the fraction coefficient $\alpha, 0 \leq \alpha \leq 1$; the first portion $\alpha\tau$ is used for energy harvesting while the remaining $\tau(1 - \alpha)$ is for transmission and reception. During each time slot, the relay and the source node is selected to transmit data based on the buffer state and the available links that can guarantee the successful transmission or reception of one packet.

If S is selected, it will generate a transmission frame of size $2r_0\tau$ bits intended to the two destination nodes D_1 and D_2 and send it to the relay node, where r_0 is the target transmission rate of the system. Each frame contains two parts, the first one contains data symbols to be sent to D_1 and the second to D_2. The relay buffer has $L \geq 2$ storage units, each can store $2r_0\tau(1 - \alpha)$ bits. The relay node decodes the received frame and stores into the storage device. Each storage device is split into two parts of the same length. The first part is used to store the information symbol intended to D_1. The second one is used for D_2.

2.3   Signal Model

During each time slot, if the source node is selected to transmit, it combines two symbols x_1 and x_2 to form a transmission packet, $x_S = x_1 + x_2$ where $\mathbb{E}\{|x_1|^2\} = P_1$ and $\mathbb{E}\{|x_2|^2\} = P_2$. Based on the NOMA principle [13], in order for the relay node to decode the information symbols x_1 and x_2, the source node sets different power levels to x_1 and x_2. Assume that $P_1 \neq P_2$ and denote the total transmit power at the source by $P_S = P_1 + P_2$. Then, the received signal at the relay is given by

$$y_R = h_1 \sum_{k=1}^{2} \sqrt{P_k} x_k + w_R. \tag{1}$$

The instantaneous SNR of the source-to-relay link is given by

$$\gamma_R = \frac{P_S |h_1|^2}{N_0}. \tag{2}$$

Recall that τ is the block duration of an entire communication period in which the information is transmitted from S to D_i. For each period τ, the first duration, $\alpha\tau$, is used for energy harvesting at R, while the remaining duration, $(1 - \alpha)\tau$, is used for transmitting and receiving the information. Therefore, the amount of harvested energy at the relay in one time slot is given by [14, 15].

$$E_h = \alpha\tau\eta P_S |h_1|^2, \tag{3}$$

where η denotes the energy conversion efficiency coefficient whose values range from 0 to 1, depending on the harvesting electric circuitry. From (3), the transmission power of the relay is given by

$$P_R = \frac{E_h}{(1 - \alpha)\tau/2} = \frac{2\alpha\eta P_S |h_1|^2}{(1 - \alpha)}. \tag{4}$$

In a specific time slot, if selected the relay R transmits a superposition modulated symbol $x_R = \sqrt{a_1 P_R} x_1 + \sqrt{(1 - a_1) P_R} x_2$ stored in the buffer, where x_1 and x_2 denote information symbols intended for D_1 and D_2, respectively; a_1 is the power allocation coefficient for D_1 and $(1 - a_1)$ represents the power allocation coefficient for D_2. At the end of each time slot, the received signal at the destinations is given by

$$y_{D_i} = \sqrt{P_R} g_i (\sqrt{a_1} x_1 + \sqrt{1 - a_1} x_2) + w_{D_i}. \tag{5}$$

When $|g_1|^2 > |g_2|^2$, based on the NOMA principle, the relay node allocates more power for D_2 in order to balance the fairness of the system performance. Due to the broadcast nature of the wireless environment, the instantaneous signal-to-interference-and noise ratio (SINR) of the $R \to D_2$ link given by

$$\gamma_{D_2}^{x_2} = \frac{(1 - a_1) P_R |g_2|^2}{a_1 P_R |g_2|^2 + N_0}, \tag{6}$$

where the information symbol x_1 is treated as the interference at D_2. D_1 needs to decode its own information symbol x_1. Using the ideal SIC estimation [16], D_1 can remove the estimated information symbol x_2 from the received signal. From (5), the instantaneous SNR and SINR of the $R \to D_1$ link is given by

$$\gamma_{D_1}^{x_2 \to x_1} = \frac{(1 - a_1) P_R |g_1|^2}{a_1 P_R |g_1|^2 + N_0}. \tag{7}$$

$$\gamma_{D_1}^{x_1} = \frac{a_1 P_R |g_1|^2}{N_0}. \tag{8}$$

3   Outage Probability Analysis for Buffer-Aided Relay Systems

In this section, we analyze the outage performance of the downlink SWIPT NOMA relay system for two cases, namely, the end-to-end outage probability and the individual outage probability for each destination node.

3.1   Overall Outage Probability

The overall outage probability (OOP) of the system is defined as the probability that neither the $S \to R$ link nor the $R \to D_i$ links is available for transmission so that the target predefined transmission rate is achieved. In wireless communication links, outage occurs when the output SNR, γ_{e2e}, falls below a certain threshold, $\gamma_{th} = 2^{\frac{2r_0}{1-\alpha}} - 1$. Using theoretical analysis we have derived the exact and approximated closed-form expressions for the OOP of the downlink SWIPT-NOMA relay system. The proof of the Theorem is omitted here due to page limit.

Theorem 1. *The overall outage probability of the downlink SWIPT-NOMA relay system when the relay knows both g_1 and g_2 is given by:*

$$\text{OOP} = 1 - \frac{1}{\Omega_1} \sum_{k=0}^{\infty} \frac{(-1)^k}{k!} \left(\frac{\Psi_{\min}}{\Omega_2} \right)^k \left[\frac{(-1)^k}{(k-1)!} \left(\frac{1}{\Omega_1} \right)^{k-1} \text{Ei} \left(-\frac{\gamma_{th}}{\Omega_1 P_S} \right) + \Delta_1 \right], \tag{9}$$

where $\mathrm{Ei}(x)$ denotes the exponential integral function [17], $\phi = \frac{2\alpha\eta}{1-\alpha}$, $a_1 < \frac{1}{1+\gamma_{th}}$,

$$\Psi_{min} = \min\left\{\frac{\gamma_{th}}{a_1\phi P_S}, \frac{\gamma_{th}}{\phi P_S(1-a_1(1+\gamma_{th}))}\right\}, \quad \Delta_1 = \frac{\exp\left(-\frac{\gamma_{th}}{\Omega_1 P_S}\right)}{\left(\frac{\gamma_{th}}{P_S}\right)^{k-1}}\sum_{j=0}^{k-2}\frac{(-1)^j\left(\frac{\gamma_{th}}{\Omega_1 P_S}\right)^j}{\prod\limits_{\ell=0}^{j}(k-1-\ell)}.$$

In the high SNR regime, the approximated OOP is given by

$$\mathrm{OOP} \approx 2 - \exp\left(-\frac{\gamma_{th}}{\Omega_1 P_S}\right) - \sqrt{\frac{4\Psi_{min}}{\Omega_1\Omega_2}}\mathcal{K}_1\left(\sqrt{\frac{4\Psi_{min}}{\Omega_1\Omega_2}}\right), \tag{10}$$

where $\mathcal{K}_1(\cdot)$ is the first-order modified Bessel function of the second kind.

3.2   Outage Probability at Destination Nodes

In this section, we derive the closed-form expression for the outage probability at the destination nodes. When a destination node suffers from outage, the other can detect its own information symbols. The system may switch to using an orthogonal multiple access scheme such as the code division multiple access (CDMA). After a straighforward mathematical analysis, we have found the closed-form expression for the outage probability of the destination nodes given in Theorem 2. Due to page limit, the proof of the theorem is omitted here.

Theorem 2. *The outage probability at* D_1 *and* D_2 *are given respectively in* (11) *and* (12), $a_1 \le \frac{1}{1+\xi_i}$, $i \in \{1,2\}$.

$$\mathrm{OP}_{D_1} = 1 - \frac{1}{\Omega_1}\sum_{t=0}^{\infty}\frac{(-1)^t}{t!}\left(\frac{\mathcal{Q}_{max}}{\Omega_2}\right)^t\left[\frac{(-1)^t}{(t-1)!}\left(\frac{1}{\Omega_1}\right)^{t-1}\mathrm{Ei}\left(-\frac{\xi_1}{\Omega_1 P_S}\right) + \Delta_2\right], \tag{11}$$

$$\mathrm{OP}_{D_2} = 1 - \frac{1}{\Omega_1}\sum_{m=0}^{\infty}\frac{(-1)^m}{m!}\left(\frac{b}{\Omega_2}\right)^m\left[\frac{(-1)^m}{(m-1)!}\left(\frac{1}{\Omega_1}\right)^{m-1}\mathrm{Ei}\left(-\frac{\xi_2}{\Omega_1 P_S}\right) + \Delta_3\right], \tag{12}$$

where $\mathcal{Q}_{max} = \max\left\{\frac{\xi_1}{a_1\phi P_S}, \frac{\xi_1}{\phi P_S(1-a_1(1+\xi_1))}\right\}$, $b = \frac{\xi_2}{\phi P_S(1-a_1(1+\xi_2))}$, $\xi_1 = 2^{\frac{2r_1}{1-\alpha}} - 1$, $\xi_2 = 2^{\frac{2r_2}{1-\alpha}} - 1$, r_1 *and* r_2 *are the target trans-mission rates at* D_1 *and* D_2, $\Delta_2 = \frac{\exp\left(-\frac{\xi_1}{\Omega_1 P_S}\right)}{\left(\frac{\xi_1}{P_S}\right)^{t-1}}\sum_{k=0}^{t-2}\frac{(-1)^k\left(\frac{1}{\Omega_1}\right)^k\left(\frac{\xi_1}{P_S}\right)^k}{\prod\limits_{\ell=0}^{k}(t-1-\ell)}$,

$$\Delta_3 = \frac{\exp\left(-\frac{\xi_2}{\Omega_1 P_S}\right)}{\left(\frac{\xi_2}{P_S}\right)^{m-1}}\sum_{q=0}^{m-2}\frac{(-1)^q\left(\frac{1}{\Omega_1}\right)^q\left(\frac{\xi_2}{P_S}\right)^q}{\prod\limits_{v=0}^{q}(m-1-v)}.$$

4   Outage Probability with Buffer Aided Relay

In this section, we investigate the outage performance of the downlink SWIPT-NOMA relay system with a buffer-aided relay. For convenience of analysis, we assume that the source node always has data to transmit. We also assume that each packet contains one transmit symbol. During each time slot, either the source or the relay node is selected to transmit a packet. In order for this selection to be possible, the information on the outage states of the links S → R and R → D is required. The system can use one bit for the feedback information from the destination to the relay node. This feedback helps the relay node R to know whether the link R → D is in outage. One more bit, fed back from the relay to the source node, is used to control the state of the source node, i.e. in the transmit or silent mode. The source node transmits data packets to the relay which decodes the packets and stores them in the buffer. The relay node then forwards the packets to the destination node. If the source node is selected to transmit but the link S → R is in outage, the source remains silent. Similarly, if the relay node is selected to transmit but the link R → D is in outage, the relay also stays silent. Since the outage could be avoided the system performance will be improved significantly. Unlike the case without a buffer-aided relay, the outage event in this case is defined as the probability that the relay does not receive and transmit data, i.e. the relay remains silent. To describe the state transition of the buffer-aided relay operation, we denote the outage events of S → R link and R → D link by $\mathcal{O}_{SR}$ and $\mathcal{O}_{RD}$, respectively. The probabilities that these links are not in outage are given respectively by $1 - \mathcal{O}_{SR} = \bar{\mathcal{O}}_{SR}$ and $1 - \mathcal{O}_{RD} = \bar{\mathcal{O}}_{RD}$. Table 1 shows different states of the relay nodes and associated outage probabilities.

Table 1. The relay decision scheme

Case	SR	RD	l	Relay	$\mathcal{OP}$
A	0	0		Silent	$\mathcal{O}_{SR}\mathcal{O}_{RD}$
B	0		$l = 0$	Silent	$\mathcal{O}_{SR}$
C		0	$l = L$	Silent	$\mathcal{O}_{RD}$
D	1	0	$l < L$	Receive	$\bar{\mathcal{O}}_{SR}\mathcal{O}_{RD}$
E	0	1	$l > 0$	Transmit	$\mathcal{O}_{SR}\bar{\mathcal{O}}_{RD}$
F	1	1	$l \geqslant 2$	Transmit	$\bar{\mathcal{O}}_{SR}\bar{\mathcal{O}}_{RD}$
G	1	1	$l \leqslant 1$	Receive	$\bar{\mathcal{O}}_{SR}\bar{\mathcal{O}}_{RD}$

In Table 1, "SR" denotes the link from the source to the relay node, "RD" refers to the link from the relay to the destination node; 'l' and 'L' respectively represent the packets stored in the buffer and the buffer size at the relay node. 'Relay' denotes the state decision of the relay node (silent, receive or transmit), '$\mathcal{OP}$' is the system outage probability. The outage and non-outage links are indicated by '0' and '1', respectively.

To calculate $\mathcal{OP}$ of the system, based on Table 1 we create the Markov chain. We start at the initial state $l = 0$ (i.e. when the buffer is empty). If the link SR is in outage which means the source does not transmit, then the buffer will be empty. In other words, the buffer state moves from $l = 0$ to $l = 0$ with probability of $\mathcal{O}_{\mathrm{SR}}$ (Case B). When the link SR is not in outage, we consider two cases. The first case is when the link RD is in outage (Case D). Consequently, the relay receives the signal, making the buffer state move from $l = 0$ to $l = 1$ with probability $(1 - \mathcal{O}_{\mathrm{SR}})\mathcal{O}_{\mathrm{RD}}$. The second case is when the link RD is not in outage (Case G). The relay receives the signal, making the buffer state move from $l = 0$ to $l = 1$ with probability $(1 - \mathcal{O}_{\mathrm{SR}})(1 - \mathcal{O}_{\mathrm{RD}})$. Combining these two cases shows that the buffer state moves from $l = 0$ to $l = 1$ with probability $1 - \mathcal{O}_{\mathrm{SR}}$. Similarly, we can obtain the probability of moving to the next state. From here, we have the Markov chain showing the state transitions as depicted in Fig. 2. When the buffer is empty ($l = 0$), it stays in the empty state with probability of $\mathcal{O}_{\mathrm{SR}}$ (case B) and receives a packet with probability $1 - \mathcal{O}_{\mathrm{SR}}$ (case D, G). When the buffer has one packet ($l = 1$), it stays in the current state with probability of $\mathcal{O}_{\mathrm{SR}}\mathcal{O}_{\mathrm{RD}}$ if the relay does not receive and transmit (case A). If the relay receives one packet, it moves to the new state ($l = 2$) with probability of $1 - \mathcal{O}_{\mathrm{SR}}$ (case D, G) and returns to the initial state ($l = 0$) with probability $\mathcal{O}_{\mathrm{SR}}(1 - \mathcal{O}_{\mathrm{RD}})$ (case E). When the buffer has l packets ($2 \leqslant l \leqslant L - 1$), it stays in this state with probability of $\mathcal{O}_{\mathrm{SR}}\mathcal{O}_{\mathrm{RD}}$ (case A), receives one packet with probability $(1 - \mathcal{O}_{\mathrm{SR}})\mathcal{O}_{\mathrm{RD}}$ (case D), and transmits one packet with probability $1 - \mathcal{O}_{\mathrm{RD}}$ (case E, F). If the buffer is full, which means that it has L packets, it remains in the same state with probability $\mathcal{O}_{\mathrm{RD}}$ (case C) and transmits one packet with probability $1 - \mathcal{O}_{\mathrm{RD}}$ (case E, F).

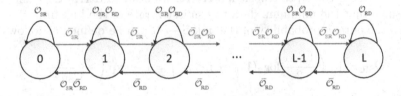

Fig. 2. The Markov chain diagram of buffer states at the relay node.

Using Table 1 and the given Markov chain, the system outage probability is calculated as

$$\mathcal{OP} = \mathcal{O}_{\mathrm{SR}} \Pr\{l = 0\} + \mathcal{O}_{\mathrm{RD}} \Pr\{l = L\} + \mathcal{O}_{\mathrm{SR}}\mathcal{O}_{\mathrm{RD}}(1 - \Pr\{l = 0\} - \Pr\{l = L\}), \quad (13)$$

where $\Pr\{l = 0\}$ and $\Pr\{l = L\}$ are the probabilities of the events that the buffer is empty and full, respectively. To derive $\mathcal{OP}$ of the system in (13), we define a state transition matrix $\mathbf{A}$ with size of $(L + 1) \times (L + 1)$ of the Markov chain, where $\mathbf{A}_{ij}$ denotes the element of the ith row and jth column of the matrix $\mathbf{A}$ which refers to the probability of moving from state i at time t to state j at time $t + 1$, i.e.,

$$\mathbf{A}_{ij} = \Pr\{l_{t+1} = j | l_t = i\}. \quad (14)$$

For the case of $L = 5$, matrix $\mathbf{A}$ is expressed as follows

$$
\mathbf{A} = \begin{pmatrix}
\mathcal{O}_{SR} & \bar{\mathcal{O}}_{SR} & 0 & 0 & 0 & 0 \\
\mathcal{O}_{SR}\bar{\mathcal{O}}_{RD} & \mathcal{O}_{SR}\mathcal{O}_{RD} & \bar{\mathcal{O}}_{SR} & 0 & 0 & 0 \\
0 & \bar{\mathcal{O}}_{RD} & \mathcal{O}_{SR}\mathcal{O}_{RD} & \bar{\mathcal{O}}_{SR}\mathcal{O}_{RD} & 0 & 0 \\
0 & 0 & \bar{\mathcal{O}}_{RD} & \mathcal{O}_{SR}\mathcal{O}_{RD} & \bar{\mathcal{O}}_{SR}\mathcal{O}_{RD} & 0 \\
0 & 0 & 0 & \bar{\mathcal{O}}_{RD} & \mathcal{O}_{SR}\mathcal{O}_{RD} & \bar{\mathcal{O}}_{SR}\mathcal{O}_{RD} \\
0 & 0 & 0 & 0 & \bar{\mathcal{O}}_{RD} & \mathcal{O}_{RD}
\end{pmatrix}. \tag{15}
$$

Note that matrix $\mathbf{A}$ is asymmetric because the states are not symmetric and the number of links to other states is not the same, leading to different transition probabilities. The stationary distribution $\boldsymbol{\pi}$ of the Markov chain is expressed as

$$
\boldsymbol{\pi} = (\mathbf{A} - \mathbf{I} + \mathbf{B})^{-1}\mathbf{b}, \tag{16}
$$

where $\mathbf{I}$ is an identity matrix, $\mathbf{B}$ is an $(L+1) \times (L+1)$ matrix with all elements equal to 1, and $\mathbf{b} = [1\ 1\ ...\ 1]^T$.

Theorem 3. *With the buffer-aided relaying, the outage probability of the down-link SWIPT-NOMA relay system is given by*

$$
\mathcal{OP} = \sum_{i=1}^{L+1} \pi_i \mathbf{A}_{ii}. \tag{17}
$$

To determine the state transit matrix $\mathbf{A}$, we need to derive $\mathcal{O}_{SR}$ and $\mathcal{O}_{RD}$. We assume that the minimum data transmission rate of the link from $S \rightarrow R$ is r_0, then the outage probability of the link $S \rightarrow R$ can be defined as follows

$$
\mathcal{O}_{SR} = \Pr\left(\frac{1-\alpha}{2}\log_2(1+\gamma_R) < r_0\right) = 1 - \exp\left(-\frac{\gamma_{th}}{\Omega_{SR}P_S}\right).
$$

Based on the SIC principle, if D_1 is able to remove x_2 from its received signal, the outage probability of the link $R \rightarrow D_i$ is given by

$$
\mathcal{O}_{RD} = \Pr\left(\frac{1-\alpha}{2}\log_2\left(1+\max\{\gamma_{D_1}^{x_1}, \gamma_{D_2}^{x_2}\}\right) < r_0\right). \tag{18}
$$

After some mathematical manipulations, we obtain

$$
\mathcal{O}_{RD} = 1 - \sqrt{\frac{4\mathcal{A}}{\Omega_{SR}}}\mathcal{K}_1\left(\sqrt{\frac{4\mathcal{A}}{\Omega_{SR}}}\right) - \sqrt{\frac{4\mathcal{B}}{\Omega_{SR}}}\mathcal{K}_1\left(\sqrt{\frac{4\mathcal{B}}{\Omega_{SR}}}\right) + \sqrt{\frac{4\mathcal{C}}{\Omega_{SR}}}\mathcal{K}_1\left(\sqrt{\frac{4\mathcal{C}}{\Omega_{SR}}}\right),
\tag{19}
$$

where $\mathcal{A} = \frac{\gamma_{th}}{\Omega_{RD_1}a_1\phi P_S}$ and $\mathcal{B} = \frac{\gamma_{th}}{\Omega_{RD_2}\phi P_S(1-a_1(1+\gamma_{th}))}$ with $\mathcal{C} = \mathcal{A} + \mathcal{B}$ and $\mathcal{K}_1(\cdot)$ is the first-order modified Bessel function of the second kind.

5   Numerical Results

In this section, numerical results are analyzed to illustrate the outage performance of the downlink SWIPT-NOMA relay system. For comparison, we also provide the performance of the SWIPT-OMA relay system with the same parameters. Configurations and parameters of the systems are explained as follows. D_1 is closer to R than D_2. The power allocation coefficient for D_1 is fixed at $a_1 = 0.3$ and that for D_2 is $1 - a_1$. The energy harvesting fraction is $\alpha = 0.3$ and the energy conversion efficiency is $\eta = 0.95$. The system data rate is $r_1 = 1$ and $r_2 = r_0 = 0.5$ [b/s/Hz].

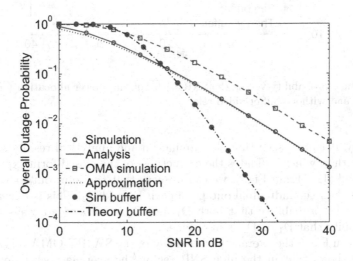

Fig. 3. Overall outage probability versus average SNRs for the case with buffer and without buffer-aided relay.

Figure 3 illustrates the overall outage probability versus the average transmit power of S in two cases, i.e., with and without buffer aided relay. As observed from the figure, the system with buffer-aided relay outperforms that without buffer-aided relay significantly. The outage curves clearly show that the system with a buffer-aided relay achieves diversity order 2 while that without a buffer-aided relay only diversity 1. Furthermore, the approximated results obtained using (10) match well with those using the exact calculation in (9), especially at the high SNR region. Therefore, the approximation in (10) can be used for convenient calculation of OOP. We can also see that the outage performance of the NOMA system is much better than that of the OMA system.

Figure 4 compares the overall outage probability of the system and the outage probability of D_1 and D_2, respectively. We can see that the outage performance of D_1 is better than D_2. This is because the distance from R to D_2 is longer than that from R to D_1 [18]. The overall outage probability is calculated from the probability of the event that both D_1 and D_2 cannot decode their symbols

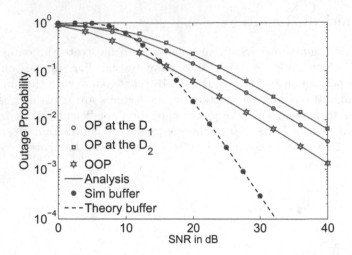

Fig. 4. Outage probability versus the SNR with optimal power allocation for the case with buffer and without buffer-aided relay.

successfully. It can be seen that the simulation and analytical results agree well with each other, which validates the correctness of the closed-form expression of (9), (11) and (12). From Fig. 4, we can observe that the joint outage of D_1 and D_2 are less than the individual outage of each D_1 and D_2. This is relevant since in practice the probability that both D_1 and D_2 in outage is always less than the probability that D_1 or D_2 is in outage.

We plot in Fig. 5 the average packet delay of the SWIPT-OMA relay system. The figure shows that in the high SNR region the average packet delay equal 4 packets. This delay is acceptable when considering the fairness of the outage performance. As a result, depending on the specific requirement of the system we can choose to the relay node with or without the buffer-aid. More specifically, when the system requires better performance, the buffer-aided relay node is used, otherwise, it can use a relay without buffer if the destination node needs to maintain low delay reception. Figure 6 depicts the effect of the power allocation on the outage probability. In order to get the results we derived the coefficient for D_2, the coefficient for D_1 is then given by $1 - a_1$. As shown in the figure different data rates, r_2, exhibit different minimum values of the outage probability. We can see from the figure that when the transmission rate r_2 reduces, the power allocation coefficient for D_2 also decreases to get better system performance for the fairness of D_1 and D_2.

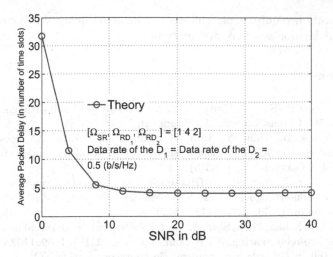

Fig. 5. The average packet delay versus SNR.

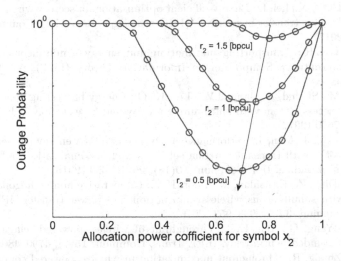

Fig. 6. Effect of power allocation on the outage probability for different data rates; $E_b/N_0 = 10\,\mathrm{dB}$.

6   Conclusion

In this paper, we investigated performance of a downlink NOMA-SWIPT relay system under the assumption that the relay is equipped and not equipped with a buffer for storing data if transmission links are in outage. We derived closed-form expressions for the exact and approximated outage probability of the system. We showed that using a buffer at the relay can increase the outage performance of the system significantly while keeping the incurred delay at an acceptable level. The system then can choose whether to use the buffer at the relay depending on the application requirement. Due to the incurred delay the system is more suitable for delay non-sensitive applications.

References

1. Wang, Y., Ren, B., Sun, S., Kang, S., Yue, X.: Analysis of non-orthogonal multiple access for 5G. China Commun. **13**(Suppl. 2), 52–66 (2016)
2. Dai, L., Wang, B., Yuan, Y., Han, S., Chih-Lin, I., Wang, Z.: Non-orthogonal multiple access for 5G: solutions, challenges, opportunities, and future research trends. IEEE Commun. Mag. **53**(9), 74–81 (2015)
3. Yang, Z., Ding, Z., Fan, P., Al-Dhahir, N.: The impact of power allocation on cooperative non-orthogonal multiple access networks with SWIPT. IEEE Trans. Wirel. Commun. **16**(7), 4332–4343 (2017)
4. Kader, M.F., Shahab, M.B., Shin, S.-Y.: Exploiting non-orthogonal multiple access in cooperative relay sharing. IEEE Commun. Lett. **21**(5), 1159–1162 (2017)
5. Luo, S., Teh, K.C.: Adaptive transmission for cooperative NOMA system with buffer-aided relaying. IEEE Commun. Lett. **21**(4), 937–940 (2017)
6. Du, C., Chen, X., Lei, L.: Energy-efficient optimisation for secrecy wireless information and power transfer in massive MIMO relaying systems. IET Commun. **11**(1), 10–16 (2017)
7. Varshney, L.R.: Transporting information and energy simultaneously. In: 2008 IEEE International Symposium on Information Theory (ISIT), pp. 1612–1616, July 2008
8. Ashraf, M., Shahid, A., Jang, J.W., Lee, K.-G.: Energy harvesting non-orthogonal multiple access system with multi-antenna relay and base station. IEEE Access **5**, 17660–17670 (2017)
9. Han, W., Ge, J., Men, J.: Performance analysis for NOMA energy harvesting relaying networks with transmit antenna selection and maximal-ratio combining over Nakagami-m fading. IET Commun. **10**(18), 2687–2693 (2016)
10. Liu, Y., Ding, Z., Elkashlan, M., Poor, H.V.: Cooperative non-orthogonal multiple access with simultaneous wireless information and power transfer. IEEE J. Sel. Areas Commun. **34**(4), 938–953 (2016)
11. Luo, S., Yang, G., Teh, K.C.: Throughput of wireless-powered relaying systems with buffer-aided hybrid relay. IEEE Trans. Commun. **15**(7), 4790–4801 (2016)
12. Ju, H., Zhang, R.: Throughput maximization in wireless powered communication networks. IEEE Trans. Wirel. Commun. **13**(1), 418–428 (2014)
13. Benjebbour, A., Saito, K., Li, A., Kishiyama, Y., Nakamura, T.: Non-orthogonal multiple access (NOMA): concept and design. In: Signal Processing for 5G: Algorithms and Implementations, pp. 143–168, August 2016
14. Luo, S., Teh, K.C., Wang, W.: Wireless-powered cooperative communications with buffer-aided relay. In: Proceedings of 2016 International Conference on Communications (ICC), pp. 1–6. IEEE, May 2016
15. Michalopoulos, D.S., Suraweera, H.A., Schober, R.: Relay selection for simultaneous information transmission and wireless energy transfer: a tradeoff perspective. IEEE J. Sel. Areas Commun. **33**(8), 1578–1594 (2015)
16. Pedersen, K.I., Kolding, T.E., Seskar, I., Holtzman, J.M.: Practical implementation of successive interference cancellation in DS/CDMA systems. In: 5th IEEE International Conference on Universal Personal Communications, vol. 1, pp. 321–325. IEEE (1996)

17. Zwillinger, D.: Table of Integrals, Series, and Products. Elsevier, Amsterdam (2014)
18. Lu, X., Wang, P., Niyato, D., Kim, D.I., Han, Z.: Wireless networks with RF energy harvesting: a contemporary survey. IEEE Commun. Tutor. **17**(2), 757–789 (2015)

Analyzing Seismic Signal Using Support Vector Machine for Vehicle Motion Detection

Thang Duong Nhat[1]([✉]) and Mai Nguyen Thi Phuong[2]

[1] Center for Training of Excellent Students,
Hanoi University of Science and Technology, Hanoi, Vietnam
thang.dn120885@sis.hust.edu.vn
[2] Department of Precision Mechanical and Optical Engineering,
Hanoi University of Science and Technology, Hanoi, Vietnam
mai.nguyenthiphuong@hust.edu.vn

Abstract. A system to process seismic signals of vehicles passing between two sensor stations had been developed and experimented. To evaluate the feasibility of the system before field test with a real vehicle and to support the classification model with artificial data later, the input seismic data were simulated from Green's method function that accounts only for Rayleigh surface wave. The system using the Machine Learning Classification method SVM to classify data collected from two stations at any time have the state of passed or not. By processing the signal, the system could detect whether the vehicle had passed the crossing line or not with the accuracy of 99.10% for *simulated* data and 94.22% for *experiment* data. The experiment and results suggested that processing seismic signals to monitor control lines is feasible.

Keywords: Machine Learning · Seismic signal · Motion detection

1 Introduction

The monitoring tasks over an area for vehicle detection have many applications, most commonly for security over an interested area. For example, a chemical weapons production facility needs to detect illegal or suspicious movements quickly, so an official can take suitable actions in response. For a wide surveillance zone, it would be costly and difficult to only use personnel to cover the area completely. For alternative technology solutions, many motion detectors are developed with unsuitable characteristic such as the Active Sensors with high energy consumption that needed power supplement regularly are impractical in many cases.

For peace-keeping and security tasks, the system is required to have a medium - large working range, can sustain for a long period of time and does not emit signal that notify any unwanted party. Thus, the author's approach turns to the seismic wave analyzing. This is a passive detection technique, so the equipment

© ICST Institute for Computer Sciences, Social Informatics and Telecommunications Engineering 2019
Published by Springer Nature Switzerland AG 2019. All Rights Reserved
T. Q. Duong and N.-S. Vo (Eds.): INISCOM 2018, LNICST 257, pp. 178–190, 2019.
https://doi.org/10.1007/978-3-030-05873-9_15

would not emit any signal to the environment. At the same time, it is also cheaper, smaller and has a wider range of working than some other sensors.

For the passive detection methods, a large proportion of research is based on processing acoustic signal, image and infrared signal [3,5,9]. Seismic signal, on the other hand, is being studied less because it is more complicated. It consists of different types of wave, propagates in different forms, with different speeds and directions, and are dependent on the geology of the interested environment.

Despise all the difficulty, this is still an attractive research approach for the stated problem. The reasons for that was the seismic waves are less sensitive to Doppler effects, noises introduced by the moving vehicle and atmosphere compare to sound, image, and infrared signals. Seismic wave also holds the possibility for non-line of sight detection at significant range (Table 1).

Table 1. Capabilities of the seismic detection method [10]

Target type	Detection range (m)
Vehicles-wheeled (light)	200
Vehicles-wheeled (heavy)	400
Vehicles-tracked (light)	500
Vehicles- tracked (heavy)	1000

The seismic wave is categorized into two main type: body waves (e.g. compressional (P) wave, shear (S) wave) and surface waves (e.g. Rayleigh wave, Love wave). At the measuring point, the signal collected is the sum of both body waves that propagate in three dimensions through the interior of the earth and the surface waves that propagate in two dimensions through the surface of the earth. This property tells us that the diminishing rate of a signal for surface waves is R^2, much less than R^3 of the body waves. Hence, most of the signal gather at sampling point come from surface waves.

When considering surface waves, Rayleigh wave holds the largest proportion of impact energy, 67%, while that of the shear wave and the compressional wave are 26% and 7% [8]

The seismic wave propagation to the surrounding in a spherical surface for body waves and in a circle for surface waves. Though there are multiple kinds of waves, when the faraway target are detected, only the Rayleigh wave, which accounted the most part of the energy in the seismic wave, can be relied on to analyze. Hence, in this research, the main focused were simulating and analyzing Rayleigh wave for vehicle detection.

2   Experiment

The experiment took place on the site of Hanoi University of Science and Technology, Hanoi, Vietnam in 28 Apr. 2018. The target of the experiment was to collect seismic signal generated from the vehicle motion to test the function of the detection system.

In the experiment, the following equipments was used: two sensor stations each with a geophones LGT-20D10 and a circuit with opamp OPA2134PA for collecting data; 3 marking stations each with one motion sensor module HC-SR501 for marking the begin, the end and the threshold; two notebook computers, two Arduino Uno, battery, relay SRD-12VDC-SL-C, IC LM7805CV TO220 and connecting cables.

The experiment use the default Arduino 10-bit ADC port with internal reference voltage of 5 V. To eliminate the 50 Hz 'hum' noise from the power grid, the Arduinos are power by the Notebook's battery through the connecting cables. The circuit with opamp OPA2134PA was built to offset signal with 2.5 V base and 0 gain. The experiment setup with two Sensor stations S0, S1 and three Marking stations M0, M1, M2 is described in the figure below. When there is motion at M0 or M2, the signal collected from the geophones transmitted to the offset circuit, converted to digital and recorded in a Arduino for each geophone with Sampling frequency of 400 Hz. The M1 sensor marking the threshold would notify both Arduino when the vehicle moved past it and saved labeling data. To synchronize the starting, threshold and stopping moment between both Arduinos, each of the three Marking stations is connected to both Arduinos by connecting cables. After collected, the data were manually transfered into the notebook for later processing (Fig. 1).

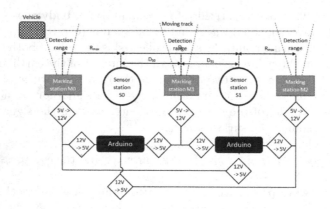

Fig. 1. Experiment setup

The geophones were placed 0.15 m underground and 0.5 m away from roadside of the testing common road made of asphalt. The Marking stations were placed at roadside with the height of 0.25 m from the road surface. The geophone testing range is R_{max} and the distance between the geophones are D. The vehicle were measured at three different speeds V, with three repetitions for each. The speeds were controlled by the driver watching the speedometer so that any acceleration and deceleration happened outside M0-M2 range and the speed was remained constant between them. Recording was started and stopped automatically when the sensors at Marking station detected vehicle motion in the road. The vehicle used in the experiment was a KIA Morning 2015 car.

3   Method

3.1   Simulated Data Generation

Firstly, a simple quarter car model (QCM) had been used by the author to represent the forces exerted to the ground as the vehicle move over irregular surfaces. For wheeled vehicles moving over the perfectly flat ground, the irregularity forces were still presented due to the small gaps in tire treads. To justify QCM as a valid vehicle simplification, several assumptions had been made [7]:

- A point contact patch assumption is deemed sufficient as typical wavelengths of generated Rayleigh waves are greater than the characteristic dimensions of a vehicle.
- Total vehicle mass is distributed evenly to all wheel stations at all time.
- The road surface is rigid.
- Freezing the low frequency 'body bounce' vibration (around 1–2 Hz). This assumption can be made since the generated ground vibrations are usually at high frequency.

In Fig. 2, the F_t represented the force exerted by the compression of the tire spring due to the vertical displacement of the wheel z_r. Thus, the vertical displacement of the wheel could be represented by a Frequency response function (FRF) with the input $z_r(t)$.

The input $z_r(t)$ as shown in Fig. 2 is the elevation changes caused by the tire tread and the irregularity of the surface road. For simplicity, the variation in surface profile over which the wheel (modeled as a point contact) traverses could be estimated as a finite series of a half sine wave pulses.

$$\begin{cases} z_r(t) = z_{r_{max}} \sin(2\pi f_{tr} t) & for \ z_r(t) \geq 0 \\ z_r(t) = 0 & otherwise \end{cases} \tag{1}$$

The frequency of the input f_{tr} could be calculated by the corresponded moving velocity V of the vehicle over the tread pitch a. Using a simple Fourier Integration, the input in the frequency domain had the expression:

$$z_r(\omega) = \int_{-\infty}^{\infty} z_r(t) e^{-i\omega t} dt \tag{2}$$

According to the QCM, the displacement of the wheel from its static position are described in the equation:

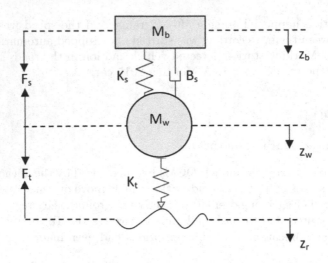

Fig. 2. A quarter car vehicle model

$$M_w \frac{\partial^2 z_w}{\partial^2 t} + B_s \frac{\partial z_w}{\partial t} + (K_t + K_s) z_w = K_t z_w (vt) \qquad (3)$$

where K_t and K_s are the Tire compliance and the Suppressing spring stiffness individually. Using Fourier transform to solve Eq. (3), the elevation of the wheel in the frequency domain is:

$$z_w(\omega) =$$
$$\frac{\omega_1^2 \, z_r(\omega)}{\sqrt{(\omega_0^2 - \omega^2)^2 + (2\omega\alpha)^2}} exp\left[-i \, tan^{-1}\left(\frac{2\omega\alpha}{\omega_0^2 - \omega^2}\right)\right] \qquad (4)$$

where $\omega_0 = ((K_t + K_s)/M_w)^{1/2}$ is the Hop resonance frequency, $\omega_1 = (K_t/M_w)^{1/2}$ is the Tire "bouncing" resonance frequency and $\alpha = B_s/2M_w$ is the Normalized damping coefficient.

Having both the displacement of the wheel and the input signal in the frequency domain, the equation of the force spectrum for a single wheel axle could be established:

$$F_t(\omega) = K_t[(z_w(\omega) - z_r(\omega)] \qquad (5)$$

The QCM described above is only valid for modeling a single axle wheel displacement. Considering the effects of multiple axles, a simple superposition of all-wheel hop displacement responses should be taken to establish the ground force spectra. The wheel hop response differently at each axle differs by a phase shift that depended on the distance from it to the front axle E_{1n} divided by the vehicle forward speed V

$$F_z^{mw}(\omega) = F_z(\omega) \cdot \left(1 + \exp\left(i\omega\frac{E_{12}}{V}\right) + \exp\left(i\omega\frac{E_{13}}{V}\right) \cdots\right.$$

$$\left. + \exp\left(i\omega\frac{E_{1N}}{V}\right)\right) \quad (6)$$

where $F_z(\omega)$ is the force spectrum for a single wheel axle.

Another thing needed to be calculate is the Rayleigh determinant and its derivative:

$$F(k) = \left(2k^2 - k_s^2\right)^2 - 4k^2 v_l v_s \quad (7)$$

where k is the projected distance onto the $z = 0$ plane of the current wave vector, $v_l, s = (k^2 - k_l, s^2)^{1/2}$ are no specified expressions, $k_{l,s} = \omega/c_{l,s}$ are the wavenumber of bulk longitudinal and shear acoustic waves. $c_l = [(\lambda + 2\mu)/\rho]^{1/2}$ is the phase velocities of bulk longitudinal acoustic waves, $c_s = (\mu/\rho)^{1/2}$ is the phase velocities of shear acoustic waves, where $\lambda = 2\mu\sigma/(1-\sigma)$ is the Lame first parameter, σ is the Poisson's ratio, μ is the shear modulus and ρ is the soil mass density.

Note that in the consideration case, the contribution of bulk waves to the ground vibration field generated on the surface are proportional to $(k_l \cdot r)^{-2}$ and $(k_s \cdot r)^{-2}$ respectively for longitudinal and shear waves, where r is the distance from the vibration source to the observation point. For comparison, the Rayleigh waves contribution is proportional only to $(k_R r)^{-1/2}$. Thus, further calculations will take into account only the contribution of Rayleigh surface waves (in Eq. (7): $k = k_R$, where k_R is the wave number of a Rayleigh wave).

Solving equation $F(k) = 0$ [6], the k_R was achieved as the real root of this equation, thus also determined the velocity of Rayleigh waves $c_R = \omega/k_R$. Taking account of attenuation of generated ground vibrations in the ground result in $k_R = (\omega/c_R)(1 + i\gamma)$, where $0 < \gamma \ll 1$ is the Loss factor which describes the linear dependence of a Rayleigh wave attenuation coefficient on frequency ω.

Next, the vibration spectra generated by the vehicle-induced ground forces using Green's function method (taking into account only generated Rayleigh surface waves) were expressed:

$$v_z(\omega) =$$

$$\left(\frac{2\pi}{k_R.r}\right)^{\frac{1}{2}} \frac{(-i\omega)\,k_R k_S v_s v_l}{2\pi\mu F'(k_R)} F_z^{mw}(\omega) \cdot e^{-k_R \gamma.r} \cdot e^{ik_R r - \frac{3\pi}{4}} \quad (8)$$

where $F(k_R)$ is the derivative of $F(k)$ taken at $k = k_R$ and R is the distance from the vibration source to the observation point.

To develop a more robust system, a family of NON-LINEAR moving tracks, which still satisfy the requirement that the moving velocity is a constant, needed to be generated. Thus, the experiment setup was developed and described in Fig. 3.

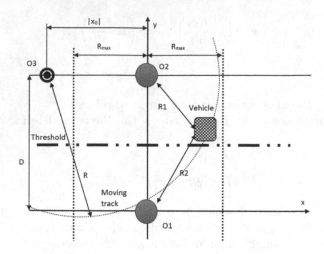

Fig. 3. Simulated experiment setup

where the two sensors were placed at O1 and O2 which have the distance between two points is D and R_{max} is the working radius of the geophone.

There were infinite number of Nonlinear moving tracks from line $y = D$ to line $y = 0$, but to satisfy the requirement of constant moving velocity, the solutions were limited than before. Thus, the moving track of vehicle was chosen as a function of a circle in Descartes coordinate system.

$$(x - x_0)^2 + (y - y_0)^2 = R^2 \tag{9}$$

$$\begin{cases} x = x_0 + R sin \left(\frac{V}{R}t + \varphi\right) \\ y = y_0 + R cos \left(\frac{V}{R}t + \varphi\right) \end{cases} \tag{10}$$

To applied the simulated location r to Eq. (8), notice that the distance r from the vehicle to the observation point here is a constant. Since the vehicle was moving continuously while collecting data, the distance from the vehicle to the observation point was also varied. To use the Eq. (8), some assumption is required to simplify the problem.

The traveling distance of the vehicle were divided into n small bins. An assumption had been made that when the vehicle is inside a bin, the distance between it and the observation point remained unchanged and equaled to the distance of the center of the bin to the observation point. Thus, the Eq. (8) could be used to obtain a vertical displacement spectrum for each bin with its corresponding R_n (Fig. 4).

After acquiring a set of vertical displacement of the wheel in the frequency domain with respect to each Rn to the observation point, performed Inverse Discrete Fourier Transform (Inverse Fast Fourier Transform to be precise) and

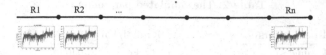

Fig. 4. Vehicle travel distance is divided into small bins in which all data inside a bin have the same distance R_n to the observation point.

merged all data with respect to each bin defined before. The result simulated signal has the form in Fig. 5.

The parameters used in the simulation process were shown in Table 2.

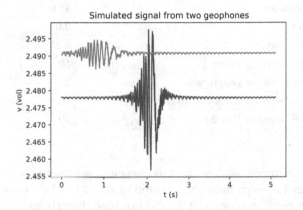

Fig. 5. Simulated signals of 2 sensors. Each peak at the time when the vehicle is closest to the sensors.

3.2 Data Classification with Support Vector Machines (SVM)

SVM is a supervised learning model with associated learning algorithms that analyze data used for classification and regression analysis. It was first proposed by Cortes and Vapnik [4]. In 1992, a method to create a nonlinear classifier is proposed by Boser, Guyon and Vapnik [2]. This method applied Kernel trick, originally proposed by Aizerman et al. [1], to maximum the margin hyperplane. Since there are available library that had been optimized, the Sklearn library had been chosen to implement SVM.

After acquiring the labeled data, an attempt was made to train the model directly with the Raw data by SVM algorithm. The result was very poor and highly sensitive to the skewness of the data. Looking back into Green's function method to find the vertical displacement in Eq. (8), it shown that the displacement was dependence on the distance in the Frequency Domain but not in the Time Domain. Thus, the main features to train the model was chosen to be the transformed data in the frequency domain.

Table 2. The simulated parameters

QCM parameters	Symbol/Unit	Kia Morning 2015
Total vehicle mass	M_v/kg	940
Mass of wheel	M_w/kg	~15
Number of wheel	N_w	4
Vehicle forward velocity	V/kmh^{-1}	5, 10, 15
Tread pitch	a/m	1.409^{-3}
Magnitude of discontinuity	$z_{r_{max}}$/m	0.005
Wheelbases	$E_1 2$/m	2.385
Soil mass density	ρ/kgm^{-3}	1800
Shear modulus	μ/Nm^{-2}	$4x10^7$
Loss factor	γ	0.05
Poisson's ratio	σ	0.25
Geophone testing range	R_{max}/m	10
Distance between geophones	D/m	20
Distance between threshold to S0	D_{S0}/m	10
Distance between threshold to S1	D_{S1}/m	10

The hyperparameter *window size* was chosen to represent the number of data taken to analyzed at each individual sampling moment. As a common practice, the Discrete Fourier Transform (DFT) was not used directly to convert data from the time domain to the frequency domain. Instead, the Fast Fourier Transform (FFT) was chosen for the increase in preprocessing speed. Thus, the hyperparameter *window size* is chosen as $windowsize = 2^n$ with n is a positive integer and *window size* is bound by $windowsize \leq number\ of\ data$. After that, the data was padded with the size according to the *window size*.

Because each feature of a data was not completely independent, the ground velocity varies in different ranges [Min, Max] of different datasets, performing Min Max Scaling or Normalization were needed on all Fourier transformed features. After that, the Min and Max values was added as two new features for each data.

Next, the transformed data set were randomly divided into the Training set, the Cross-Validation set and the Test set with ratio 8:1:1. Then, the Standardization procedure were performed on each feature so that they have $mean = 0$ and $standard\ deviation = 1$. That was the final step of the preprocessing procedure.

After that, the model was trained with the preprocessed Training set using the Sklearn SVM library.

Finally, the hyperparameter *window size* was tested and evaluated on the classification model. Since hyperparameters depend on the characteristic of the dataset, the *window size* was chosen by testing with different value on the dataset to see which would bring the highest accuracy. To test the effect of the different window sizes with different dataset size, a test on multiple simulated datasets were made and the result are shown in Sect. 4.

4   Result and Discussion

4.1   Result

Simulated Data Generation. After choosing the size of the data set as 2048 (2^{11}) number of data, 40 simulated datasets were generated with the randomized starting point, velocity, Non-linear moving tracks as presented in Sect. 3.1. A results of the moving track and simulated signals in the time domain were shown in Fig. 6.

A result of experiment data were shown in Fig. 7.

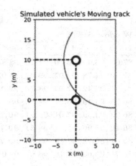

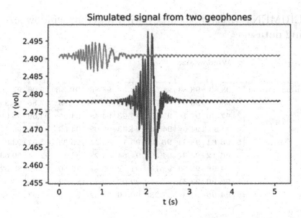

Fig. 6. A sample of track and simulated signal in the time domain.

Data Classification. The effect of the different *window sizes* with different number of SIMULATED dataset can be seen in Table 3:

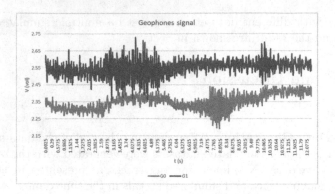

Fig. 7. A sample of track and simulated signal in the time domain.

Table 3. (SIMULATED DATA) Accuracy with different *window sizes* and different number of training datasets

		Window size (2^n)								
		1	2	3	4	5	6	7	8	9
No. of training dataset	5	87.60	93.46	96.00	97.75	98.83	**99.02**	96.88	93.36	88.28
	10	87.60	94.78	97.07	99.02	99.02	**99.17**	98.29	95.85	91.11
	15	87.01	93.42	97.20	98.70	**99.19**	99.12	98.57	96.58	92.84
	20	86.35	94.19	96.97	98.56	98.80	**99.39**	98.93	97.56	94.53
	25	86.99	94.39	97.01	98.81	**99.18**	99.02	**99.18**	97.38	95.02
	30	86.10	93.96	97.51	98.31	**99.19**	**99.19**	98.96	97.84	94.84
	35	85.81	93.71	97.15	98.63	**99.26**	**99.26**	99.20	97.81	94.53
	40	86.01	93.63	97.16	98.68	99.05	**99.28**	99.13	97.96	95.47

Table 4. (EXPERIMENT DATA) Accuracy with different window sizes and different number of training datasets

		Window size (2^n)								
		1	2	3	4	5	6	7	8	9
No. of training dataset	1	98.83	98.83	98.59	98.59	98.83	**99.53**	99.06	97.66	96.96
	2	98.71	98.61	98.71	97.64	**99.04**	98.93	97.86	93.35	93.89
	3	**97.30**	97.18	95.92	95.23	94.48	94.25	94.66	95.00	96.84
	4	93.83	94.68	94.60	94.43	95.02	94.09	94.85	96.38	**97.32**
	5	**95.61**	95.34	**95.61**	95.14	95.47	95.20	93.51	93.95	95.03
	6	**95.12**	93.97	94.59	94.64	94.28	94.19	92.51	90.63	92.15
	7	**94.96**	93.69	93.11	91.12	90.87	89.60	89.80	90.52	90.44
	8	**93.90**	93.81	92.11	90.07	89.35	88.42	87.02	87.67	88.24
	9	**94.90**	93.87	93.23	90.76	90.32	88.03	87.97	88.82	88.84

The effect of the different *window sizes* with different number of EXPERIMENT dataset can be seen in Table 4:

The predicting accuracy of the system trained with all nine EXPERIMENT datasets are shown in Fig. 8 in which the dash-line shown training accuracy while the other shown cross-validation accuracy.

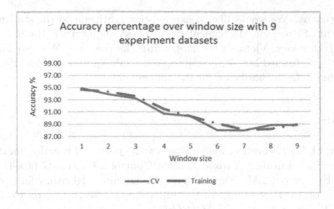

Fig. 8. Accuracy over *window size* with nine datasets..

Choosing the *window size* that brought the highest accuracy in a large size dataset, the *window size = 6* was chosen for SIMULATED data and *window size = 1* for EXPERIMENT data. To not be biased by the cross-validation data set, a final analysis on a Blind Test Set that the model has never seen before was used to evaluate the system and result in the accuracy of 99.10% for SIMULATED data and 94.22% for EXPERIMENT data.

Some conclusions can be made in Fig. 8 and Table 4. Overall, the predicting accuracy of the system are very high. Beside, there were a slight decrease in the accuracy when more datasets was added and the more datasets were used, the smaller the *window size* would bring the highest accuracy. This result implied that the current datasets were still not large enough to generalize the model to reach its best potential. Thus, acquiring more data and combining with the artificial datasets to train the model would much likely increase the predicting accuracy.

4.2 Discussion

At present state, a working system have been constructed to collect and analyze seismic signal generate from a moving vehicle. Preprocessing and Machine Learning techniques are used to classify acquired data thus determined whether the vehicle have passed the threshold or not.

The tests has been conducted only on a common road at HUST university using a 2015 KIA Morning car. Further tests are needed for integrating the artificial data and acquiring more data with different vehicle types to develop more robust analyzing system. With a small modification, the system can be made to work in real-time that satisfy the requirements of a long-lasting, unnoticeable, robust vehicle detection system with large monitoring range.

Acknowledgment. We would like to express our gratitude to the staff members in the Department of Precision Mechanical and Optical Engineering, Hanoi University of Science and Technology, Vietnam for supporting the research. We also would like to show our gratitude toward Mr. Anh Nguyen, Mr. Binh Nguyen and Mr. Phuong Le for their help in building the hardware for the experiment.

References

1. Aizerman, M.A.: Theoretical foundations of the potential function method in pattern recognition learning. Autom. Remote Control **25**, 821–837 (1964)
2. Boser, B.E., Guyon, I.M., Vapnik, V.N.: A training algorithm for optimal margin classifiers. In: Proceedings of the Fifth Annual Workshop on Computational Learning Theory, pp. 144–152. ACM (1992)
3. Choe, H.C., Karlsen, R.E., Gerhart, G.R., Meitzler, T.J.: Wavelet-based ground vehicle recognition using acoustic signals. In: Wavelet Applications III, vol. 2762, pp. 434–446. International Society for Optics and Photonics (1996)
4. Cortes, C., Vapnik, V.: Support vector networks. Mach. Learn. **20**, 273–297 (1995)
5. Estrin, D., Girod, L., Pottie, G., Srivastava, M.: Instrumenting the world with wireless sensor networks. In: Proceedings of 2001 IEEE International Conference on Acoustics, Speech, and Signal Processing, (ICASSP 2001), vol. 4, pp. 2033–2036. IEEE (2001)
6. Krylov, V.V.: Computation of ground vibrations generated by accelerating and braking road vehicles. Modal Anal. **2**(3), 299–321 (1996)
7. Krylov, V.V.: Generation of ground elastic waves by road vehicles. J. Comput. Acoust. **9**(03), 919–933 (2001)
8. Lan, J., Nahavandi, S., Lan, T., Yin, Y.: Recognition of moving ground targets by measuring and processing seismic signal. Measurement **37**(2), 189–199 (2005)
9. Sabatier, J.M., Xiang, N.: An investigation of acoustic-to-seismic coupling to detect buried antitank landmines. IEEE Trans. Geosci. Remote Sens. **39**(6), 1146–1154 (2001)
10. Stotts, L.B.: Unattended-ground-sensor-related technologies: an army perspective. In: Unattended Ground Sensor Technologies and Applications II, vol. 4040, pp. 2–11. International Society for Optics and Photonics (2000)

Smart-IoUT 1.0: A Smart Aquatic Monitoring Network Based on Internet of Underwater Things (IoUT)

Anand Nayyar[1,2(✉)], Cuong Huynh Ba[1,2],
Nguyen Pham Cong Duc[1,2], and Ha Dac Binh[1,2]

[1] Graduate School, Duy Tan University, Da Nang, Vietnam
{anandnayyar, hadacbinh}@duytan.edu.vn,
cuonghuynh18590@gmail.com, congducdtu@gmail.com
[2] Faculty of Electronics and Electrical Engineering,
Duy Tan University, Da Nang, Vietnam

Abstract. Internet of Underwater Things (IoUT) is defined as a network of smart interconnected underwater objects for monitoring/carrying out underwater operations. IoUT enables a system of autonomous underwater vehicles (AUV) communicating with each other, sensing, collecting and transmitting data to control centers above the surface at regular Internet speeds. The information could be a great source to carry out a wide range of tasks like crash surveying, shipwrecks discovery, detection of tsunamis early signs, animal health monitoring as well as collecting real-time aquatic information, archaeological expeditions etc. This paper analyzes the complete terminology of IoUT - Concept, Architecture and challenges. In addition to this, a novel IoUT based working prototype, i.e. Smart IoUT 1.0 is proposed for sensing and collecting underwater information. Smart IoUT 1.0 is equipped with 2 sensing nodes containing 4 different sensors - EZO Dissolved Oxygen Sensor, DS18B20 Temperature sensor, pH analog and Water Turbidity Sensor for acquiring live data and access it anywhere via Internet using Thingspeak.com. The paper provides strong base ground for researchers to tackle serious challenges posed by IoUT as extending the concept of IoT to underwater is completely different.

Keywords: Internet of Underwater Things (IoUT) · Internet of Things (IoT)
Architecture · Challenges · Underwater Sensor Networks (UWSN)
Smart IoUT 1.0 · Node MCU ESP8266 development board
Underwater monitoring · Sensors

1 Introduction

The planet earth is also referred as *"Majestic Blue Marble"* due to prevalence of water on the earth's surface. In other terms, 71% of the earth's surface is covered with water and rest 29% is continents and islands. 96.5% of earth's water is ocean's salt water and only 3.5% remaining is fresh water and other lakes and glaciers. Wireless transmission

© ICST Institute for Computer Sciences, Social Informatics and Telecommunications Engineering 2019
Published by Springer Nature Switzerland AG 2019. All Rights Reserved
T. Q. Duong and N.-S. Vo (Eds.): INISCOM 2018, LNICST 257, pp. 191–207, 2019.
https://doi.org/10.1007/978-3-030-05873-9_16

of information via oceans is one of the enabling technologies, laying strong foundation for the development of ocean-observation systems, smart underwater sensor network (UWSN) and even future oriented Internet of Things enabled UWSN i.e. IoUT. New technologies have enabled new ways for monitoring and sensing aquatic environments via sensors. Underwater sensing has led to diverse applications ranging from simple aquaculture to the oil industry and also includes monitoring of river and sea pollution, oceanographic data collection, natural disturbances prediction, conducting search-survey-rescue missions, marine life study and surveillance. Accordingly, Underwater Sensor Network [1, 8, 13] is highly emerging technology for carrying out all under-water operations. UWSN is regarded as a fusion of Wireless Technology and Smart Sensors with MEMS capability having smart sensing, computing and communication capability. UWSN is defined as, *"Network of Autonomous Sensor Nodes operating Underwater spatially distributed to sense water related parameters like Temperature, Pressure, Oxygen Level and other underwater monitoring"* [3]. The sensed data is communicated back to base station utilized by human beings for diverse research, analysis and other sorts of operations. The sensor nodes operating underwater are either fixed or mobile and connected via wireless communication modules to transfer data. Underwater Sensor Network operates in a scenario where the set of wireless nodes after acquiring the data transmit the data back to buoyant gateway nodes and gateway nodes relay the data back to the control and monitoring station called Remote Station. Underwater Sensor Networks are classified into four different architectures which lays foundation for designing UWSN applications: 1D-UWSN, 2D-UWSN, 3D-UWSN and 4D-UWSN [2, 15, 16]. 1D-UWSN architecture defines autonomous deployment of underwater nodes where every node operates in standalone position and solely responsible for performing all tasks of sensing, processing and transmission back to the remote station. Example: AUV (Autonomous Vehicles) diving underwater, performs sensing and collecting information and transmits the information back to the remote station. 2D-UWSN architecture, defines UWSN sensor nodes deployed underwater in the form of clusters. Every cluster consists of cluster head, i.e. Anchor Node. All nodes collect data and relays back to anchor node via Horizontal link communication and anchor node, in turn, relays, data back to surface buoyant node via Vertical link communication. 3D-UWSN, sensor nodes are deployed like 2D-UWSN in a cluster manner at varying depths and communication happens in three forms - InterCluster - Nodes communication; IntraCluster - Nodes and Anchor Node Communication; Anchor-Buoyant communication. 4D-UWSN architecture is a combination of 1D, 2D, 3D UWSN architecture and comprise of Remotely operated Underwater vehicles. Figure 1 demonstrates 1D, 2D, 3D and 4D-UWSN architectures.

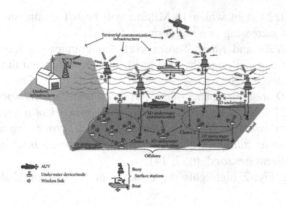

Fig. 1. Underwater wireless sensor networks - architecture

In recent years, lots of extensive interest with regard to modification and enhancement of Underwater Wireless Sensor Network, Smart Cities and Internet of Things (IoT) is carried out and lots of researchers are working towards making Underwater Sensor Network compliant with Internet of Things (IoT). With the use of IoT, UWSN and Smart Sensors cum Tracking Technologies, IoUT is developed. The objective of this research paper is to propose Smart IoUT - A smart IoUT based aquatic monitoring network equipped with lots of UWSN sensors for live monitoring. To the best of our knowledge, this is the first Smart IoUT 1.0 proposed for Live Underwater Sensor Monitoring with IoT capabilities.

Structure of Paper
The paper is structured as follows: Section 2 discusses of Internet of Underwater Things (IoUT) - Evolution, Introduction, Architecture and Challenges surrounding IoUT. Section 3 highlights Smart IoUT 1.0 - A Novel IoUT based solution designed for monitoring underwater environment - Overview, Components - hardware and software used and sheds light on Architecture cum Circuit and Working of sensor nodes. Section 4 gives preview of Smart IoUT 1.0 prototype's real time results captured from live underwater environment via thingspeak.com. Section 5 concludes the paper with future scope.

2   Internet of Underwater Things (IoUT)

2.1   Evolution and Introduction

Internet of Things (IoT) [4] was coined by Kevin Ashton to lay foundation of how IoT can be developed by "adding RF-Identification and other sensors to everyday objects". With the passage of time, IoT has gone to the next level and consists of network of entities, i.e. Physical objects, home appliances, watches and any other devices embedded via electronics, software, sensors, actuators and connectors enabling data exchange. As per the latest report by Ericsson, 29 Billion devices will be connected to the Internet by 2022 out of which 18 Million will be IoT complaint and every year since 2016, IoT is increasing at a staggering rate of 21%.

With IoT, UWSN and Smart Sensor cum Communication technology [9–11], underwater monitoring can go to the next level. IoT has facilitated the design of next-generation underwater wireless sensor network termed as IoUT. The term "IoUT" was first discussed by Domigo in a paper titled "An Overview of the Internet of Underwater Things" in 2012 [5]. IoUT is defined as *"World-Wide Network of Interconnected Smart Underwater objects with digital entity capable of sensing, processing and transmitting information to remote stations with a combination of smart tracking technologies, Internet and Intelligent Sensors"* [6, 7, 14].

The following Fig. 2 highlights the concept of Internet of Underwater Things (IoUT):

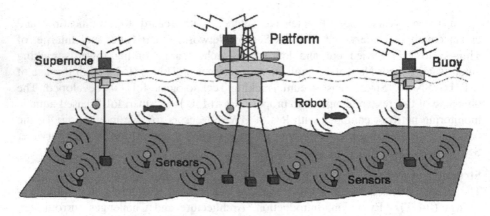

Fig. 2. IoUT (Internet of Underwater Things)

Internet of Underwater Things (IoUT) consists of following components:

IoUT = Underwater Wireless Sensor Networks + Internet + Tracking Technologies + Smart/Intelligent Sensors

With Internet of Underwater Things (IoUT), underwater monitoring becomes more efficient and promising. The following three reasons highlight the importance of IoUT in real-world:

- With 71% of the planet earth covered with only water, researchers, ocean scientists and marine biologists believe that still, till date more than 60% of the ocean's area is unexplored. For diverse explorations and wide area monitoring, IoUT is required.
- IoUT lays a strong foundation for deploying smart and autonomous sensors, especially mobile sensors for wide coverage, sensing lots of underwater properties and even marine life precision monitoring.

- IoUT can also be utilized by military to defend enemy attacks, and in addition to this IoUT can pose a strong advantage towards underwater natural disaster monitoring, unlocking lots of ocean secrets and aquatic life new species discovery.

2.2 Architecture

Figure 3 outlines a standard Internet of Underwater Things (IoUT) architecture.

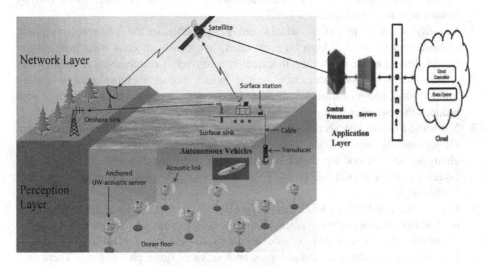

Fig. 3. Standard IoUT architecture

The architecture of IoUT comprise of three layers:

- Perception Layer: Perception layer comprises of all Underwater sensor nodes, Underwater autonomous vehicles (UAVs), surface links and all sorts of monitoring stations. The primary objective of this layer is to collect diverse information regarding underwater objects, aquatic life and all other water properties.
- Network Layer: The task of network layer is to acquire and process information received from the perception layer. This layer comprises of all wired as well as wireless links, cloud platforms, internet and even remote connected servers for storing information.
- Application Layer: It consists of all GUI-based front end applications to analyze the sensed information in the form of data.

2.3 Challenges of Internet of Underwater Things (IoUT)

The established era of IoT cannot be simply extended to underwater as conditions in water are highly opposite as compared to land. Internet of Underwater Things (IoUT) being mix of both Underwater Wireless Sensor Network and Internet of Things (IoT) technology poses serious challenges for scientists, researchers, engineers and even marine technicians to tackle before IoUT becomes reality. Some practical challenges surrounding IoUT are enlisted below [6, 7, 12]:

1. Energy Efficiency: Energy utilization is one of the first and foremost challenging issue surrounding all types of sensor networks and even IoT objects. In order to operate underwater, communication systems and sensor nodes require more power as transmission distance in IoUT is very large. In order to keep sensor nodes operational for prolonged period of time, energy efficient algorithms are required, especially designed for IoUT as traditional and existing energy solutions for UWSN cannot be suitably adapted for IoUT.
2. Security: Lots of malicious attacks can be possible in the underwater aquatic environment because of high bit error rates, low bandwidth and high propagation delays, this in turns, impact the link quality and overall connectivity in underwater communications. As underwater sensor nodes consume more energy during transmission, energy exhaustion attacks can even impact sensor nodes which in turn reduces the overall lifetime of IoUT network.
3. Mobility: For any UWSN network, all the sensor nodes operate in mobile condition. Underwater is impacted via water currents at short durations repeatedly, which changes the network topology of underwater operational nodes. Same mobility issues can even impact IoUT. IoUT demands specially designed mobility models for break-free operation.
4. Lack of Standardization: IoUT has a stringent requirement of standard architectures to address interoperability issues between heterogeneous underwater systems. Currently, there exists NO de jure or de facto standard for IoUT. This makes heterogeneity of devices, technologies and services quite challenging. There is a strong requirement for IoUT protocols to provide interoperability between heterogeneous underwater objects. In addition to this, gateways are required to facilitate communication between underwater sensors and IP-based networks.
5. Transmission Speeds & Propagation Speeds: Like in UWSN, where the transmission rate is near to 10 Kbps and bandwidth wastage is usually reported. Same, is the case with IoUT, less transmission speed impact overall performance. Considering the propagation speed of UWSN which is near to about 1500 m/s, is quite low, IoUT face serious issues with regard to End-to-End delay.

3   Smart IoUT 1.0 - A Smart Aquatic Monitoring Network Based on Internet of Underwater Things (IoUT)

3.1   Overview

Smart IoUT 1.0 is regarded as Smart Aquatic Monitoring Internet of Underwater Things (IoUT) based solution for monitoring various pedagogies of water like Dissolved Oxygen, Temperature, Water Turbidity and pH level. Smart IoUT 1.0 integrates the concept of Underwater Wireless Sensor Network (UWSN) and Internet of Things (IoT) with the objective to measure the water quality of lakes, rivers etc. with smart sensor nodes. Smart IoUT 1.0 is based on the concept of "Plug-Play-Sense" where the entire prototype can just be deployed to the water by starting up and sensors will start sensing, collecting and transmitting real-time data to Internet. Smart IoUT 1.0 is highly cheap, efficient, lightweight and powered via solar power technology for long-term operation. Smart IoUT 1.0 prototype is well tested at MyKhe Beach and Da Nang beach during early morning and late evening and the prototype has given a whopping accuracy of data sensing near to about 99%. The results are best as compared to existing high cost IoUT devices readily available in the market.

3.2 Components Used

This section enlists various Hardware and Software components that make up Smart IoUT 1.0 prototype for carrying out underwater sensing operations.

a. Hardware Components
1. NodeMCU ESP 8266 Development Board: The NodeMCU is SoC-Based ESP8266 Wi-Fi chip development kit with easy interface design and makes use of Arduino compilers for programming and coding. The board integrates GPIO, PWM, IIC, 1-Wire, ADC, USB TTL and PCB Antenna (Fig. 4).

Fig. 4. NodeMCU ESP8266 development board

2. TP4056 Charger: TP4056 is an efficient constant current charger for LiPo/Li-ion batteries. Highly efficient in utilization, especially portal applications because of low external component count and small outline package. Technical Specifications: Automatic recharge, under voltage lockout, current monitor and 2 LED's for charging mode and termination signal. Uses 5 V-input voltage, 4.2 V-charging voltage and 1A-charge current (Fig. 5).

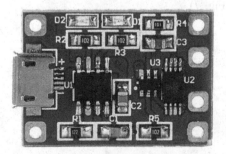

Fig. 5. TP4056 charger

3. ESP8266: ESP8266 Wi-Fi Module integrates Software on Chip (SoC) with TCP/IP protocol stack to facilitate Wi-Fi access to any microcontroller.
Technical Specifications: 802.11b/g/n; Wi-Fi Direct, 1 MB Flash Memory, SDIO 1.1/2.0, SPI, UART.

3.2.1   Sensors

1. Atlas EO Dissolved Oxygen Kit: The Dissolved Oxygen Kit acquires data and give the results in the form of Mg/L. It bundles everything to take precision, full range D. O. readings for environmental monitoring, wine making, fish monitoring etc.

Technical Specifications: Full oxygen readings range from 0.01 to +35.99 mg/l; precision dissolved oxygen reading with accuracy ±0.2; Temperature compensation; Salinity Compensation; Pressure compensation; UART; I2C; HDPE dissolved oxygen probe and operating voltage 3.3 V to 5 V (Fig. 6).

Fig. 6. Atlas EO dissolved oxygen kit

2. DS18B20 Temperature Sensor

The DS18B20 Temperature sensor provides Celsius temperature measurements with 9 to 12-bit precision. The DS18B20 has 64-bit serial code which allows multiple DS18B20s to function on same 1-wire bus.

Technical Specifications: No external components; converts temperature to 12-bit digital word in 750 ms; operational in different form factors like Body Temperature and even underwater (Fig. 7).

Fig. 7. DS18B20 temperature sensor

3. Analog Turbidity Sensor

Analog Turbidity Sensor is capable to detect water quality by measuring turbidity level. The sensor detects suspended particles in water by measuring light transmittance and scattering rate, which changes relative to the amount of total suspended solids (TSS) in the water. With the increase in TSS, the liquid turbidity level increases. The sensor is highly efficient to determine the water quality in swimming pools, rivers and other laboratory-based measurements.

Technical Specifications: Response Time: <500 ms; Analog Output – 0–4.5 V, 100 M Insulation Resistance, Operating Temperature: 5 °C to 90 °C, Operating Voltage: 5 V DC (Fig. 8).

Fig. 8. Analog turbidity sensor

4. pH Sensor: pH sensor is used to determine the hydrogen level of the water. It consists of an LED as Power Indicator, BNC Connector and is well connected to the Arduino controller via analog input port.

Technical Specifications - Measuring range: 0–60 °C, Response time <1 min; measuring range 0–14 pH; 5 V power (Fig. 9).

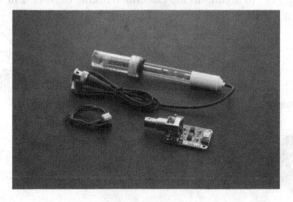

Fig. 9. pH sensor

3.2.2   Software

1. Arduino IDE Environment [17, 18]: Arduino IDE is a platform independent base for Arduino hardware and can run on multiple operating system platforms. It consists of a text editor for code writing, message area, the text console and a toolbar with common functions and menus. It can connect to all sorts of Arduino Boards for uploading programs and communication with boards. Programs in Arduino IDE are written as sketches with file extension. ino and IDE provides a strong interface by displaying all error messages and other information.
2. Thingspeak.com: It is an open source IoT Application and API to store data received from sensors and makes use of the HTTP protocol to display the results in the form of graphs. It is a powerful IoT platform to provide aggregation, visualization and analysis of live data in the cloud. It enables device configuration to send data to thingspeak.com using REST API or MQTT.

3.2.3   Architecture

Smart IoUT 1.0 consists of two sensor nodes: Node 1 and Node 2. This section highlights the electronics circuit and working of Node 1 and Node 2.

Node 1

Node 1 consists of two sensors: Temperature Sensor: DSB1820 and EZO Dissolved Oxygen Sensor. The following Table 1 highlights technical specifications of Node 1.

Table 1. Node 1 technical specifications

Node	Hardware	Unit	Amplitude
Node 1	Solar panel	V	0–9 V
	Tp4056 charger circuit		
	NodeMCU ESP8266 development board		
	Atlas scientific DO kit		
	Battery 5 v 2000 mAh	V	2000 mAh
	EZO dissolved oxygen sensor	mg/L	0.01–100 mg/L
	DS18B20 temperature sensor	°C	−55 °C to +125 °C

The following Fig. 10 highlights the General Working of Node 1 in Smart IoUT 1.0.

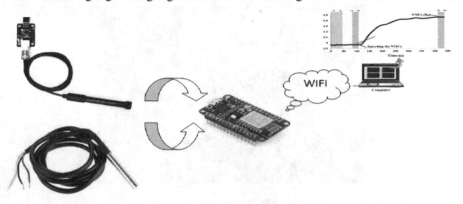

Fig. 10. Node 1 - smart IoUT 1.0

The following Fig. 11 highlights the Schematic/Circuit Diagram of Node 1.

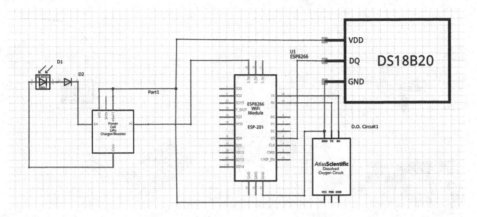

Fig. 11. Node 1 circuit diagram

Circuit Explanation

The node 1 uses the NodeMCU ESP8266 as the central processor. It reads the temperature sensor and oxygen concentration in water for 2 min and sends the data to Thingspeak.com. Node 1 is powered using Solar Panel, to power TP4056 module to charge the battery.

Node 2

Node 2 consists of two sensors: Water Turbidity Sensor and pH Sensor.

The following Table 2 highlights in details of Node 2.

Table 2. Node 2 technical specifications

Node	Sensor	Unit	Amplitude
Node 2	Water turbidity sensor	NTU	0–3000
	pH analog	pH	0–14
	Solar panel	V	0–9 V
	Tp4056 charger circuit		
	NodeMCU ESP8266 development board		
	Battery 5 V 2000 mAh	mAh	2000

The following Fig. 12 highlights the General Working of Node 2 in Smart IoUT 1.0.

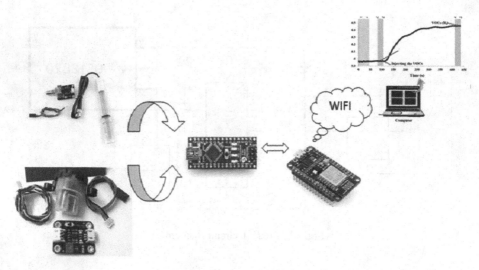

Fig. 12. Node 2 - smart IoUT 1.0

The Following Fig. 13 highlights the Schematic/Circuit Diagram of Node 2.

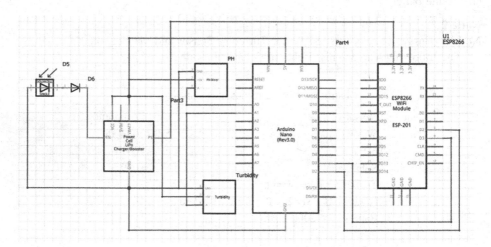

Fig. 13. Node 2 circuit diagram

Circuit Explanation

In Node 2, Arduino Nano reads the sensor values and sends the sensor data to ESP8266. The data is sent online to Thingspeak.com every 2 min for 1 time. The data transmission system uses solar panel to power TP4056 module which in turn charges the battery.

4   Smart IoUT 1.0 - A Smart Aquatic Monitoring Network Based on Internet of Underwater Things (IoUT) - Working Prototype and Results

This section highlights the complete working of Smart IoUT 1.0 and highlights results of the data captured from real-world operating underwater.

4.1   Flowchart and Working

The complete working of Smart IoUT 1.0 can be depicted via Flowchart in Fig. 14.

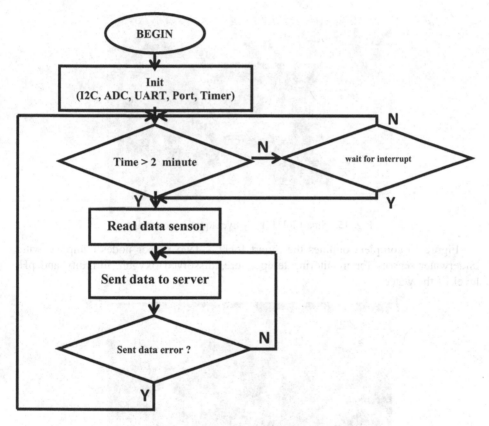

Fig. 14. Smart IoUT 1.0 flowchart

Smart IoUT 1.0 Working:

- First, the main board initializes the initial value such as: I2C, ADC, UART, Timer, the port that will use.
- Second, after initialization completed, the timer value is set for the node time sensor data acquisition. Here we use the duration of 2 min to get the data once. We use interrupts so that when the signal is returned we will send data to the server

immediately. If there is no interrupt or setting time is not timeout, the main board into sleep mode to save energy.

- Next, the collected data is sent to the SERVER over Wi-Fi with the device ESP8266 and communicate with the virtual server https://thingspeak.com/ via http protocols HTTP protocol for the purpose of testing the interface to the server is stable and efficient.

4.2   Live Smart IoUT 1.0 - Prototype - Highlight and Live Demonstration

The following Figs. 15 and 16 highlights the complete ready to use Smart IoUT-1.0 for live aquatic monitoring.

Fig. 15. Smart IoUT 1.0 - live aquatic monitoring

Figure 15 complete outlines the Smart IoUT - Two sensor nodes equipped with underwater sensors for monitoring temperature, dissolved oxygen, turbidity and pH level of the water.

Fig. 16. (Live implementation of smart IoUT 1.0) operational in lake in Da Nang City capturing live water properties

The Fig. 16 outlines the live working of Smart IoUT 1.0 device on MyKhe Beach monitoring the live parameters of underwater.

4.3 Results

The following Fig. 17a–d demonstrates the Live results acquired underwater by Smart IoUT 1.0 in terms of Oxygen Level, Temperature, pH Level and Turbidity Level from water and results are acquired online via Thingspeak.com

The Fig. 17(a) demonstrates the Dissolved oxygen level monitored underwater. The graph shows the variations of mg/L under water. It is observed that pH level variates from 11.1 to 11.14 underwater and the results are quite impressive in sensor detection.

Figure 17(b) gives the temperature readings underwater. It is observed that underwater temperature at the time of measure ranges from 32 °C to 33.5 °C. Observation of temperature like this, helps to study underwater currents.

Figure 17c demonstrates Turbidity level underwater. This measurement is highly important to have a good correlation to the amount of suspended matter in water and use to monitor the level of sediment, algae and particle pollution. The sensor has detected the turbidity level underwater from 2996 to 2998 (NTU). The readings show the presence of suspended solids in the water which are at OK rate. But sometimes, the suspended solids increases which marks change in the graph.

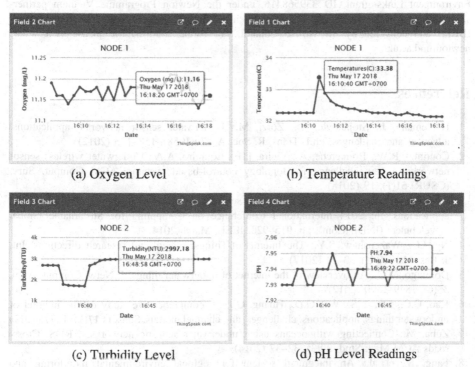

(a) Oxygen Level   (b) Temperature Readings

(c) Turbidity Level   (d) pH Level Readings

Fig. 17. Live Thingspeak.com results - IoT based results - Oxygen, Temperature, Turbidity, pH level - Node 1 and Node 2

Figure 17(d) determines the pH level of the water i.e. Acidity and Alkalinity of water. The value ranges from 0 to 14. The value 7–8 represents neutrality. In the graph, it is observed that value of pH ranges from 7.93 to 7.94 which marks that the water has neutral amount of acidity and alkalinity. This determines that there is less underwater pollution.

5 Conclusion and Future Scope

In this paper, we propose a novel IoUT based prototype "Smart IoUT 1.0" i.e. Smart Internet of Underwater Things 1.0. The research paper provides comprehensive information with regard to IoUT in terms of evolution, architecture, challenges. Smart IoUT 1.0 is the first and one of its kind IoT based device for capturing underwater data and sending it live on Internet using Thingspeak.com. The prototype comprises of two nodes and both the nodes have capability to sense Temperature, Oxygen Level, Turbidity Level as well as pH level underwater.

In the near future, the device would be enhanced with more sensor nodes equipped with more underwater sensors with widened coverage and more work would be done in order to enhance the overall lifetime of the sensor nodes.

Acknowledgement. This work was supported by Newton Prize 2017 and by a Research Environment Links grant, ID 339568416, under the Newton Programme Vietnam partnership. The grant is funded by the UK Department of Business, Energy and Industrial Strategy (BEIS) and delivered by the British Council. For further information, please visit www.newtonfund.ac.uk/.

References

1. Heidemann, J., Stojanovic, M., Zorzi, M.: Underwater sensor networks: applications, advances and challenges. Phil. Trans. R. Soc. A **370**(1958), 158–175 (2012)
2. Coutinho, R.W., Boukerche, A., Vieira, L.F., Loureiro, A.A.: Underwater wireless sensor networks: a new challenge for topology control-based systems. ACM Comput. Surv. (CSUR) **51**(1), 19 (2018)
3. Bhambri, H., Swaroop, A.: Underwater sensor network: architectures, challenges and applications. In: 2014 International Conference on Computing for Sustainable Global Development (INDIACom), pp. 915–920. IEEE, March 2014
4. Ng, I.C., Wakenshaw, S.Y.: The Internet-of-Things: review and research directions. Int. J. Res. Mark. **34**(1), 3–21 (2017)
5. Domingo, M.C.: An overview of the internet of underwater things. J. Netw. Comput. Appl. **35**(6), 1879–1890 (2012)
6. Kao, C.C., Lin, Y.S., Wu, G.D., Huang, C.J.: A comprehensive study on the internet of underwater things: applications, challenges, and channel models. Sensors **17**(7), 1477 (2017)
7. Zuba, M.: Connecting with oceans using underwater acoustic networks. XRDS: Crossroads ACM Mag. Stud. **20**(3), 32–37 (2014)
8. Fang, S., et al.: An integrated system for regional environmental monitoring and management based on internet of things. IEEE Trans. Ind. Inf. **10**(2), 1596–1605 (2014)

9. Xu, M., Liu, L.: Sender-receiver role-based energy-aware scheduling for Internet of underwater things. IEEE Trans. Emerg. Top. Comput. (99), 1 (2016)
10. Jiang, S.: On the marine internet and its potential applications for underwater inter-networking. In: Proceedings of the Eighth ACM International Conference on Underwater Networks and Systems, p. 13. ACM, November 2013
11. Jiang, L., Da Xu, L., Cai, H., Jiang, Z., Bu, F., Xu, B.: An IoT-oriented data storage framework in cloud computing platform. IEEE Trans. Ind. Inf. **10**(2), 1443–1451 (2014)
12. Li, S., Da Xu, L., Zhao, S.: The Internet of Things: a survey. Inf. Syst. Front. **17**(2), 243–259 (2015)
13. Xu, G., Shen, W., Wang, X.: Applications of wireless sensor networks in marine environment monitoring: a survey. Sensors **14**(9), 16932–16954 (2014)
14. Derawi, M., Zhang, H.: Internet of Things in real-life—a great understanding. In: Zeng, Q.-A. (ed.) Wireless Communications, Networking and Applications. LNEE, vol. 348, pp. 337–350. Springer, New Delhi (2016). https://doi.org/10.1007/978-81-322-2580-5_32
15. Nayyar, A., Puri, V., Le, D.N.: Comprehensive analysis of routing protocols surrounding underwater sensor networks (UWSNs). In: Balas, V., Sharma, N., Chakrabarti, A. (eds.) Data Management, Analytics and Innovation. AISC, vol. 808, pp. 435–450. Springer, Singapore (2019). https://doi.org/10.1007/978-981-13-1402-5_33
16. Nayyar, A., Balas, V.E.: Analysis of simulation tools for underwater sensor networks (UWSNs). In: Bhattacharyya, S., Hassanien, A., Gupta, D., Khanna, A., Pan, I. (eds.) International Conference on Innovative Computing and Communications. LNNS, vol. 55, pp. 165–180. Springer, Singapore (2019). https://doi.org/10.1007/978-981-13-2324-9_17
17. Nayyar, A., Puri, V.: A review of Arduino board's, Lilypad's & Arduino shields. In: 2016 3rd International Conference on Computing for Sustainable Global Development (INDIACom), pp. 1485–1492. IEEE, March 2016
18. Nayyar, A.: An encyclopedia coverage of compiler's, programmer's & simulator's for 8051, PIC, AVR, ARM, Arduino embedded technologies. Int. J. Reconfigurable Embed. Syst. **5**(1) (2016)

Wireless Power Transfer Under Secure Communication with Multiple Antennas and Eavesdroppers

Duc-Dung Tran[1], Dac-Binh Ha[1(✉)], and Anand Nayyar[2]

[1] Faculty of Electrical and Electronics Engineering, Duy Tan University, Danang, Vietnam
dung.td.1227@gmail.com, hadacbinh@duytan.edu.vn
[2] Graduate School, Duy Tan University, Danang, Vietnam
anandnayyar@duytan.edu.vn

Abstract. In this paper, we analyze the physical layer secrecy performance of a 5G radio frequency energy harvesting (RF-EH) network in the presence of multiple passive eavesdroppers. In this system, the source is considered as an energy-limited node, hence it harvests energy from RF signals generated by a power transfer station to use for information transmission. Additionally, in order to enhance the energy harvesting and system performance, the source is equipped with multiple antennas and employs maximal ratio combining (MRC) and transmit antenna selection (TAS) techniques to exploit the benefits of spatial diversity. Given these settings, the exact close-form expressions of existence probability of secrecy capacity and secrecy outage probability are derived. Furthermore, the obtained results indicate that multiple antennas technique applied at the source not only facilitates energy harvesting but also improves secrecy performance of the investigated network. Finally, Monte-Carlo simulation is provided to confirm our analytical results.

Keywords: Energy harvesting · Maximal ratio combining
Secrecy capacity · Transmit antenna selection · Wireless power transfer

1 Introduction

In recent years, a new solution, namely Radio Frequency (RF) energy harvesting (EH) has ben proposed for powering wireless devices in 5G network [1–3]. This technology helps to prolong the lifetime of the devices without recharging or replacing the batteries which is usually inconvenient or even impossible (e.g., medical devices embedded inside human bodies). More importantly, it can bring energy-constrained communication nodes the possibility of collecting energy and receiving information simultaneously.

On the other hand, many recent research works have been devoted to physical layer security (PLS) [4–6]. Unlike the traditional secure methods (e.g., upper-layer cryptographic protocols), which provide security under the assumptions

© ICST Institute for Computer Sciences, Social Informatics and Telecommunications Engineering 2019
Published by Springer Nature Switzerland AG 2019. All Rights Reserved
T. Q. Duong and N.-S. Vo (Eds.): INISCOM 2018, LNICST 257, pp. 208–220, 2019.
https://doi.org/10.1007/978-3-030-05873-9_17

of limited computational capability at the eavesdroppers and error-free physical link, PLS regarded as a complement to the cryptographic technology, improves secrecy performance of wireless networks by exploiting the nature of wireless channels [7,8].

As to the RF-EH network, PLS has been studied to evaluate the information security ability of this network [8–14]. Specifically, the works in [8,10] have addressed secure cooperative communication in RF-EH networks. He et al. [11] has investigated the optimal placement of the EH node with PLS considerations by solving two optimization problems: maximizing the average EH power subject to a secrecy outage constraint and minimizing the secrecy outage probability subject to an EH constraint. In [12], PLS in single-input single-output (SISO) simultaneous wireless information and power transfer (SWIPT) system has been evaluated by analyzing the optimization problems of transmit power allocations and power splitting ratios. Moreover, to utilize the advantages of spatial diversity, multiple antennas technique has been studied in [13,14]. In particular, the work in [13] has investigated a MIMO information-energy broadcast system. The authors have supposed that a multi-antenna source can transmit information and energy simultaneously to a multi-antenna information receiver and a multi-antenna energy receiver, where the energy receiver is considered as an eavesdropper. Also, they has proposed global optimal solutions to the secrecy rate maximization (SRM) problem. Furthermore, the authors in [14] have formulated the resource allocation algorithm design for secure MISO communication systems with RF energy harvesting receivers as a non-convex optimization problem. Given these studies, it can be concluded that besides reducing the received signal at eavesdroppers, multiple antennas technique can improve the received signals at both IRs and ERs. As a result, it can enhance the secrecy performance and RF-EH. However, to the best of our knowledge, the secrecy performance analysis for multi-antenna information source considered as energy harvesting nodes has not been studied.

In this paper, we focus on anlyzing the secrecy performance of a wireless power transfer system with multiple antennas at the source over Rayleigh fading channels. In particular, the network consists of one power transfer station, one energy-constrained source and one destination in the presence of multiple passive eavesdroppers. The source is equipped with multiple antennas to facilitate the energy harvesting and information transmission processes, meanwhile all the remaining nodes have a single antenna. Our contributions can be summarized as follows. First, we propose a communication protocol which helps to improve energy collecting capability of the source and secrecy performance of the system. Second, we derive the exact close-form expressions of existence probability of secrecy capacity and secrecy outage probability. Third, we evaluate the secrecy performance of the considered system under the influence of various system parameters, namely energy harvesting time, energy conversion efficiency, source location, and average transmit signal-to-noise ratio (SNR). Finally, we confirm the analytical results by using Monte Carlo simulation.

The remainder of this paper is organized as follows. System and channel model is presented in Sect. 2. Secrecy performance of the considered system is analyzed in Sect. 3. The numerical results are discussed in Sect. 4. We conclude our work with future scope in Sect. 5.

Notation: boldface lowercase letters are used to denote vectors. $\mathcal{CN}(0, N_0)$ is a scalar complex Gaussian distribution with zero mean and variance N_0. $\mathbb{E}[\cdot]$ indicates the expectation operator. $|\cdot|$ and $\|\cdot\|$ denote the absolute value and the Euclidean norm, respectively. $\mathcal{K}_\nu(\cdot)$ stands for the νth – order modified Bessel function of the second kind.

2   System and Channel Model

In this paper, we investigate a secure communication system with energy harvesting, as illustrated in Fig. 1. The network consists of one power transfer station denoted by P, one energy-constrained information source denoted by S, one destination denoted by D and K passive eavesdroppers denoted by $E_k(1 \leq k \leq K)$. It is assumed that P, D, and E_k are equipped with a single antenna, meanwhile S has N_S antennas. Our proposed protocol divides the overall communication process into two phases. In the first phase, source S harvests energy from the power transfer station P by using MRC scheme to combine the received signals from N_S antennas. In the second phase, source S transmits information to destination D by employing TAS scheme to reduce the signal processing cost and choose the best antenna which maximizes the instantaneous SNR at the destination (i.e., maximize channel coefficient of $S \rightarrow D$ link) [7]. Moreover, we assume that all the channels subject to Rayleigh distribution.

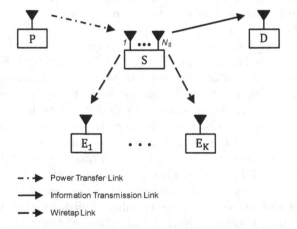

Fig. 1. System model for secure communication with energy harvesting.

First, the source S harvests energy from the power station P in the time duration of by using MRC technique. The power used for the second phase is given by [15]

$$P_S = \frac{E_h}{(1-\alpha)T} = \frac{\eta\alpha P_0\|\mathbf{h}_{PS}\|^2}{(1-\alpha)d_S^{\theta_S}} = \frac{\eta\alpha P_0}{1-\alpha}\gamma_S, \tag{1}$$

where, $1 < \eta \leq 1$ is the energy conversion efficiency which depends on the rectification process and the energy harvesting circuitry; P_0 is the transmit power of the power station; T is the block time in which a certain block of information is conveyed from the source node to the destination node; α is the fraction of the block time in which source S collects energy from the RF signal produced by the power station P, $0 < \alpha < 1$. Considering the link from the power station to the source, $\|\mathbf{h}_{PS}\|^2$ is the channel power gain following exponential distribution, with $\mathbf{h}_{PS}$ is $1 \times N_S$ channel vector between an antenna at the power station P and N_S antennas at the source S, $\gamma_S = \frac{\|\mathbf{h}_{PS}\|^2}{d_S^{\theta_S}}$, d_S is the distance, θ_S is the path-loss exponent. The probability density function (PDF) of γ_S has the following form [7]

$$f_{\gamma_S}(x) = \frac{x^{N_S-1}e^{-\frac{x}{\lambda_S}}}{\Gamma(N_S)\lambda_S^{N_S}}, \tag{2}$$

where, $\Gamma(\cdot)$ is the Gamma function, $\lambda_S = \frac{[\|\mathbf{h}_{PS}\|^2]}{d_S^{\theta_S}}$.

In the remaining duration of $(1-\alpha)T$, the source transmits signal $x(t)$ to the destination by applying TAS scheme and the signal received at D can be presented as

$$y(t) = \frac{\sqrt{P_S}h_{SD}}{\sqrt{d_D^{\theta_D}}}x(t) + n_D, \tag{3}$$

where, h_{SD} is the channel coefficient of $S \rightarrow D$ link with $|h_{SD}|^2 = \max_{1\leq i\leq N_S}\left\{|h_{SD,i}|^2\right\}$, d_D and θ_D are the distance and path-loss exponent of $S \rightarrow D$ link, respectively, n_D is white complex Gaussian noise, $n_D \sim \mathcal{CN}(0, N_0)$.

The eavesdroppers try to extract the transmitted information at S without active attack. The signal received at E_k $(1 \leq k \leq K)$ is given by

$$z_k(t) = \sqrt{\frac{P_S}{d_{E,k}^{\theta_{E,k}}}}h_{SE,k}y(t) + n_{E,k}, \tag{4}$$

where, $h_{SE,k}$ is the channel coefficient of $S \rightarrow E_k$ link, $d_{E,k}$ and $\theta_{E,k}$ are the distance and path-loss exponent from source S to the eavesdropper E_k, respectively. $n_{E,k}$ is white complex Gaussian noise, $n_{E,k} \sim \mathcal{CN}(0, N_0)$.

The instantaneous received SNR at the destination D and the eavesdropper E_k are, respectively, presented as

$$\gamma_{SD} = \frac{\eta\alpha P_0\|\mathbf{h}_{PS}\|^2|h_{SD}|^2}{(1-\alpha)\,N_0 d_S^{\theta_S} d_D^{\theta_D}} = a\gamma_S\gamma_D, \tag{5}$$

$$\gamma_{SE,k} = \frac{\eta\alpha P_0\|\mathbf{h}_{PS}\|^2|h_{SE,k}|^2}{(1-\alpha)\,N_0 d_S^{\theta_S} d_{E,k}^{\theta_{E,k}}} = a\gamma_S\gamma_{E,k}, \tag{6}$$

where, $a = \frac{\eta\alpha\gamma_0}{1-\alpha}$, $\gamma_S = \frac{\|\mathbf{h}_{PS}\|^2}{d_S^\theta}$, $\gamma_D = \frac{|h_{SD}|^2}{d_D^\theta}$, $\gamma_{E,k} = \frac{|h_{SE,k}|^2}{d_{E,k}^\theta}$, $\gamma_0 = \frac{P_0}{N_0}$ is the average transmit SNR. The cummulative density function (CDF) of γ_D and $\gamma_{E,k}$ are, respectively, given by

$$F_{\gamma_D}(x) = \left(1 - e^{-\frac{x}{\lambda_D}}\right)^{N_S}$$

$$\overset{(o)}{=} 1 + \sum_{p=1}^{N_S} \binom{N_S}{p} (-1)^p e^{-\frac{px}{\lambda_D}}, \tag{7}$$

and

$$F_{\gamma_{E,k}}(x) = 1 - e^{-\frac{x}{\lambda_{E,k}}}, \tag{8}$$

where (o) is obtained by applying ([16], Eq. (1.111)) for binomial expansion, $\lambda_D = \frac{(|h_{SD}|^2)}{d_D^\theta}$ and $\lambda_{E,k} = \frac{(|h_{SE,k}|^2)}{d_{E,k}^\theta}$.

3   Secrecy Performance Analysis

3.1   Preliminaries

In this subsection, secrecy capacity and joint CDF of γ_{SD} and $\gamma_{SE,k}$ $(1 \leq k \leq K)$ are presented.

The instantaneous secrecy capacity corresponding to the k-th eavesdropper is defined as [5]

$$C_{S,k} = \max\{0, C_{SD} - C_{SE,k}\}$$
$$= \begin{cases} 0, & \gamma_{SD} \leq \gamma_{SE,k} \\ \log_2\left(\frac{1+\gamma_{SD}}{1+\gamma_{SE,k}}\right), & \gamma_{SD} > \gamma_{SE,k} \end{cases}, \tag{9}$$

where, $C_{SD} = \log_2(1 + \gamma_{SD})$ and $C_{SE,k} = \log_2(1 + \gamma_{SE,k})$ are the instantaneous channel capacities of the legal and k-th eavesdropper channels, respectively.

Lemma 1. *Let X and Y be γ_{SD} and $\gamma_{SE,k}$ $(1 \leq k \leq K)$, respectively. Under Rayleigh fading, the joint CDF of X and Y is derived as*

$$F_{X,Y}^{(k)}(x,y) = 1 - \frac{t^{N_S}\mathcal{K}_{N_S}(t)}{\Gamma(N_S)2^{N_S-1}} + \sum_{p=1}^{N_S} \binom{N_S}{p} \frac{(-1)^p}{\Gamma(N_S)2^{N_S-1}} \tag{10}$$
$$\times \left[u^{N_S}\mathcal{K}_{N_S}(u) - v^{N_S}\mathcal{K}_{N_S}(v)\right],$$

where, $\binom{N_S}{p} = \frac{N_S!}{(N_S-p)!p!}$ *is the binomial coefficient,* $X = \gamma_{SD}$, $Y = \gamma_{SE,k}$, $t = 2\sqrt{\frac{y}{a\lambda_S\lambda_{E,k}}}$, $u = 2\sqrt{\frac{px}{a\lambda_S\lambda_D}}$, $v = 2\sqrt{\frac{p\lambda_{E,k}x+\lambda_D y}{a\lambda_S\lambda_D\lambda_{E,k}}}$.

Proof. See Appendix A.

3.2 Existence Probability of Secrecy Capacity

At this point, we investigate the existence probability of secrecy capacity (P_{CS}), which is one of the important measures to evaluate the secrecy performance of a wireless network. For the k-th eavesdropper, it is computed as

$$P_{CS,k} = \Pr(C_{S,k} > 0)$$

$$= \int_0^\infty \int_0^x f_{\gamma_{SD},\gamma_{SE,k}}(x,y)dydx$$

$$= \int_0^\infty \left[\frac{\partial F_{\gamma_{SD},\gamma_{SE,k}}(x,y)}{\partial x}\right]_{y=x} dx. \tag{11}$$

Proposition 1. *In the presence of K eavesdroppers, the existence probability of secrecy capacity of the considered system over Rayleigh fading channels can be derived as*

$$P_{CS} = \prod_{k=1}^K \sum_{p=1}^{N_S} \binom{N_S}{p} \frac{(-1)^{p+1}\lambda_D}{p\lambda_{E,k}+\lambda_D}. \tag{12}$$

Proof. See Appendix B.

3.3 Secrecy Outage Probability

Another prominent measure which is used to evaluate the secrecy performance is secrecy outage probability (P_{Out}). It is defined as the probability that the instantaneous secrecy capacity less than a given secrecy transmission rate, $R > 0$. Given the definition, P_{Out} is formulated by (for the k-th eavesdropper)

$$P_{Out,k} = \Pr(C_{S,k} < R)$$

$$= \int_0^\infty \int_0^{2^R(1+y)-1} f_{\gamma_{SD},\gamma_{SE,k}}(x,y)dxdy$$

$$= \int_0^\infty \left[\frac{\partial F_{\gamma_{SD},\gamma_{SE,k}}(x,y)}{\partial y}\right]_{x=2^R(1+y)-1} dx. \tag{13}$$

Proposition 2. *In the presence of K eavesdroppers, the secrecy outage probability of the considered system over Rayleigh fading channels can be derived as*

$$P_{Out} = 1 - \prod_{k=1}^K \sum_{p=1}^{N_S} \binom{N_S}{p} \frac{(-1)^{p+1}\lambda_D}{\Gamma(N_S)\,2^{N_S-1}\,(p2^R\lambda_{E,k}+\lambda_D)}$$

$$\times \left[2\sqrt{\frac{p(2^R-1)}{a\lambda_S\lambda_D}}\right]^{N_S} \mathcal{K}_{N_S}\left[2\sqrt{\frac{p(2^R-1)}{a\lambda_S\lambda_D}}\right]. \tag{14}$$

Proof. See Appendix C.

4   Numerical Results and Discussion

In this section, numerical results in terms of the existence probability of secrecy capacity (P_{CS}) and the secrecy outage probability (P_{Out}) are presented to clarify the physical layer secrecy performance of the considered system.

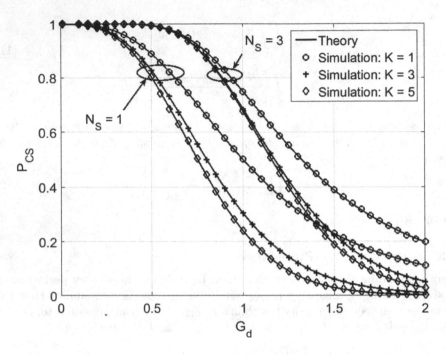

Fig. 2. P_{CS} vs. G_d with $\alpha = 0.5$, $\eta = 1$, $d_S = 2$ (m), $\theta_S = \theta_D = \theta_E = 3$, $\gamma_0 = 10$ (dB).

Accounting for the path-loss exponent of the wiretap links, we set $\theta_{E,1} = \theta_{E,2} = ... = \theta_{E,K} = \theta_E$. Furthermore, K eavesdroppers are randomly located (following uniform distribution) in a circle of radius 5 (m) from the source. To evaluate the impact of the distances d_D and $d_{E,k}$ $(1 \leq k \leq K)$ on the secrecy performance parameters $(P_{CS}$ and $P_{Out})$, we define a factor $G_d = \frac{d_D}{d_E}$, with $d_E = \min_{1 \leq k \leq K} \{d_{E,k}\}$. It can be seen that when $G_d < 1$, the position of the destination is closer to the source than that of all the K eavesdroppers. Additionally, as $G_d > 1$, some of eavesdroppers or all of them are placed closer to the source than the destination.

Figures 2 and 3 plots P_{CS} and P_{Out} versus G_d, respectively. Moreover, different values of N_S and K are considered in these figures. It can be seen that while the larger number of antennas at the source (N_S) leads to the growth of the secrecy performance $(P_{CS}$ increases and P_{Out} decreases), it decreases $(P_{CS}$ decreases and P_{Out} increases) with respect to the upward trend in the number of eavesdroppers (K). Moreover, we can see that the increase in G_d makes P_{CS}

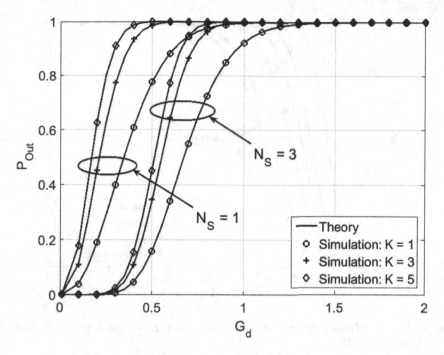

Fig. 3. P_{Out} vs. G_d with $\alpha = 0.5$, $\eta = 1$, $d_S = 2$ (m), $R = 1$ (bit/s/Hz), $\theta_S = \theta_D = \theta_E = 3$, $\gamma_0 = 10$ (dB).

degrade and P_{Out} upgrade. This can be explained that the greater G_d is, the larger $S \rightarrow D$ distance is, hence the capacity of legal channel is much smaller than that of illegal channels. When $G_d \rightarrow 0$ then $P_{CS} \rightarrow 1$ and $P_{Out} \rightarrow 0$, and when $G_d > 1$ then $P_{CS} \rightarrow 0$ and $P_{Out} \rightarrow 1$.

As can be seen from (12) and (14) that P_{CS} does not depend on α, η and γ_0 (i.e., it is not influenced by the amount of harvested energy at the source) but P_{Out} is observed as a function of these system parameters as shown in Figs. 4 and 5, respectively. Specifically, it is apparent from Fig. 4 that when α and η scale up (i.e., the amount of harvested energy increases) then P_{Out} scales down, hence the secrecy performance of the considered system is improved. The similar conclusions is also obtained when γ_0 increases as illustrated in Fig. 5.

Generally, from the aforementioned results (in Figs. 2, 3, 4, and 5), one can conclude that the secrecy performance of the considered system is better with the increase in the number of antennas as well as the amount of harvested energy at the source. Furthermore, the presence of multiple eavesdroppers makes the secrecy performance get worse. Finally, our analysis is confirmed by the excelent match between the analytical and simulation results.

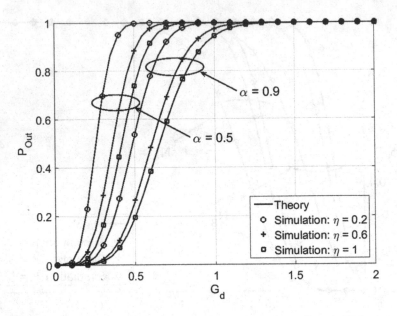

Fig. 4. P_{Out} vs. G_d with $N_S = 2$, $K = 3$, $d_S = 2$ (m), $R = 1$ (bits/s/Hz), $\theta_S = \theta_D = \theta_E = 3$, $\gamma_0 = 10$ (dB).

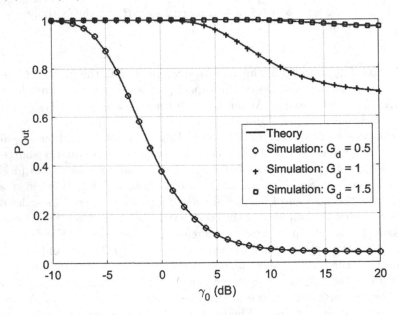

Fig. 5. P_{Out} vs. γ_0 with $\alpha = 0.5$, $\eta = 1$, $N_S = 2$, $K = 3$, $d_S = 2$ (m), $R = 1$ (bit/s/Hz), $\theta_S = \theta_D = \theta_E = 3$.

5   Conclusion and Future Scope

In this paper, we have investigated a RF-EH network under the secure communication over Rayleigh fading channels, in which the energy-constrained source is equipped with multiple antennas. Furthermore, the MRC and TAS schemes have been employed at the source to exploit the benefits of multiple antennas. To analyze the secrecy performance of the considered system, we have derived the exact close-form expressions of existence probability of secrecy capacity and secrecy outage probability. Also, by means of these analytical results, the impact of various system parameters, such as energy harvesting time, energy conversion efficiency, source location, and average transmit SNR, on the secrecy performance have been evaluated. Furthermore, our results have showed that by applying multiple antennas at the source and using MRC as well as TAS techniques, both secrecy performance of the considered system and EH capability of the source are improved.

In the future work, we will extend this system model to multiuser schedule scheme with hybrid energy harvesting architecture, i.e., the combination of time switching and power splitting, to improve the secrecy performance of wireless networks.

Acknowledgement. This work was supported by Newton Prize 2017 and by a Research Environment Links grant, ID 339568416, under the Newton Programme Vietnam partnership. The grant is funded by the UK Department of Business, Energy and Industrial Strategy (BEIS) and delivered by the British Council. For further information, please visit www.newtonfund.ac.uk/.

Appendix A: Proof of Lemma 1

According to the probability definition, the joint CDF of X and Y is given by

$$
\begin{aligned}
F_{X,Y}^{(k)}(x,y) &= \Pr\left(X < x, Y < y\right) \\
&= \Pr\left(\gamma_D < \frac{x}{a\gamma_S}, \gamma_{E,k} < \frac{y}{a\gamma_S}\right) \\
&= \int_0^\infty F_{\gamma_D}\left(\frac{x}{az}\right) F_{\gamma_{E,k}}\left(\frac{y}{az}\right) f_{\gamma_S}(z)dz.
\end{aligned}
\tag{A.1}
$$

Substituting (2), (7), and (8) into (A.1) and then using ([16], Eq. (3.471.9)), the final result is obtained as shown in (10).

Appendix B: Proof of Proposition 1

By employing ([16], Eq. (6.561.16)) and ([16], Eq. (8.486.14)), (11) is expanded as

$$
\begin{aligned}
P_{CS,k} &= \sum_{p=1}^{\infty} \binom{N_S}{p} \frac{(-1)^p 2p}{\Gamma(N_S) 2^{N_S-1} a \lambda_S \lambda_D} \left\{ \int_0^{\infty} v_x^{N_S-1} \mathcal{K}_{N_S-1}[v_x] \, dx \right. \\
&\qquad\qquad \left. - \int_0^{\infty} (u)^{N_S-1} \mathcal{K}_{N_S-1}(u) \, dx \right\} \\
&= \sum_{p=1}^{\infty} \binom{N_S}{p} \frac{(-1)^p}{\Gamma(N_S) 2^{N_S-1}} \left\{ \frac{p \lambda_{E,k}}{p \lambda_{E,k} + \lambda_D} \int_0^{\infty} t^{N_S} \mathcal{K}_{N_S-1}(t) \, dx \right. \qquad\text{(B.1)} \\
&\qquad\qquad \left. - \int_0^{\infty} t^{N_S} \mathcal{K}_{N_S-1}(t) \, dx \right\} \\
&= \sum_{p=1}^{\infty} \binom{N_S}{p} \frac{(-1)^{p+1} \lambda_D}{p \lambda_{E,k} + \lambda_D},
\end{aligned}
$$

where, $v_x = 2\sqrt{\frac{(p\lambda_{E,k}+\lambda_D)x}{a\lambda_S\lambda_D\lambda_{E,k}}}$ and $u = 2\sqrt{\frac{px}{a\lambda_S\lambda_D}}$.

In the presence of K eavesdroppers, the existence probability of secrecy capacity is given by

$$
\begin{aligned}
P_{CS} &= \Pr\left(C_{S,1} > 0, ..., C_{S,K} > 0\right) \\
&= \prod_{k=1}^{K} \Pr\left(C_{S,k} > 0\right).
\end{aligned}
\qquad\text{(B.2)}
$$

The final result of P_{CS} in (12) is obtained by substituting (B.1) into (B.2).

Appendix C: Proof of Proposition 2

Similar to the process of calculating (11) in Appendix B, the integral in (13) is solved by the help of ([16], Eq. (6.561.16)) and ([16], Eq. (8.486.14)) and the result is indicated in (C.1) as follows:

$$P_{Out,k} = \frac{2}{\Gamma(N_S) \, 2^{N_S-1} a \lambda_S \lambda_{SE,k}} \left\{ \int_0^\infty (t)^{N_S-1} \mathcal{K}_{N_S-1}(t) \, dy + \sum_{p=1}^{N_S} \binom{N_S}{p} (-1)^p \right.$$

$$\left. \times \int_0^\infty [w]^{N_S-1} \mathcal{K}_{N_S-1}[w] \, dy \right\}$$

$$= \frac{1}{\Gamma(N_S) \, 2^{N_S-1}} \left\{ \int_0^\infty t^{N_S} \mathcal{K}_{N_S-1}(t) \, dt \right.$$

$$\left. + \sum_{p=1}^{N_S} \binom{N_S}{p} \frac{(-1)^p \lambda_D}{p 2^R \lambda_{E,k} + \lambda_D} \int_u^\infty t^{N_S} \mathcal{K}_{N_S-1}(t) \, dt \right\}$$

$$= 1 - \sum_{p=1}^{N_S} \binom{N_S}{p} \frac{(-1)^{p+1} \lambda_D}{\Gamma(N_S) \, 2^{N_S-1} (p 2^R \lambda_{E,k} + \lambda_D)}$$

$$\times \left[2\sqrt{\frac{p(2^R-1)}{a\lambda_S\lambda_D}} \right]^{N_S} \mathcal{K}_{N_S} \left[2\sqrt{\frac{p(2^R-1)}{a\lambda_S\lambda_D}} \right],$$

$$\text{(C.1)}$$

where, $t = 2\sqrt{\frac{y}{a\lambda_S\lambda_{E,k}}}$ and $w = 2\sqrt{\frac{(p 2^R \lambda_{E,k} + \lambda_D) y + p\lambda_{E,k}(2^R-1)}{a\lambda_S\lambda_D\lambda_{E,k}}}$.

For K eavesdroppers, the secrecy outage probability is computed as

$$P_{Out} = 1 - \Pr(C_{S,1} \geq R, ..., C_{S,K} \geq R)$$

$$= 1 - \prod_{k=1}^{K} \Pr(C_{S,k} \geq R) \qquad \text{(C.2)}$$

$$= 1 - \prod_{k=1}^{K} [1 - \Pr(C_{S,k} < R)].$$

Substituting (C.1) into (C.2), the final result of P_{Out} in (14) is derived.

References

1. Lu, X., Wang, P., Niyato, D., Kim, D.I., Han, Z.: Wireless networks with RF energy harvesting: a contemporary survey. IEEE Commun. Surv. Tutor. **17**(2), 757–789 (2015)
2. Krikidis, I., Timotheou, S., Nikolaou, S., Zheng, G., Ng, D.W.K., Schober, R.: Simultaneous wireless information and power transfer in modern communication systems. IEEE Commun. Mag. **52**(11), 104–110 (2014)
3. Ha, D.B., Tran, D.D., Tran, H.V., Hong, E.K.: Performance of amplify-and-forward relaying with wireless power transfer over dissimilar channels. Elektronika ir Elektrotechnika J. **21**(5), 90–95 (2015)

4. Bloch, M., Barros, J., Rodrigues, M.R., McLaughlin, S.W.: Wireless information-theoretic security. IEEE Trans. Inform. Theory **54**(6), 2515–2534 (2008)
5. Tran, D.D., Ha, D.-B., Tran-Ha, V., Hong, E.K.: Secrecy analysis with MRC/SC-based eavesdropper over heterogeneous channels. IETE J. Res. **61**(4), 363–371 (2015)
6. Liu, Y., Wang, L., Duy, T.T., Elkashlan, M., Duong, T.Q.: Relay selection for security enhancement in cognitive relay networks. IEEE Wirel. Commun. Lett. **4**, 46–49 (2015)
7. Yang, N., Yeoh, P.L., Elkashlan, M.: Transmit antenna selection for security enhancement in MIMO wiretap channels. IEEE Trans. Comm. **61**(1), 144–154 (2013)
8. Li, Q., Zhang, Q., Qin, J.: Secure relay beamforming for simultaneous wireless information and power transfer in nonregenerative relay networks. IEEE Trans. Veh. Technol. **63**(5), 2462–2467 (2014)
9. Kalamkar, S.S., Banerjee, A.: Secure communication via a wireless energy harvesting untrusted relay. IEEE Trans. Veh. Technol. **66**(3), 2199–2213 (2017)
10. Chen, X., Chen, J., Liu, T.: Secure wireless information and power transfer in large-scale MIMO relaying systems with imperfect CSI. In: IEEE GLOBECOM, pp. 4131–4136 (2014)
11. He, B., Zhou, X.: On the placement of RF energy harvesting node in wireless networks with secrecy considerations. In: IEEE Globecom Workshops, pp. 1355–1360 (2014)
12. Xing, H., Liu, L., Zhang, R.: Secrecy wireless information and power transfer in fading wiretap channel. IEEE Trans. Veh. Technol. **65**(1), 180–190 (2016)
13. Shi, Q., Xu, W., Wu, J., Song, E., Wang, Y.: Secure beamforming for MIMO broadcasting with wireless information and power transfer. IEEE Trans. Wirel. Commun. **14**(5), 2841–2853 (2015)
14. Ng, D.W.K., Lo, E.S., Schober, R.: Robust beamforming for secure communication in systems with wireless information and power transfer. IEEE Trans. Wirel. Commun. **13**(8), 4599–4615 (2014)
15. Nasir, A.A., Zhou, X., Durrani, S., Kennedy, R.A.: Relaying protocols for wireless energy harvesting and information processing. IEEE Trans. Wirel. Commun. **12**(7), 3622–3636 (2013)
16. Gradshteyn, I., Ryzhik, I.: Table of Integrals, Series, and Products, 7th edn. Academic Press, Cambridge (2007)

Impact of Opportunistic Transmission on MCIK-OFDM: Diversity and Coding Gains

Thien Van Luong and Youngwook Ko$^{(\boxtimes)}$

Institute of ECIT, Queen's University Belfast, Belfast BT3 9DT, UK
{tluong01,y.ko}@qub.ac.uk

Abstract. This work proposes an opportunistic scheduling scheme for Multi-Carrier Index Keying - Orthogonal Frequency Division Multiplexing (MCIK-OFDM), which is termed as OS-MCIK-OFDM. Particularly, in every transmission, the proposed scheme allows only one machine whose worst sub-channel is the maximum among several machines' worst sub-channels, to communicate with the central device, employing MCIK-OFDM technique. As a result, OS-MCIK-OFDM can harvest the multi-user diversity gain to enhance the reliability of MCIK-OFDM, especially when the number of machines increases. For performance analysis, we derive the closed-form expression for the symbol error probability (SEP), which is then asymptotically analyzed to develop unique features that can address achievable diversity and coding gains, as well as impacts of system parameters. Finally, simulation results are presented to validate the accuracy of the derived SEP performance of OS-MCIK-OFDM and specifically its superiority over the opportunistic scheduling OFDM.

Keywords: MICK-OFDM · OFDM-IM · Index modulation · IM
Opportunistic scheduling · Symbol error probability (SEP)

1 Introduction

Multi-carrier Index Keying - Orthogonal Frequency Division Multiplexing (MCIK-OFDM) or the so-called OFDM with Index Modulation (OFDM-IM) has been recently proposed as an appealing multi-carrier modulation for future wireless communication systems [1,2]. Unlike the classical OFDM, in every MCIK-OFDM transmission, only a subset of sub-carriers are activated to carry data bits through not only the M-ary modulation symbols, but also the indices of active sub-carriers. This interestingly makes MCIK-OFDM more reliable and energy efficient than the conventional OFDM. In addition, MCIK-OFDM can provide a better trade-off between the reliability and the spectral efficiency (SE), just by changing the number of active sub-carriers.

A wide range of MCIK-OFDM concepts have been proposed to improve either the bit error rate (BER) or the SE performance, which can be found in the survey [3]. For example, in [4,5], the simple repetition code is proposed to MCIK-OFDM and the resulting scheme called as ReMO can provide the better BER

© ICST Institute for Computer Sciences, Social Informatics and Telecommunications Engineering 2019
Published by Springer Nature Switzerland AG 2019. All Rights Reserved
T. Q. Duong and N.-S. Vo (Eds.): INISCOM 2018, LNICST 257, pp. 221–231, 2019.
https://doi.org/10.1007/978-3-030-05873-9_18

than more complex IM systems in [6,7]. A tight bound on the BER is derived in [8], while the achievable rate and the outage probability of MCIK-OFDM are analyzed in [9,10], respectively. The combinations of MCIK-OFDM with diversity reception techniques and the low-complexity greedy detector are presented in [11,12]. To increase the SE, in [13], the authors apply the IM concept to both in-phase and quadrature components to double index bits. Besides, a number of methods aiming at boosting the SE can be seen in [14,15]. The application of MCIK-OFDM to the multi-input multi-output (MIMO) framework is presented in [16]. Recently, impacts of channel state information (CSI) uncertainty on the symbol error probability (SEP) and the BER of MCIK-OFDM are investigated in [17,18], respectively. In [19], a number of spreading matrices are employed to noticeably increase the diversity gain of MCIK-OFDM.

It is worth mentioning that none of existing works has explored the multi-user diversity improvement in the MCIK-OFDM framework. Thus, in this paper, we study the opportunity of employing the multi-user diversity scheme to enhance the reliability of MCIK-OFDM. More specifically, main contributions of this work are as follows:

- We first propose an opportunistic scheduling MCIK-OFDM (OS-MCIK-OFDM) scheme, in which only one machine with the worst sub-channel being maximum is selected to communicate with the central device, employing MCIK-OFDM technique in every transmission. Such the scheduling strategy can enable the proposed scheme to harvest a notable multi-machine diversity gain when the number of manchine increases.
- The average SEP is derived in closed-form, which is then asymptotically analyzed to provide insights into diversity and coding gains achieved by OS-MCIK-OFDM, and effects of system parameters on the SEP performance. We also theoretically compare the average SEPs between the proposed scheme and the opportunistic scheduling OFDM (OS-OFDM) with the same scheduling criterion.
- Simulations results are provided to validate the performance of our proposed scheme and the theoretical analysis. It is shown that OS-MCIK-OFDM significantly outperforms OS-OFDM in terms of the SEP performance.

The rest of the paper is organized as follows. Section 2 introduces the system model. The SEP performance analysis is carried out in Sect. 3. Section 4 presents simulation results, while Sect. 5 concludes the paper.[1]

2   System Model

Consider an uplink multicarrier multi-user system where Q machines communicate with a central access point (AP), using the MCIK-OFDM techique. Assume

---

[1] *Notation:* Upper-case bold and lower-case bold letters represent matrices and vectors, respectively. $(.)^T$, $\|.\|$, $C(,)$ and $\lfloor.\rfloor$ denote the transpose operator and the Frobenious norm, the binomial coefficient and the floor function, respectively. $\mathcal{CN}(0, \sigma^2)$ denotes the complex Gaussian distribution with zero mean and variance σ^2. $\mathbb{E}\{.\}$ and $\mathcal{M}(.)$ denote the average value and the moment generating function, respectively.

that all machines and the AP are equipped with a single antenna. At each times-lot, only one out of Q machines is opportunistically selected to send its data to the AP, employing MCIK-OFDM with N available sub-carriers.

In particular, denote by $h_q(\alpha)$ the channel from the q-th machine to the AP at sub-carrier α, where $q = 1, ..., Q$ and $\alpha = 1, ..., N$. Suppose that $h_q(\alpha)$ represents the Rayleigh fading channels as being independent and identically distributed (i.i.d) complex Gaussian random variable (RVs) with zero mean and unit variance, i.e., $h_q(\alpha) \sim \mathcal{CN}(0, 1)$. Notice from [17] that the error performance of MCIK-OFDM mainly depends on the worst sub-channel, we propose an opportunistic scheduling scheme for MCIK-OFDM, called as OS-MCIK-OFDM, where the AP selects the q^*-th machine to communicate, satisfying

$$q^* = \arg \left\{ \max_{q=1,...,Q} \left[\min_{\alpha=1,...,N} \left(|h_q(\alpha)|^2 \right) \right] \right\}. \tag{1}$$

Here, the AP is assumed to perfectly know the CSI from all Q machines to conduct such the scheduling strategy. It is noteworthy that by exploiting the multi-machine diversity, OS-MCIK-OFDM is expected to improve the error performance over the classical MCIK-OFDM.

As for the MCIK-OFDM transmission, the selected machine dynamically activates only K out of N sub-carriers to carry data bits via K complex M-ary symbols and indices of active sub-carriers, while the remaining $N - K$ ones are zero padded. Specifically, in every transmission, p incoming data bits are divided into two bit streams ($p = p_1 + p_2$). The first p_1 bits enters an index mapper to identify a combination of K active indices, which is denoted by $\theta = \{\alpha_1, ..., \alpha_K\}$, where $\alpha_k \in \{1, ..., N\}$ for $k = 1, ..., K$. The index set θ can be referred as an index symbol. While, the remaining p_2 bits enters an M-ary mapper to be mapped to K data symbols denoted by $\mathbf{s} = [s(\alpha_1), ..., s(\alpha_K)]$, where $s(\alpha_k) \in \mathcal{S}$ and $\mathcal{S}$ is the M-ary constellation. Combining θ and $\mathbf{s}$, the MCIK-OFDM transmitted vector of the selected machine is generated as $\mathbf{x}_{q^*} = [x_{q^*}(1), ..., x_{q^*}(N)]^T$, where $x_{q^*}(\alpha) = s(\alpha)$ for $\alpha \in \theta$ and $x_{q^*}(\alpha) = 0$ for $\alpha \notin \theta$. For each $\alpha \in \theta$, assume $\mathbb{E}\left\{ |x_{q^*}(\alpha)|^2 \right\} = \varphi E_s$, where E_s and φ are the average power per M-ary symbol and the power allocation coefficient, respectively.

For given N and K, there are $C(N, K)$ possible combinations of active sub-carrier indices, thus the number of index bits is given by $p_1 = \lfloor \log_2 C(N, K) \rfloor$. Meanwhile, the number of bits conveyed via K non-zero data symbols is $p_2 = K \log_2 M$. Consequently, the data rate of system can be given as

$$R = \frac{\lfloor \log_2 C(N, K) \rfloor + K \log_2 M}{N}. \tag{2}$$

At the AP, the received signal from the selected machine in the frequency domain, is given by

$$\mathbf{y}_{q^*} = \mathbf{H}_{q^*} \mathbf{x}_{q^*} + \mathbf{n}_{q^*}, \tag{3}$$

where $\mathbf{H}_{q^*} = \text{diag}[h_{q^*}(1), ..., h_{q^*}(N)]$ is the channel matrix, and $\mathbf{n}_{q^*}$ denotes the additive white Gaussian noise (AWGN) vector of machine q^* with its entries satisfying $n_{q^*}(\alpha) \sim \mathcal{CN}(0, N_0)$. Denote by $\bar{\gamma}$ the average signal-to-noise ratio (SNR) per active sub-carrier, which is given as $\bar{\gamma} = \varphi E_s / N_0$.

For signal detection, the maximum likelihood (ML) is employed, carrying out an exhaustive search as follows

$$\left(\hat{\theta}, \hat{\mathbf{s}}\right) = \arg\min_{\theta, \mathbf{s}} \|\mathbf{y}_{q^*} - \mathbf{H}_{q^*}\mathbf{x}_{q^*}\|^2. \tag{4}$$

Then, $\hat{\theta}$ and $\hat{\mathbf{s}}$ are used to recover p transmitted data bits. Next, we analyze the SEP performance of the proposed OS-MCIK-OFDM with the ML detection.

3   SEP Performance Analysis

To derive the closed-form expression for the SEP of OS-MCIK-OFDM using the ML, it is essential to know distribution characteristics of the instantaneous SNRs of the selected machine. Thus, we introduce the following important theorem and its corollary.

Theorem 1. *Denote* $\gamma_{q,\alpha} = \bar{\gamma}|h_q(\alpha)|^2$ *be the instantaneous SNR of the q-th machine at sub-carrier α. Based on (1), the selected machine satisfies $q^* = \arg\{\max_{q=1,\ldots,Q}[\min_{\alpha=1,\ldots,N}(\gamma_{q,\alpha})]\}$. Then, the moment generating function (MGF) of $\gamma_{q^*,\alpha}$ is approximated, at high SNRs, by*

$$\mathcal{M}_{\gamma_{q^*,\alpha}}(t) \approx \frac{N^{Q-1}Q!}{\prod_{q=1}^{Q}(qN - t\bar{\gamma})}. \tag{5}$$

Proof. See Appendix A.

Notice that although we can not find out the exact MGF of $\gamma_{q^*,\alpha}$, its approximation in (5) is still capable of providing a tight closed-form expression for the SEP of OS-MCIK-OFDM.

Corollary 1. *Denote* $\gamma_{\Sigma_{q^*}} = \gamma_{q^*,\alpha} + \gamma_{q^*,\tilde{\alpha}}$, *where* $\gamma_{q^*,\alpha} = \bar{\gamma}|h_{q^*}(\alpha)|^2$ *and* $\alpha \neq \tilde{\alpha} = 1, \ldots, N$, *then the MGF of $\gamma_{\Sigma_{q^*}}$ can be approximated by*

$$\mathcal{M}_{\gamma_{\Sigma_{q^*}}}(t) \approx \left[\frac{N^{Q-1}Q!}{\prod_{q=1}^{Q}(qN - t\bar{\gamma})}\right]^2. \tag{6}$$

Proof. Since $\gamma_{q^*,\alpha}$ and $\gamma_{q^*,\tilde{\alpha}}$ are the i.i.d RVs, we have $\mathcal{M}_{\gamma_{\Sigma_{q^*}}}(t) = \mathcal{M}_{\gamma_{q^*,\alpha}}^2(t)$, where $\mathcal{M}_{\gamma_{q^*,\alpha}}(t)$ is given in (5). This concludes the proof.

3.1   SEP Derivation

We first recall the definition of a symbol error event for MCIK-OFDM in [17] that a symbol error event occurs if any of the $K + 1$ symbols including K non-zero M-ary symbols and one index symbol θ are incorrectly decoded. As a result, there are a maximum of $K+1$ symbols in error events. Based on this definition, the average SEP of OS-MCIK-OFDM can be given by [17]

$$\overline{P}_s \leq \frac{\overline{P}_I \left(2 - \frac{1}{M}\right) + K\overline{P}_M}{K+1}, \tag{7}$$

where $\overline{P}_M$ denotes the average SEP of classical M-ary symbols and $\overline{P}_I$ is the average index error probability (IEP). Notice that both $\overline{P}_I$ and $\overline{P}_M$ depend on the channel characteristics of the selected machine. Particularly, utilizing Theorem 1, $\overline{P}_M$ of the q^*-th machine using M-ary PSK symbols can be obtained as in the following lemma.

Lemma 1. *The average SEP of classical M-ary symbols (PSK) of the selected machine q^* in (1) is approximated by*

$$\overline{P}_M \approx \frac{\xi\Psi}{12} \left[\frac{1}{\prod_{q=1}^{Q}(qN + \rho\bar{\gamma})} + \frac{3}{\prod_{q=1}^{Q}(qN + 4\rho\bar{\gamma}/3)} \right], \tag{8}$$

where $\xi = 1, 2$ for $M = 2$ and $M > 2$, respectively, $\Psi = N^{Q-1}Q!$ and $\rho = \sin^2(\pi/M)$.

Proof. See Appendix B.

The instantaneous IEP of machine q^* using the ML (denoted by P_I) can approximated by [17]

$$P_I \approx \frac{K}{N} \sum_{\alpha=1}^{N} \sum_{\tilde{\alpha}\neq\alpha=1}^{N-K} Q\left(\sqrt{\frac{\gamma_{q^*,\alpha} + \gamma_{q^*,\tilde{\alpha}}}{2}}\right), \tag{9}$$

where $\gamma_{\Sigma_{q^*}} = \gamma_{q^*,\alpha} + \gamma_{q^*,\tilde{\alpha}}$. Applying the approximation of Q-function as $Q(x) \leq \frac{1}{2}e^{-\frac{x^2}{2}}$ to (9), we obtain

$$P_I \approx \frac{K}{2N} \sum_{\alpha=1}^{N} \sum_{\tilde{\alpha}\neq\alpha=1}^{N-K} e^{-\frac{\gamma_{q^*,\alpha} + \gamma_{q^*,\tilde{\alpha}}}{4}}. \tag{10}$$

Here, utilizing the MGF approach to (9) with the MGF of $\gamma_{\Sigma_{q^*}}$ given in Corollary 1, we attain

$$\overline{P}_I \approx \frac{K(N-K)}{2} \left[\frac{4^Q N^{Q-1}Q!}{\prod_{q=1}^{Q}(4qN + \bar{\gamma})} \right]^2. \tag{11}$$

Finally, plugging (8) and (11) into (7), the average SEP of the selected machine is obtained in the closed-form as

$$\overline{P}_s \approx \frac{K(N-K)\left(2-\frac{1}{M}\right)}{2(K+1)} \left[\frac{4^Q N^{Q-1}Q!}{\prod_{q=1}^{Q}(4qN + \bar{\gamma})} \right]^2$$

$$+ \frac{\xi K N^{Q-1}Q!}{12(K+1)} \left[\frac{1}{\prod_{q=1}^{Q}(qN + \rho\bar{\gamma})} + \frac{3}{\prod_{q=1}^{Q}(qN + \frac{4\rho\bar{\gamma}}{3})} \right]. \tag{12}$$

3.2 Asymptotic Analysis

We asymptotically investigate the SEP performance of OS-MCIK-OFDM to provide insights into achievable diversity and coding gains, as well as impacts of system parameters such as N, K and Q on the performance.

Specifically, at high SNRs, the average SEP in (12) can be asymptotically approximated by

$$\overline{P}_s \approx \frac{\xi K^{Q+1} Q! \left(1 + \frac{3^{Q+1}}{4^Q}\right)}{12N\left(K+1\right)\rho^Q} \left(\frac{1}{\gamma_0^Q}\right), \tag{13}$$

where $\gamma_0 = E_s/N_0$ is the average SNR per sub-carrier. This leads to the coding gain given by

$$c = \frac{\rho}{K} \sqrt[Q]{\frac{12N\left(K+1\right)}{\xi K Q! \left(1 + \frac{3^{Q+1}}{4^Q}\right)}}. \tag{14}$$

It is shown via (13) that OS-MCIK-OFDM achieves the diversity order of Q. Thus, a larger Q can significantly enhance the error performance of systems. However, increasing Q leads to a decrease in the coding gain, due to $\lim_{Q\to\infty} c = \lim_{Q\to\infty} \frac{\rho}{K} \sqrt[Q]{\frac{1}{Q!}} = 0$. Moreover, as can be seen from (14), the coding gain can be improved if K gets smaller or N gets larger, however at the cost of the spectral efficiency.

We now compare the proposed OS-MCIK-OFDM with the opportunistic scheduling OFDM (OS-OFDM) which is also based on the criterion (1). The average SEP of OS-OFDM (denoted by $\overline{P}_{s_0}$) can be attained from (13) when K tends to N, as follows

$$\overline{P}_{s_0} \approx \frac{\xi N^{Q-1} Q! \left(1 + \frac{3^{Q+1}}{4^Q}\right)}{12\rho^Q} \left(\frac{1}{\gamma_0^Q}\right). \tag{15}$$

Using (13) and (15), the coding gain achieved by OS-MCIK-OFDM over OS-OFDM is calculated by $g = 10\log_{10}\left(\overline{P}_{s_0}/\overline{P}_s\right)^{1/Q}$, which is

$$g = 10\log_{10}\left(\frac{N}{K} \sqrt[Q]{\frac{K+1}{K}}\right) \text{ (dB)}. \tag{16}$$

As seen from (16), increasing Q makes g smaller, however g always satisfies a lower bound that $g \geq 10\log_{10}\left(\frac{N}{K}\right)$ for every Q. In addition, when $Q = 1$, (16) becomes [17, Eq. (27)], which validates the accuracy of our analysis.

4   Simulation Results

Simulation results are now presented to validate the performance of the proposed OS-MCIK-OFDM, the theoretical analysis as well as its superiority over OS-OFDM, under independent and identically distributed Rayleigh fading channels.

Figure 1 depicts the average SEP of OS-MCIK-OFDM when $(N, K, M) = (4, 1, 4)$ and $Q \in \{1, 2, 4, 6, 8\}$. Both the theoretical and asymptotic results are also provided. As observed from Fig. 1, the SEP performance of the proposed scheme is significantly improved when increasing Q. For example, when $Q = 2$ and at BER $= 10^{-4}$, our scheme achieves an SNR gain of 13 dB over MCIK-OFDM with $Q = 1$. Moreover, this gain increases to approximately 20 dB when $Q = 4$. However, if Q gets larger than 4, i.e., $Q = 6$ or 8, the performance gain attained becomes insignificant. This is caused by the decrease of the coding gain c when increasing Q as analyzed in Subsect. 3.2. In addition, Fig. 1 validates the tightness of derived theoretical and asymptotic expressions for the average SEP of OS-MCIK-OFDM, especially at higher SNRs.

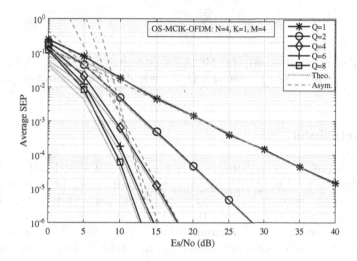

Fig. 1. SEP performance of OS-MCIK-OFDM when $(N, K, M) = (4, 1, 4)$ and $Q \in \{1, 2, 4, 6, 8\}$.

In Fig. 2, we compare the average SEPs between OS-MCIK-OFDM with $(N, K, M) = (2, 1, 2)$ and OS-OFDM with $(N, K) = (2, 2)$, at the same data rate of 1 bps/Hz and $Q \in \{1, 2, 4, 8\}$. It can be seen from Fig. 2 that our proposed scheme remarkably outperforms its benchmark, especially when Q is not too large. For instance, when $Q = 2$ and at high SNRs, the proposed scheme provides an SNR gain of more than 4 dB over OS-OFDM. This matches well with (16), which results in a similar gain of $g = 4.5$ dB. It also should be noted from Fig. 2 that when Q gets larger, the gap between two schemes becomes smaller, which confirms our analysis at the end of Subsect. 3.2.

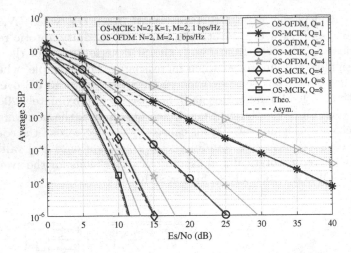

Fig. 2. SEP comparison between OS-MCIK-OFDM with $(N, K, M) = (2, 1, 2)$ and OS-OFDM with $(N, M) = (2, 2)$, when $Q \in \{1, 2, 4, 8\}$, at the same data rate of 1 bps/Hz.

5   Conclusions

We have proposed the opportunistic scheduling MCIK-OFDM (OS-MCIK-OFDM) and analyzed the SEP performance. In particular, the tight closed-form expression for the average SEP was derived. Then, the asymptotic analysis was conducted to provide insights into the SEP performance as well as impacts of system parameters such as N, K, and Q. More specifically, the proposed scheme achieves the diversity order being the number of machines, i.e., Q, and the achievable coding gain decreases when Q increases. Simulation results clearly validated the theoretical analysis, and particularly the superiority of the proposed scheme over the OS-OFDM, especially when Q is in a moderate size. In the future work, the proposed method can be extended to more realistic cases with non-identically distributed fading channels and imperfect CSI condition.

Appendix A

Proof of Theorem 1

Based on system model with the Rayleigh fading, it can be shown that $\gamma_{q,\alpha} = \bar{\gamma} |h_q(\alpha)|^2$ are the i.i.d RVs with the probability density function (PDF) and the cumulative distribution function (CDF), respectively, given by

$$f_\gamma(x) = \frac{1}{\bar{\gamma}} e^{-\frac{x}{\bar{\gamma}}}, \tag{17}$$

$$F_\gamma(x) = 1 - e^{-\frac{x}{\bar{\gamma}}}. \tag{18}$$

Relying on the scheduling method in (1), we first find the distribution of the following RV $\gamma_{q^*} = \max_{q=1,..,Q} [\min_{\alpha=1,..,N} (\gamma_{q,\alpha})]$. For this, denote $\gamma_q = \min_{\alpha=1,..,N} (\gamma_{q,\alpha})$ for given q. Using the order statistics, the CDF of γ_q can be calculated as

$$F_{\gamma_q} (x) = 1 - [1 - F_\gamma (x)]^N = 1 - e^{-\frac{Nx}{\bar{\gamma}}}. \tag{19}$$

Thus, $\gamma_{q^*} = \max_{q=1,..,Q} (\gamma_q)$ has the CDF given by

$$F_{\gamma_{q^*}} (x) = [F_{\gamma_q} (x)]^Q = \left(1 - e^{-\frac{Nx}{\bar{\gamma}}}\right)^Q. \tag{20}$$

Owing to the fact that $\gamma_{q^*} = \min_{\alpha=1,..,N} (\gamma_{q^*,\alpha})$, the CDF of $\gamma_{q^*,\alpha}$ can be obtained by the following equation

$$F_{\gamma_{q^*}} (x) = 1 - [1 - F_{\gamma_{q^*,\alpha}} (x)]^N. \tag{21}$$

Substituting (20) into (21), then solving this equation, we obtain

$$F_{\gamma_{q^*,\alpha}} (x) = 1 - \left[1 - \left(1 - e^{-\frac{Nx}{\bar{\gamma}}}\right)^Q\right]^{\frac{1}{N}}. \tag{22}$$

Here, the PDF of $\gamma_{q^*,\alpha}$ can be attained by taking the derivative of $F_{\gamma_{q^*,\alpha}} (x)$ in (22) as

$$f_{\gamma_{q^*,\alpha}} (x) = \frac{Q}{\bar{\gamma}} \left[1 - \left(1 - e^{-\frac{Nx}{\bar{\gamma}}}\right)^Q\right]^{\frac{1-N}{N}} \left(1 - e^{-\frac{Nx}{\bar{\gamma}}}\right)^{Q-1} e^{-\frac{Nx}{\bar{\gamma}}}. \tag{23}$$

As can be shown from (23), it is difficult to directly compute the MGF of $\gamma_{q^*,\alpha}$ based on its PDF in (23). Hence, this motivates us to find out an approximation of $f_{\gamma_{q^*,\alpha}} (x)$, that can facilitate the derivation of the MGF of $\gamma_{q^*,\alpha}$. For this, we consider (23) at high SNRs and discover that

$$\lim_{\bar{\gamma} \to \infty} \left[1 - \left(1 - e^{-\frac{Nx}{\bar{\gamma}}}\right)^Q\right]^{\frac{1-N}{N}} = 1. \tag{24}$$

Consequently, the PDF of $\gamma_{q^*,\alpha}$ can be approximated, in high SNR regions, by

$$f_{\gamma_{q^*,\alpha}} (x) \approx \frac{Q}{\bar{\gamma}} \left(1 - e^{-\frac{Nx}{\bar{\gamma}}}\right)^{Q-1} e^{-\frac{Nx}{\bar{\gamma}}}. \tag{25}$$

Finally, the MGF of $\gamma_{q^*,\alpha}$ can be calculated, by performing the inverse Laplace transform in (25), as follows

$$\mathcal{M}_{\gamma_{q^*},\alpha}(z) \approx \mathcal{L}^{-1}\left\{\frac{Q}{\bar{\gamma}}\left(1 - e^{-\frac{Nx}{\bar{\gamma}}}\right)^{Q-1}e^{-\frac{Nx}{\bar{\gamma}}}\right\}$$

$$= \mathcal{L}^{-1}\left\{\frac{Q}{\bar{\gamma}}\sum_{q=0}^{Q-1}C(Q-1,q)(-1)^q e^{-\frac{(q+1)Nx}{\bar{\gamma}}}\right\}$$

$$= Q\sum_{q=0}^{Q-1}\frac{C(Q-1,q)(-1)^q}{(q+1)N - t\bar{\gamma}}$$

$$= \frac{N^{Q-1}Q!}{\prod_{q=1}^{Q}(qN - t\bar{\gamma})}. \tag{26}$$

The theorem is completely proved.

Appendix B

Proof of Lemma 1

According to [20], the instantaneous SEP of M-ary PSK symbols can be approximated by $P_M(\alpha) \approx \xi Q\left(\sqrt{2\gamma_{q^*},\alpha}\sin(\pi/M)\right)$, where we recall γ_{q^*},α being the instantaneous SNR of the selected machine at sub-carrier α, and $\xi = 1$ for $M = 2$ and $\xi > 1$ for $M > 2$. Using the approximation of Q-function as $Q(x) \approx \frac{1}{12}e^{-\frac{x^2}{2}} + \frac{1}{4}e^{-\frac{2x^2}{3}}$, we obtain

$$P_M(\alpha) \approx \frac{\xi}{12}\left(e^{-\gamma_{q^*},\alpha\rho} + 3e^{-\frac{4\gamma_{q^*},\alpha\rho}{3}}\right), \tag{27}$$

where $\rho = \sin^2(\pi/M)$.

Next, applying the MGF approach in (27), with the MGF of γ_{q^*},α given in Theorem 1, the average SEP of the classical M-ary PSK symbols of machine q^* can be computed as

$$\overline{P}_M \approx \mathbb{E}_{\gamma_{q^*},\alpha}\left\{\frac{\xi}{12}\left(e^{-\gamma_{q^*},\alpha\rho} + 3e^{-\frac{4\gamma_{q^*},\alpha\rho}{3}}\right)\right\}$$

$$= \frac{\xi}{12}\left[\mathcal{M}_{\gamma_{q^*},\alpha}(\rho) + \mathcal{M}_{\gamma_{q^*},\alpha}(-4\rho/3)\right]$$

$$= \frac{\xi N^{Q-1}Q!}{12}\left[\frac{1}{\prod_{q=1}^{Q}(qN + \rho\bar{\gamma})} + \frac{3}{\prod_{q=1}^{Q}(qN + \frac{4\rho\bar{\gamma}}{3})}\right].$$

This concludes the proof.

References

1. Abu-alhiga, R., Haas, H.: Subcarrier-index modulation OFDM. In: Proceedings of IEEE International Symposium on Personal, Indoor and Mobile Radio Communications, pp. 177–181, September 2009

2. Basar, E., Aygolu, U., Panayirci, E., Poor, H.V.: Orthogonal frequency division multiplexing with index modulation. IEEE Trans. Signal Process. **61**(22), 5536–5549 (2013)
3. Basar, E., Wen, M., Mesleh, R., Renzo, M.D., Xiao, Y., Haas, H.: Index modulation techniques for next-generation wireless networks. IEEE Access **5**, 16693–16746 (2017)
4. Luong, T.V., Ko, Y., Choi, J.: Repeated MCIK-OFDM with enhanced transmit diversity under CSI uncertainty. IEEE Trans. Wireless Commun. **17**(6), 4079–4088 (2018)
5. Luong, T.V., Ko, Y.: A closed-form symbol error probability for MCIK-OFDM with frequency diversity. In: Proceedings of IEEE SPAWC, July 2017, pp. 1–5 (2017)
6. Basar, E.: OFDM with index modulation using coordinate interleaving. IEEE Wireless Commun. Lett. **4**(4), 381–384 (2015)
7. Zheng, J., Chen, R.: Achieving transmit diversity in OFDM-IM by utilizing multiple signal constellations. IEEE Access **5**(99), 8978–8988 (2017)
8. Ko, Y.: A tight upper bound on bit error rate of joint OFDM and multi-carrier index keying. IEEE Commun. Lett. **18**(10), 1763–1766 (2014)
9. Wen, M., Cheng, X., Ma, M., Jiao, B., Poor, H.V.: On the achievable rate of OFDM with index modulation. IEEE Trans. Signal Process. **64**(8), 1919–1932 (2016)
10. Luong, T.V., Ko, Y.: Symbol error outage performance analysis of MCIK-OFDM over complex TWDP fading. In: Proceedings of European Wireless, May 2017, pp. 1–5 (2017)
11. Crawford, J., Chatziantoniou, E., Ko, Y.: On the SEP analysis of OFDM index modulation with hybrid low complexity greedy detection and diversity reception. IEEE Trans. Veh. Technol. **66**(9), 8103–8118 (2017)
12. Luong, T.V., Ko, Y.: The BER analysis of MRC-aided greedy detection for OFDM-IM in presence of uncertain CSI. IEEE Wireless Commun. Lett., to be published
13. Zheng, B., Chen, F., Wen, M., Ji, F., Yu, H., Liu, Y.: Low-complexity ML detector and performance analysis for OFDM with in-phase/quadrature index modulation. IEEE Commun. Lett. **19**(11), 1893–1896 (2015)
14. Mao, T., Wang, Z., Wang, Q., Chen, S., Hanzo, L.: Dual-mode index modulation aided OFDM. IEEE Access **5**, 50–60 (2017)
15. Wen, M., Basar, E., Li, Q., Zheng, B., Zhang, M.: Multiple-mode orthogonal frequency division multiplexing with index modulation. IEEE Trans. Commun. **65**(9), 3892–3906 (2017)
16. Basar, E.: On multiple-input multiple-output OFDM with index modulation for next generation wireless networks. IEEE Trans. Signal Process. **64**(15), 3868–3878 (2016)
17. Luong, T.V., Ko, Y.: Impact of CSI uncertainty on MCIK-OFDM: tight, closed-form symbol error probability analysis. IEEE Trans. Veh. Technol. **67**(2), 1272–1279 (2018)
18. Van Luong, T., Ko, Y.: A tight bound on BER of MCIK-OFDM with greedy detection and imperfect CSI. IEEE Commun. Lett. **21**(12), 2594–2597 (2017)
19. Luong, T.V., Ko, Y., Choi, J.: Precoding for spread OFDM IM. In: Proceedings of IEEE 87th Vehicular Technology Conference (VTC Spring), pp. 1–5 (2018)
20. Proakis, J.: Digital Communications. McGraw-Hill, New York City (2001)

An Improved Occlusion Detection with Constraints Approach for Video Processing

Tuan-Anh Vu, Hung Ngoc Phan, Tu Kha Huynh, and Synh Viet-Uyen Ha[(✉)]

School of Computer Science and Engineering, International University,
Vietnam National University, Ho Chi Minh city, Vietnam
{anhvt14,hungpn17}@mp.hcmiu.edu.vn, {hktu,hvusynh}@hcmiu.edu.vn

Abstract. The accurate understanding of occlusion region is critical for trustworthy estimation of optical flow to prevent the negative influence of occluded pixels on disocclusion regions. However, occlusion is the result of motion. In contrast, estimating accurate optical flow is necessary to locate reliable occlusions. Hence, one of the key challenges that required further exploration and research is the accuracy at the boundaries of the moving objects. This paper presents the work in process approach that can detect occlusion regions by using some constraints such as pixel-wise coherence, segment-wise confidence and edge-motion coherence. Comparing to the previous methods, our method achieves the same efficiency by solving only one Partial Differential Equation (PDE) problem. The proposed method is faster and provides better coverage rates for occlusion regions than variation techniques in various numbers of benchmark datasets. With these improved results, we can apply and extend our approach to a wider range of applications in computer vision, such as: motion estimation, object detection and tracking, robot navigation, 3D reconstruction, image registration.

Keywords: Optical flow · Unstable region · Object boundaries
Occlusion detection · Video object extraction
Video object segmentation

1 Introduction

Occlusions detection is a famous problem in optical flow in particular and the field of image processing in general [1–3]. Moreover, most problems in video processing such as object tracking, 3D object reconstruction, motion blurring, and unexpected objects removing are difficult problems to optimize due to the lack of information of the motion vector at the pixels in the occlusion regions between two consecutive frames.

This research is funded by Vietnam National University Ho Chi Minh City (VNU-HCM) under grant number C2016-28-11/HD-KHCN.

T. Q. Duong and N.-S. Vo (Eds.): INISCOM 2018, LNICST 257, pp. 232–242, 2019.
https://doi.org/10.1007/978-3-030-05873-9_19

Suppose an object moves from one position in the frame at time t to another position in the frame at time $t+1$ through the motion vectors. Occlusion regions will focus on two main regions: Region U', at the head of the motion vectors, and Region U", at the tail end of the motion vectors. There are two groups of pixels that create overlap in Region U'. The first group contains the moving pixels out of the next frame. Estimations will be wrong because they do not identify these pixels' disappearance when objects move out of the frame or change their shape. The second group contains the overlapping pixels in the next frame. A part of the background image/object in the first frame disappears in the next frame due to obscured or morphing objects. Estimations will be inaccurate because the constraint of pixels is not preserved. In Region U", a part of the background image/object is not shown in the first frame will appear in the second frame. Therefore, these pixels will be inaccurate because the initial position of these pixels cannot be determined. If the object changes shape, overlapping pixels will be created.

One of the ways to improve the quality of video processing applications is to detect occlusion regions accurately. In reality, there are many methods are used to detect occlusion regions. These methods can be divided into two main groups: the computation of two PDEs optical flow problems and the combination of forward optical flow and image segmentation. In the first group [2–9], the optical flow between two frame: the I_t frame at time t and the I_{t+1} frame at time $(t + 1)$ is evaluated by solving one PDE problem. Then, the backward and forward optical flow between the frames at time $(t + 1)$ and the frame at time t is evaluated. Occlusion regions were determined by analyzing the results of the two assessments above. This group will minimize the energy function of the optimal problem by the continuous iterative interaction between occlusion detection and estimation of motion vector. In the second group [10–20], an image segmentation algorithm is used to divide the I_t frame into small pieces, and then connect the pieces together based on the information of the motion vectors. In this approach, using an image segmentation algorithm, another PDE problem must be solved in addition to the original PDE problem. However, the use of image segments to obtain groups of motion vector never gives an ideal result because small segments may not contain enough information and large segments are unreliable. Generally, to detect occlusion regions, two PDE problems must be solved: the backward and forward optical flow problems or the combine forward optical flow and image segmentation problems. Occlusion regions are determined by analyzing the results of these two PDE problems.

In this paper, we present a method that can detect occlusion regions by computing only one PDE problem and using some constraints such as pixel-wise coherence, segment-wise confidence and edge-motion coherence. It means that the estimation of optical flow procedure is only called once. Our method achieved better accuracy, faster processing time and a higher coverage rate of almost occlusion regions than other variation techniques: Alvarez et al. [2], Ayvaci et al. [6] and Estellers et al. [22]. The rest of this paper is organized as follows: Sect. 2 introduces the proposed method. Section 3 presents the experimental results and discussion to illustrate the application and usefulness of the proposed algorithm. Section 4 concludes this study with a discussion.

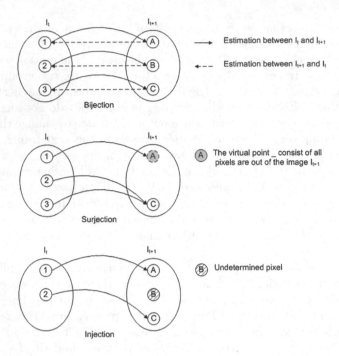

Fig. 1. Bijection pixel, surjection pixel and injection pixel.

2   The Proposed Method

When an object moves from time t to time $(t+1)$, it creates two instability areas: Region U', at the head of motion vectors and Region U", at the tail of motion vectors. There are two groups of pixels creating instability in Region U'. The first group contains pixels in the first frame that disappear or move outside the next frame. The motion estimation is disturbed between these pixels that move out the image and changes in shape of the objects. The second group contains pixels in the first frame that are overlapped together in next frame. A part of the background/object (when objects overlap) in the first frame move outside or disappears in the next frame because the objects move or change shape. The motion estimation around the brightness constancy assumption is incorrect for these pixels. In Region U", a part of the background/object absent from the previous frame appears in the next frame. The smoothness assumption among pixels is broken. Thus, their estimation is incorrect. If objects change shape, unstable pixels can result. We do not have enough information for estimating motion vectors in these regions. So, they are unstable regions. In another way, these pixels in these regions will be assessed approximately. Both Region U' and Region U" create the occlusion boundary of the moving object.

Some definitions are used in this method is defined as follow:

Definition 1 (Bijection pixel). *A bijection pixel is a pixel in the first frame satisfies two conditions:*

- *In optical flow estimation from the first frame to the next frame, it points to a pixel x in the next frame.*
- *In optical flow estimation from the next frame to the first frame, the pixel x points to it (Fig. 1).*

Definition 2 (Surjection pixel). *A pixel in the first frame points to a pixel which was pointed by at least another pixel. In this case, we have a virtual pixel consisting of all pixels are out of the next frame (Fig. 1).*

Definition 3 (Overlapping pixel). *An overlapping pixel is a pixel in the next frame that was pointed by two and more pixels in the first frame (Fig. 1).*

Definition 4 (Injection pixel). *A pixel in the first frame points to a pixel which is adjacent to an undetermined pixel. An undetermined pixel in the next frame has no pixel in the first frame pointing to (Fig. 1).*

Definition 5 (Stable pixel and unstable pixel). *A stable pixel is a bijection pixel. An unstable pixel is a surjection pixel or an injection pixel.*

Definition 6 (Unstable region). *A set of unstable pixels are connected.*

Assume that, we have a set of positions P_1 in a local region I_t and a set of positions P_2 in a local region I_{t+1}, the function f is defined on P_1 and take positions in P_2. A pixel $\mathbf{x}$ is an unstable pixel if $f(\mathbf{x})$ is surjection and injection. In other words, $\mathbf{x}$ is a stable pixel. In our algorithm, we determine unstable pixels directly by finding surjection pixels and injection pixels. When we convert the values of a motion vector (u, v) from real numbers into integers in unstable regions that are determined by using the conditions for undermined pixels and overlap pixels, we found some calculation error. Therefore, to reduce the error's effect, we restored good features by using the method in [21] and extended two ideas presented in [15].

Pixel-Wise Coherence (C_p). We denote $\mathbf{w}$ is the motion vector at pixel $\mathbf{x}$. C_p only focused and assess stableness of the data constraint of optical flow estimation. Thus, we determine a pixel is stable or unstable based on the deviation between the grey value of pixel at position (i, j) in I_t and the one at position $(i + u, j + v)$ in I_{t+1} (Eq. 1). If it is large, the pixel is unstable and vice versa.

$$C_p(\mathbf{w}, \mathbf{x}) \triangleq \exp\left(-\frac{|I_{t+1}(\mathbf{x} + \mathbf{w}) - I_t(\mathbf{x})|^2}{\sigma_I^2}\right) \quad (1)$$

Segment-Wise Confidence (C_s). We denote Nb is a neighbor region of pixel $\mathbf{x}_c$, $\mathbf{w}_\mathbf{x}^0$ is the motion vector at pixel $\mathbf{x}$ in the time (t), $\mathbf{w}_\mathbf{x}^f$ is the motion vector

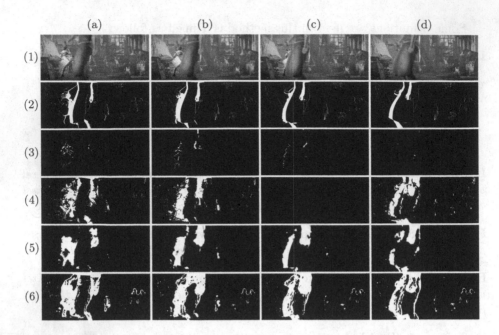

Fig. 2. Visual qualitative comparison on dataset market2: (a) frame 05 (b) frame 06, (c) frame 07, (d) frame 08. From top to bottom present (1) input image frame, (2) ground truth, (3) Ayvaci *et al.* [6], (4) Alvarez *et al.* [2], (5) Estellers *et al.* [22], (6) proposed method, respectively.

at pixel $\mathbf{x}$ in the time $(t+1)$, and $\kappa(\mathbf{x}, \mathbf{x}_c)$ constrains that two adjacent pixels must have similar color. We focused on the processing pixel and the pixels in the neighbor region of it. If they have similar color and the motion vector at a pixel changes sharply, it can be unstable (Eq. 2).

$$C_s(\mathbf{w}^f, \mathbf{w}^0, \mathbf{x}_c) = \frac{\sum\limits_{\mathbf{x} \in Nb} \exp\left(-\left\|\mathbf{w}_{\mathbf{x}}^f - \mathbf{w}_{\mathbf{x}}^0\right\|^2 \frac{C_p(\mathbf{w}_{\mathbf{x}}^0, \mathbf{x})}{\sigma_A^2}\right) \kappa(\mathbf{x}, \mathbf{x}_c)}{\sum\limits_{\mathbf{x} \in Nb} \kappa(\mathbf{x}, \mathbf{x}_c)} \tag{2}$$

$$\kappa(\mathbf{x}, \mathbf{x}_c) = (1 - M(\mathbf{x})) \exp(-\frac{|I_t(\mathbf{x}) - I_t(\mathbf{x}_c)|^2}{\sigma_s^2}) \tag{3}$$

where $M(\mathbf{x}) = 1$ if pixel x is unstable and $M(\mathbf{x}) = 0$ if pixel x is stable.

Then, the unstable regions is computing as below:

$$Com(\mathbf{w}^f, \mathbf{w}^0, \mathbf{x}_c) = \begin{cases} \varsigma.C_s(\mathbf{w}^f, \mathbf{w}^0, \mathbf{x}_c), & if \ M(\mathbf{x}) = 0 \\ C_p(\mathbf{w}_{\mathbf{x}_c}^f, \mathbf{x}_c).C_s(\mathbf{w}^f, \mathbf{w}^0, \mathbf{x}_c), & otherwise \end{cases} \tag{4}$$

where ς is a constant to penalize the unstable pixel.

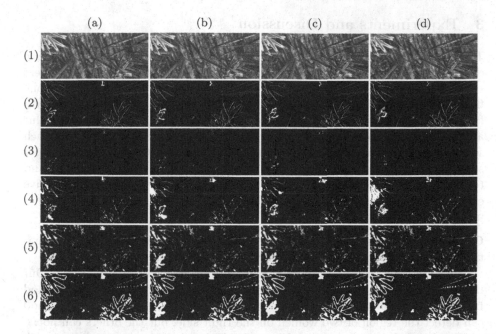

Fig. 3. Visual qualitative comparison on dataset bamboo1: (a) frame 01 (b) frame 02, (c) frame 03, (d) frame 04. From top to bottom present (1) input image frame, (2) ground truth, (3) Ayvaci *et al.* [6], (4) Alvarez *et al.* [2], (5) Estellers *et al.* [22], (6) proposed method, respectively.

Edge-Motion Coherence (C_e). We focus on the pixels on edge and the motion vector of it. We realized that the unstable pixels are usually in the marginal regions or edge of the objects. Based on that point, we will distinguish whether this pixel is an unstable pixel or texture of the object (Eq. 5).

$$C_e\left(\mathbf{w}_\mathbf{x}^f, \mathbf{w}_\mathbf{x}^0, E_t(\mathbf{x})\right) = \begin{cases} 1 & if\ E_t(\mathbf{x}).(\mathbf{w}_\mathbf{x}^f - \mathbf{w}_\mathbf{x}^0) > 0 \\ 0 & otherwise \end{cases} \tag{5}$$

where $E_t(\mathbf{x})$ is the value of edge at pixel $\mathbf{x}$.

However, the pixels near the boundary of the unstable regions after Eqs. 4 and 5 will create aliasing along spatial dimensions. Such aliasing effects can cause sudden changes in the computation of feature vector. To avoid this, we use bilinear interpolation of the pixel weight into the spatial orientation.

The final unstable regions are computed as:

$$F(\mathbf{w}^f, \mathbf{x}_c) = \begin{cases} 1 & if\ Com(\mathbf{w}^f, \mathbf{w}^0, \mathbf{x}_c)\ \&\ C_e \geq \xi \\ 0 & otherwise \end{cases} \tag{6}$$

where ξ is a threshold coefficient.

3 Experiments and Discussion

To evaluate the proposed method, we compare the proposed method to three different variational models: Alvarez *et al.* [2], Ayvaci *et al.* [6] and Estellers *et al.* [22]. We performed a number of experiments using the MPI-Sintel public flow dataset [23]. This dataset is generated by computer and has ground-truth for occlusions. We use the final pass of the data set, including shading, blur and atmospheric shading in the video sequence to cover all the cases. Each sequence has 50 frames with the resolution of 1024×436, with 24 frames per second. Experiments have been conducted to test the system performance and to measure the accuracy of the proposed method. Qualitative and quantitative evaluations of the proposed algorithm have been carried out.

Qualitative Evaluation. In this section, we will show our Qualitative evaluation on the five sequences of the MPI-Sintel dataset.

The Fig. 2 present detected occlusion regions for sequences market2. Our method has better results (approximately 30%) than the runner-up method because our method has the least false alarm rate(TPR). Especially, our method can detect the region of two women on the right side, but the others can not.

The Fig. 3 shows an unsuccessful case in the bamboo1 sequence of our model. The combination of the bamboo oscillations and the non-smooth displacements of the camera create some noises when we perform to detect occluded regions. Therefore, these things reduce the accuracy of our model. However, our model can detect occluded regions of object on the left, but the boundaries of the object is a bit thicker than ground-truth. In conclusion, our results are slightly lower than the best results. For a more detailed analysis of these models, we use a quantitative measure in the next session.

Quantitative Evaluation. To evaluate Quantitative of proposed method, we use the theory by [24–27] to calculate precision or positive predictive value (PPV), recall or true positive rate (TPR), F-measure and Jaccard index of our proposed method. If we consider each pixel is an independent object, precision is the probability of classified occluded pixel is true, while recall is the probability of detected occluded pixel is true. F-measure is combined of PPV and TPR with the weight balance between them. We also measure the number of true positive (TP), true negative (TN), false positive (FP), false negative (FN) pixels to calculate these metrics above. The formula for calculating precision and recall(sensitivity) of each class are:

$$PPV = \frac{TP}{TP+FN} \qquad TPR \doteq \frac{TP}{TP+FN} \qquad (7)$$

The F1-score is conducted from the following formula:

$$F_\beta = (1 + \beta^2) \cdot \frac{PPV.TPR}{\beta^2.PPV + TPR} \tag{8}$$

where, β is the parameter that controls the weighing balance between PPV and TPR. In this step, we use $\beta = 1$ to make no extra weight to either PPV or TPR.

And the Jaccard index is defined as the Intersection over Union (IoU) between the results and the ground-truth:

$$J = \frac{|TP|}{|TP| + |FN| + |FP|} = \frac{F}{2 - F} \tag{9}$$

Table 1 compare our method to variation techniques Alvarez et al. [2], Ayvaci et al. [6] and Estellers et al. [22]. Both Alvarez and Ayvaci compute the optical flow and occlusions between two consecutive frames, but Alvarez uses the mismatch of forward and backward flows to compute the occlusion, while Ayvaci only uses the forward flow. Therefore, Ayvaci and our method only solve one PDE problem and Alvarez must solve two PDE problems. Estellers proposes a hybrid method that integrates information from multiple consecutive frames to compute the occlusion detection such as the formulation of single-frame from the previous frames, the spatial and temporal adjustment removes isolated faulty detection incompatible in time or space.

Table 1. Comparison between our method and existing occlusion detection methods Alvarez et al. [2], Ayvaci et al. [6] and Estellers et al. [22] optimized for J-index

Sequence	Our method				Alvarez et al. [2]				Ayvaci et al. [6]				Estellers et al. [22]			
	PPV	TPR	F1	J-index	PPV	TPR	F1	J-index	PPV	TPR	F1	J-index	PPV	TPR	F1	J-index
alley1	0.11	0.54	0.18	0.10	0.24	0.28	0.26	0.15	0.02	0.43	0.03	0.02	0.43	0.23	0.29	**0.17**
bamboo1	0.20	0.76	0.32	0.19	0.38	0.34	0.36	**0.22**	0.02	0.46	0.04	0.02	0.45	0.28	0.34	0.20
bandage1	0.16	0.86	0.27	0.16	0.17	0.33	0.22	0.12	0.01	0.39	0.01	0.01	0.45	0.28	0.34	**0.20**
bandage2	0.12	0.69	0.20	**0.11**	0.1	0.28	0.15	0.08	0.01	0.21	0.02	0.01	0.42	0.12	0.19	0.10
market2	0.33	0.87	0.48	**0.31**	0.46	0.34	0.38	0.23	0.02	0.47	0.04	0.02	0.32	0.37	0.32	0.19
Overall J-index	0.17[1]				0.16[2]				0.01[3]				0.17[1]			

* Bold value denotes the best result for each sequence.
** (1), (2), (3) denotes the rank of each method in J-index.

Processing Time. Finally, we tested the processing times of our method and other methods in Fig. 4. Compared to the literature, our model is slightly slower than Ayvaci by 10%, but faster than the other two models Alvarez and Estellers by two times and 18% respectively. Despite having a slower processing time, our results are more accurate than Ayvaci.

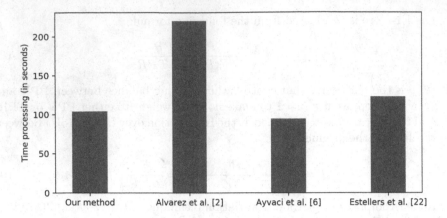

Fig. 4. The average time processing of our method and other methods.

4   Conclusion

In this paper, we had presented the results of the work-in-process that could detect occlusion regions by using pixel-wise coherence, segment-wise confidence and edge-motion coherence. Our method achieved better accuracy, faster processing time and a higher coverage rate of almost occlusion regions than other variation techniques Alvarez *et al.* [2], Ayvaci *et al.* [6] and Estellers *et al.* [22] in most cases. Our method was a method which could get the same result as usual methods by solving only one PDE problem. Compared with other occlusion detection algorithm, our algorithm got a better result. Compared against boundary algorithms, our algorithm was approximately two times faster because it halved the number of PDE problems that needed to be solved. Based on these results, future studies can extend our proposed method to detect occlusion regions better and faster.

References

1. Perez-Rua, J.-M., Crivelli, T., Bouthemy, P., Perez, P.: Determining occlusions from space and time image reconstructions. In: 2016 IEEE Conference on Computer Vision and Pattern Recognition (CVPR), pp. 1382–1391. IEEE (2016)
2. Alvarez, L., Deriche, R., Papadopoulo, T., Sánchez, J.: Symmetrical dense optical flow estimation with occlusions detection. Int. J. Comput. Vis. **75**, 371–385 (2007)
3. Strecha, C., Fransens, R., Van Gool, L.: A probabilistic approach to large displacement optical flow and occlusion detection. In: Comaniciu, D., Mester, R., Kanatani, K., Suter, D. (eds.) SMVP 2004. LNCS, vol. 3247, pp. 71–82. Springer, Heidelberg (2004). https://doi.org/10.1007/978-3-540-30212-4_7
4. Proesmans, M., Van Gool, L., Pauwels, E., Oosterlinck, A.: Determination of optical flow and its discontinuities using non-linear diffusion. In: Eklundh, J.-O. (ed.) ECCV 1994. LNCS, vol. 801, pp. 294–304. Springer, Heidelberg (1994). https://doi.org/10.1007/BFb0028362

5. Xiao, J., Cheng, H., Sawhney, H., Rao, C., Isnardi, M.: Bilateral filtering-based optical flow estimation with occlusion detection. In: Leonardis, A., Bischof, H., Pinz, A. (eds.) ECCV 2006. LNCS, vol. 3951, pp. 211–224. Springer, Heidelberg (2006). https://doi.org/10.1007/11744023_17
6. Ayvaci, A., Raptis, M., Soatto, S.: Sparse occlusion detection with optical flow. Int. J. Comput. Vis. **97**, 322–338 (2012)
7. Sadek, R., Ballester, C., Garrido, L., Meinhardt, E., Caselles, V.: Frame interpolation with occlusion detection using a time coherent segmentation. In: Proceedings of the International Conference on Computer Vision Theory and Applications, pp. 367–372. SciTePress - Science and and Technology Publications (2012)
8. Ballester, C., Garrido, L., Lazcano, V., Caselles, V.: A TV-L1 optical flow method with occlusion detection. In: Pinz, A., Pock, T., Bischof, H., Leberl, F. (eds.) DAGM/OAGM 2012. LNCS, vol. 7476, pp. 31–40. Springer, Heidelberg (2012). https://doi.org/10.1007/978-3-642-32717-9_4
9. Tao, M., Bai, J., Kohli, P., Paris, S.: SimpleFlow: a non-iterative, sublinear optical flow algorithm. Comput. Graph. Forum. **31**, 345–353 (2012)
10. Lim, K.P., Das, A., Chong, M.N.: Estimation of occlusion and dense motion fields in a bidirectional Bayesian framework. IEEE Trans. Pattern Anal. Mach. Intell. **24**, 712–718 (2002)
11. Kolmogorov, V., Zabih, R.: Computing visual correspondence with occlusions using graph cuts. In: Proceedings Eighth IEEE International Conference on Computer Vision. ICCV 2001, pp. 508–515. IEEE Computer Society (2001)
12. Sun, J., Li, Y., Kang, S.B., Shum, H.-Y.: Symmetric stereo matching for occlusion handling. In: 2005 IEEE Computer Society Conference on Computer Vision and Pattern Recognition (CVPR 2005), pp. 399–406. IEEE (2005)
13. Ince, S., Konrad, J.: Occlusion-aware optical flow estimation. IEEE Trans. Image Process. **17**, 1443–51 (2008)
14. Ben-Ari, R., Sochen, N.: Variational stereo vision with sharp discontinuities and occlusion handling. In: 2007 IEEE 11th International Conference on Computer Vision, pp. 1–7. IEEE (2007)
15. Xu, L., Chen, J., Jia, J.: A segmentation based variational model for accurate optical flow estimation. In: Forsyth, D., Torr, P., Zisserman, A. (eds.) ECCV 2008. LNCS, vol. 5302, pp. 671–684. Springer, Heidelberg (2008). https://doi.org/10.1007/978-3-540-88682-2_51
16. Stein, A.N., Hebert, M.: Occlusion boundaries from motion: low-level detection and mid-level reasoning. Int. J. Comput. Vis. **82**, 325–357 (2009)
17. He, X., Yuille, A.: Occlusion boundary detection using pseudo-depth. In: Daniilidis, K., Maragos, P., Paragios, N. (eds.) ECCV 2010. LNCS, vol. 6314, pp. 539–552. Springer, Heidelberg (2010). https://doi.org/10.1007/978-3-642-15561-1_39
18. Gao, T., Packer, B., Koller, D.: A segmentation-aware object detection model with occlusion handling. In: CVPR 2011, pp. 1361–1368. IEEE (2011)
19. Humayun, A., Mac Aodha, O., Brostow, G.J.: Learning to find occlusion regions. In: CVPR 2011, pp. 2161–2168. IEEE (2011)
20. Sundberg, P., Brox, T., Maire, M., Arbelaez, P., Malik, J.: Occlusion boundary detection and figure/ground assignment from optical flow. In: CVPR 2011, pp. 2233–2240. IEEE, Washington, DC, USA (2011)
21. Shi, J., Tomasi, C.: Good features to track. In: Proceedings of IEEE Conference on Computer Vision and Pattern Recognition CVPR 1994, pp. 593–600. IEEE Computer Society Press (1994)
22. Estellers, V., Soatto, S.: Detecting occlusions as an inverse problem. J. Math. Imaging Vis. **54**, 181–198 (2016)

23. Butler, D.J., Wulff, J., Stanley, G.B., Black, M.J.: A naturalistic open source movie for optical flow evaluation. In: Fitzgibbon, A., Lazebnik, S., Perona, P., Sato, Y., Schmid, C. (eds.) ECCV 2012. LNCS, vol. 7577, pp. 611–625. Springer, Heidelberg (2012). https://doi.org/10.1007/978-3-642-33783-3_44
24. Sokolova, M., Lapalme, G.: A systematic analysis of performance measures for classification tasks. Inf. Process. Manag. **45**, 427–437 (2009)
25. Fawcett, T.: An introduction to ROC analysis. Pattern Recognit. Lett. **27**, 861–874 (2006)
26. Powers, D.M.W.: Evaluation: from precision, recall and F-measure To Roc, informedness, markedness & correlation. J. Mach. Learn. Technol. **2**, 37–63 (2011)
27. Pont-Tuset, J.: Image segmentation evaluation and its application to object detection, Universitat Politècnica de Catalunya, UPC BarcelonaTech (2014)

Development and Deployment of an IoT-Based Reconfigurable System: A Case Study for Smart Garden

Dang Huynh-Van, Ngan Le-Thi-Chau, Khoa Ngo-Khanh,
and Quan Le-Trung[✉]

UiTiOt Research Group, Department of Computer Networks,
University of Information Technology, Vietnam National University,
Ho Chi Minh City, Vietnam
{danghv, quanlt}@uit.edu.vn,
{14520574, 14520423}@gm.uit.edu.vn

Abstract. Nowadays, Internet of Things (IoT) is not only a hot research topic but also plays a vital role in the development and deployment of many different domains, e.g., industry, transportation, education, health and agriculture as well. However, the development and deployment of real-time IoT-based systems requires stringent technical challenges on the scalability, the availability and the dynamic adaptability of the IoTs systems and applications. In this paper, we present an IoT-based architecture with a reconfigurable approach to deal with those technical challenges, and the proposed IoT-based reconfigurable system is applied for the environmental monitoring in a Smart Garden application as a case study. By implementing the load-balanced infrastructure and the over-the-air (OTA) programming, our system fulfils the large deployment of many wireless sensor devices. In addition, we have also presented and discussed the empirical results related to the resource usage, the packet transmission and the energy consumption during the reconfiguration and lay out tentative approaches on the application of our proposed system to different application domains.

Keywords: Internet of Things · Over-the-air programming · Reconfiguration Wireless sensor networks · TinyOS

1 Introduction

The explosion of Industrial Revolution 4.0 has made the phrase Internet of Things (IoT) become no stranger to many people. According to Consumers International's 2016 report cited Cisco's estimate that 50 billion Internet of Things Device will be connected by 2020. That figure is three times the amount we currently have today. The development of the Internet of Things has led to enormous IoT-based systems such as Intelligent Transportation Systems (ITS) [1], Smart Agriculture [2], Smart Shopping Systems [3], and Smart Home System [4]. Among various IoT applications, the design of smart farm has attracted attentions from both academic and industrial as the systems are closely related to country's economy. When deploying these systems, besides the usability and convenience for users, long-term maintenance and dynamic adaptability are also

© ICST Institute for Computer Sciences, Social Informatics and Telecommunications Engineering 2019
Published by Springer Nature Switzerland AG 2019. All Rights Reserved
T. Q. Duong and N.-S. Vo (Eds.): INISCOM 2018, LNICST 257, pp. 243–252, 2019.
https://doi.org/10.1007/978-3-030-05873-9_20

significant because the replacement of new devices or manually install new application costs not only a lot of money but also time and effort. Especially, when wireless sensor devices located in places with hard-to-reach areas e.g., on high mountains, deep forest.

In this paper, we try to resolve some of those problems by implementing the OTA programming in the wireless sensor network. We proposed an IoT-based Smart Garden system. This system includes three main characteristics: scalability, availability, and portability by using a load-balanced network architecture on the Open Stack [5] as well as OTA programming. The reconfiguration and remote reprogramming solution will help to change behaviors, functions, parameters in the application on sensor devices through a set of wireless sensor network protocols. So that these applications run stably in the real environment and users do not need to install/upgrade devices manually. The middleware reconfiguration in the system can create a dynamic behavior IoT-based system for the modern agriculture in the future.

There are five sections in this paper that you can follow. Section 1 introduces the research topology, mentions the current challenges of IoTs and emphasizes the importance of reconfiguring in IoT middleware as well as our paper's contribution. Section 2 is the overview of research situation in Vietnam and other countries about the IoT application and middleware reconfiguration on WSN. Section 3 clarifies our system implementation in detail. The following is the results of our system when running in many practical experiments. The final section ends this paper by summarizing our contribution and disclose some future directions of our research roadmap.

2   Related Work

IoT-Based Application. The Internet of Things (IoT) is dramatically impacting the world we live and attracting many researchers, as well as investigators. In agriculture domain, IoT is a driving force behind increased agricultural production at a lower cost. IoT applications in agriculture domain include environmental parameter monitoring (e.g., temperature, humidity, light, pH, etc.), plant monitoring, livestock monitoring, and much more. For example, CropX's[1] soil monitoring system is the hardware and software that measures moisture, temperature, and electrical conductivity in the soil. The system tells farmers when and how much to water via smartphone or PC. With transportation field, nowadays, there are many applications which provide the abilities to locate your cars, response to voice controls, monitor the fuel level, feel the connecting with other vehicles, and much more. Some commercial products include Vinli[2], Zubie[3], Voyo[4], Tesla[5], etc. Besides, there are also a variety of different IoT applications, each application has its own features, most of which are intended to serve and advance the development of society.

---

[1] https://www.cropx.com.

[2] https://www.vin.li.

[3] http://zubie.com.

[4] https://www.voyomotive.com.

[5] https://www.tesla.com.

OTA Programming/Reconfiguration. The complexity of software running on wireless sensor network has increased over the years, and the need for an over-the-air (OTA) programming tool has become noticeable. OTA programming or reconfiguration can be observed as a method for programming computer code on a remote platform included fixing bugs, updating code, updating the firmware, and managing application changes. Especially, it is important in facilitating the management and maintenance of Wireless Sensor Network (WSN) more easily. The WSNs operating system is one of the most important parts of the platform. There are a lot of operating systems available, and it is hard to determine which one is better suited for the deployed system. To deal with many constraints on low memory as well as data transmission with low power, numerous of operating systems in Contiki, Riot OS, Free OS, Tiny OS, etc. have been developed. All of them also have strengths and weaknesses but they are necessary for various requirements of WSN. In this list, Tiny OS is better suited for the IoT applications with limited resources. It supports complex programs with very low memory requirements (many applications fit within 16 KB of memory) and efficient, low-power operation [6, 7]. For the efficiency of software maintenance in the large deployed area, many OTA programming protocols have been designed for TinyOS in the past few years, the summary of various protocols run on Tiny OS is presented in Table 1 below:

Table 1. The summary of various protocols run on TinyOS

Protocol/middleware	Supported hardware	Network architecture	Features
XNP [8]	Mica, Mica2DOT	Single hop	Do not scale to a large sensor network and incremental update Have a lower bandwidth The programming time is small
Deluge [9]	Mica2, Telosb, TOSSIM Simulator	Multihop	Use optimization techniques like adjusting the packet transmit rate and spatial multiplexing Use a fixed size page as a unit of buffer management and retransmission
MNP [10]	Mica2, XSM	Multihop	Energy efficient because it reduces the active radio time of a sensor node by putting the node into "sleep" state when its neighbors are transmitting a segment that is not of interest
Sprinkler [11]	Mica2	Multihop	A reliable data dissemination service for wireless embedded devices which are constrained in energy, processing speed, and memory
TinyCubus [12]	Telosb	Multihop	A flexible, adaptive cross-layer framework for sensor networks Included 3 part: - Tiny data management framework - Tiny cross-layer framework - Tiny configuration engine

In summary, IoT-based applications have been developed and deployed in many different domains to enhance the quality of people's live. Most of these systems are based on WSNs so that OTA programming the wireless sensor devices to update/upgrade firmware remotely is significant and attracted lots of scientists. In next section, we present the implementation of our IoT-based system with the reconfigurable approach that can be applied not only in the agricultural areas, e.g., Smart Garden, Smart Farm, but also in different environmental monitoring application domains.

3   Implementation

In this section, we clearly explain our implementation by going details on each component of the framework. Our IoT-based system includes three essential components: wireless sensor networks, server system and web interface. These parts are obviously illustrated in Fig. 1 below.

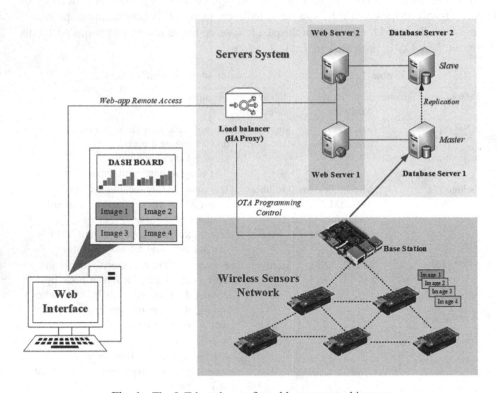

Fig. 1. The IoT-based reconfigurable system architecture

3.1   Wireless Sensor Networks

All devices used in our experiments formulate a wireless sensor network. In particular, Telosb motes were used as wireless sensors nodes, and Raspberry Pi 2 Model B worked as Base Station. Telosb motes run TinyOS and communicate with others over IEEE 802.15.4 (CC2420) network based on Active Messages, which are small (36-byte) packets associated with a 1-byte node ID. Whereas Raspberry Pi is a mini computer, which is pre-installed TinyOS developing environment so that we can attach Telosb motes to Raspberry's USB ports and this mini-computer can work as a Base Station, collect sensor data, control wireless devices as well as talk to external networks in specific applications.

Figure 2 below illustrates the network protocol stack which runs on TinyOS WSNs node in our implementation.

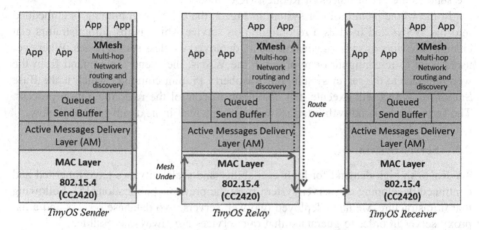

Fig. 2. The network protocol stack run on TinyOS application

As the figure shows, in sender nodes, data from Application Layer (e.g., sensor data, control values…) will be encapsulated by adding the header of under layers and delivered to other nodes by the Physical Layer (802.15.4). When one node in the network receives a message, there are two options for next operation: forwarding directly (mesh under) or routing required (route over). Mesh under allows incoming messages sent out by Mac and Physical Layer, while Route Over needs more actions with higher layers (XMesh – Network Layer). In case of the receiver is the destination, the packet will be delivered from Physical to Application and de-encapsulated to extract data.

3.2 The Over-the-Air Programming Mechanism

In our implementation, we used Deluge middleware running on TinyOS to support as the re-programming mechanism. By using Deluge, we can distribute new images, configuration settings to wireless sensor devices in order to change their behaviors and meet the high demand in real life.

Deluge is a reliable protocol for distributing large data object from one or more sources to many other nodes over multi-hop wireless sensor networks [9]. With Deluge, big object data can be disseminated through the wireless link by splitting into many small pages, and each page will be continued breaking into smaller packets so that TinyOS radio can transmit to other nodes.

When a wireless sensor device runs Deluge service, this node working in one of three states followed: Maintenance, Request (RX) and Transmit (TX). In Maintenance sate, running nodes have the newest version and ready for sending object data to others. When some nodes are working as senders, these nodes are running in Transmit (TX) mode. Otherwise, receivers, which request new version and collect packet from the senders are the instances of Request (RX) mode.

In the whole picture of our system in Fig. 2, the OTA programming is embedded into the WSNs and used as a reconfiguration service which helps administrators can change behaviors or swap images run on all deployed wireless sensor nodes. Whenever users want to re-configure or re-program the WSNs, they send a command from the web interface to the server system, our Raspberry Pi mini-computer which is the Base Station of WSNs will execute that command to control the re-configuration process. The server system and web interface will be explained in next sub-sections below.

3.3   Server System

To deal with high demand for high availability and scalability, we have inherited and continued developing our server system from the previous publications [5]. Following that infrastructure, we have deployed two web servers, two database servers and a ha-proxy server in order to guarantee that our services are always accessible.

In particular, to ensure that all necessary files of the web app running on two different servers consistent with the other, we have deployed a well-known file synchronizer solution – Unison on web servers. Unison is a file-synchronization [13] tool for OSX, Unix, and Windows, which allows replicating files, directory stored not only on different disks on the same host but also on different hosts, all modification and changes in the replica will be propagated and updated to the other.

For the purpose of storing collected data from wireless sensor networks, we set up a duplicated database system. Technically, we installed MySQL on two Centos instances and turned on the replication feature on these servers. This system works follow a Master-Slave model, everything changes on the Master server will be replicated to the Slave that keep stored data safe and highly available.

3.4   Web Interface

In the previous version [5], our web interface was designed for visualizing the collected sensor data from WSNs which deployed in a Garden or a Farm. These measured data were processed and analyzed by the fuzzy logic controller [5] then generated control signal or helpful information so that the gardeners can take care of their plants easily and effectively.

In this version, we have improved our web interface for implementing the OTA reprogramming feature. In the admin side, we've created the re-configuration options. By using this sub-menu, the administrator can easily manage and control the applications of images running on wireless sensor nodes that they deployed in their gardens. There are three different command types which can be used in this feature, namely, check version, disseminate new image, re-programming new image. Whenever the admin clicks to the corresponding command, then this signal will be sent to the base station in a secure channel. After that, the base station will control all wireless sensor devices in WSNs by executing the received command under the OTA programming mechanism supported by Deluge middleware.

4 Result and Evaluation

In this section, we illustrate our experimental outcomes after deploying the IoT-based reconfigurable system described in above section. The results give information about many different metrics related to resources usage, code disseminating time, packet transmission analysis as well as power consumption of the wireless sensor nodes.

4.1 Code Distribution Evaluation

To figure out the code distribution process in our system, we have conducted many experiments on updating new images to wireless sensor devices. Technically, we developed a simple TinyOS application, called *SensorCollection* run on Telosb motes. That application was pre-embedded Deluge service for OTA programming. To analyze packet transmission between wireless sensor nodes, we used Open Sniffer, an analyzer device which can capture and analyze communication over the wireless connection based on IEEE 802.15.4, Zigbee, 6LoWPAN, Wireless Hart and ISA 100.11a standard [14]. By using Open Sniffer device, the object data distributing process between Telosb motes was clarified with some helpful information such as code size, disseminating time, number of transmitted packets as well as the average of packet size. Besides the packet transmission analysis, we also examined power consumption for data propagating process. The Table 2 below shows the average of outcome value after examining the OTA programming with *SensorCollection* application, which is 40128 bytes in image size in 5 times.

Table 2. The experimental results of Deluge code distribution

Metric	1st time	2nd time	3rd time	4th time	5th time	Avg.
Disseminating time (s)	60.570	63.642	64.730	65.583	64.330	63.771
No. packets	2134	2145	2161	2167	2158	2153
Avg. packet/s	29.372	29.958	30.084	29.385	30.879	29.9356
Avg. packet size	113	113	113	113	113	113
Bytes	241908	243022	244739	245315	244461	243889
Avg. bytes/s	3329.568	3394.134	3407.037	3326.548	3497.976	3391.053

4.2   WSNs Application in Smart Garden Evaluation

The OTA programming or re-configuration the WSNs play a vital role in this version of our IoT-based Smart Garden system. This feature is significantly useful when a lot of wireless sensor devices are deployed in large areas. For example, in a smart garden or smart farm, loads of sensors are distributed around for environmental monitoring. In that scenarios, after the harvest time or on different development stages of a specific plant species we need to install and update new images/applications for that wireless sensors devices to adapt to realistic requirements. Our deployment system comes from that idea. In order to prove the enhancement with the previous version [5], we have used the OTA update feature to distribute new images to the activating wireless sensor node and examined the differences between these versions on Telosb motes.

Technical Description. In this investigation, we have generated three versions of SensorCollection application. In particular, in these applications, there is a Telosb mote plugged into the Raspberry Pi and works as a Base Station (node_id = 0) which collects sensor data from others follow a tree topology. These versions are distinguished by the data collecting frequency or number of used sensors. In detail, version 1 and version 2 are SensorCollection which collect and update sensor data to base station every 30 s and 3 min respectively. While version 3 is SensorCollection with just 2 sensors (temperature and humidity) and update measured data with the period of 3 min. We also compare some popular dimensions follow above experiment, e.g., resource usage, packet transmission as well as power consumption.

Empirical results. The Fig. 3 below shows experimental results. It is clears that Version 1 uses most resources, bandwidth and energy because this application uses all three sensors and collect data in high frequency (every 30 s). On the other hand, version 2 and version 3 are consume less resources. So that OTA programming for updating new images into the WSNs to meet new the requirements of current situations is really important that will save lots of cost, energy keep our system working perfectly.

4.3   Summary and Discussion

With two sub-sections above, we have presented some results of our deployed system in two key features: OTA programming and environmental monitoring applications. By using TinyOS platform, our system working with a low require on memory and power consumption. The measured data showed can be used as a dataset to research community to compare in other environments. In that research direction, we have also conducted some initial research on comparing our selected platform TinyOS to Contiki with the same purposes: re-programming and sensor collection for environmental monitoring. Some early results are shows in Table 3.

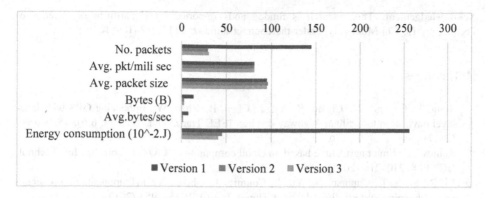

Fig. 3. The empirical comparison of 3 images version run on WSNs for the Smart Garden

Table 3. The comparison results between TinyOS and Contiki on *SensorCollection*

Comparative aspects	TinyOS	Contiki
Hardware	Telosb	Telosb
Software feature	Sensor collection	Sensor collection
Re-programing method	Deluge	Shell upload, shell-exec
Re-programming mechanism	Full-image	Module-based
Code size of updated element	40128 bytes	1702 bytes
Code size of images	40128 bytes	45478 bytes
Traffic on *SensorCollection* application in 30 min	145 packets Avg pkt size: 95 bytes Bytes: 13743 Avg. bytes/s: 7.711	11172 packets Avg pkt size: 123 bytes Bytes: 1371443 Avg. bytes/s: 762.429

5   Conclusion and Future Work

In conclusion, we have implemented the IoT-based system with the reconfigurable approach, which can be deployed in the Smart Garden or Smart Farm to deal with high demands on adaptability and portability. By using the load-balanced infrastructure of the previous version, our system keeps the scalability and availability when deploying in a large area with lots of wireless sensor devices. We also conducted many experiments to examine the operation of the proposed system in various dimensions, e.g., power consumption, memory usage and some summary of packet transmission.

A promising area of our future research would probably be integrated the context awareness on IoT reconfiguration so that the applications run on WSNs can be updated automatically to adapt more realistic requirements. On the other hand, developing a secure code distribution as well as optimizing the data disseminating process is also open issues attracting not only our group but also many other researchers from over the world.

Acknowledgement. This research is funded and supported by Department of Science & Technology Ho Chi Minh City, under the contract number: 211/2017/HĐ-SKHCN.

References

1. Song, T., Capurso, N., Cheng, X., Yu, J., Chen, B., Zhao, W.: Enhancing GPS with lane-level navigation to facilitate highway driving. IEEE Trans. Veh. Technol. **6**(6), 4579–4591 (2017)
2. TongKe, F.: Smart agriculture based on cloud computing and IOT. J. Converg. Inf. Technol. (JCTT) **8**, 210–216 (2013)
3. Li, R., Song, T., Capurso, N., Yu, J., Couture, J., Cheng, X.: IoT applications on secure smart shopping system. IEEE Internet Things J. **4**(6), 1945–1954 (2017)
4. Jie, Y., Pei, J.Y., Jun, L., Yun, G., Wei, X.: Smart home system based on IOT technologies. In: 2013 International Conference on Computational and Information Sciences, Shiyang (2013)
5. Huynh-Van, D., Tran-Quoc, K., Le-Trung, Q.: Toward a real-time development and deployment of IoTs application for smart garden on OpenStack cloud. In: Chen, Y., Duong, T.Q. (eds.) INISCOM 2017. LNICST, vol. 221, pp. 121–130. Springer, Cham (2018). https://doi.org/10.1007/978-3-319-74176-5_12
6. Levis, P., Madden, S., Polastre, J., Szewczyk, R., Whitehouse, K., Woo, A.: TinyOS: an operating system for sensor networks. In: Weber, W., Rabaey, J.M., Aarts, E. (eds.) Ambient Intelligence, pp. 115–148. Springer, Heidelberg (2005). https://doi.org/10.1007/3-540-27139-2_7
7. Amjad, M., Sharif, M., Afzal, M.K., Kim, S.W.: TinyOS-new trends, comparative views and supported sensing applications: a review. IEEE Sens. J. **16**(9), 2865–2889 (2016)
8. Jeong, J., David, C.: Incremental network programming for wireless sensors. In: 2004 First Annual IEEE Communications Society Conference on Sensor and Ad Hoc Communications and Networks (2004)
9. Hui, J.W., Culler, D.: The dynamic behavior of a data dissemination protocol for network programming at scale. In: SenSys 2004 Proceedings of the 2nd International Conference on Embedded Networked Sensor Systems, Baltimore, MD, USA (2004)
10. Sandeep, S.K., Limin, W.: MNP: multihop network reprogramming service for sensor networks. In: The 25th IEEE International Conference on Distributed Computing Systems (2005)
11. Naik, V., Arora, A., Sinha, P., Zhang, H.: Sprinkler: a reliable and energy efficient data dissemination service for wireless embedded devices. In: 26th IEEE International Real-Time Systems Symposium (RTSS 2005) (2005)
12. Marron, P.J., Lachenmann, A., Minder, D., Hahner, J., Sauter, R., Rothermel, K.: TinyCubus: a flexible and adaptive framework for sensor networks. In: Proceedings of the Second European Workshop on Wireless Sensor Networks (2005)
13. Unison file synchronizer. https://www.cis.upenn.edu/~bcpierce/unison/
14. Open Sniffer For 802.15.4, Zigbee, 6LOWPAN. https://www.sewio.net

Secure Cooperative Systems with Jamming and Unreliable Backhaul over Nakagami-m Fading Channels

Michael Stewart[1], Long D. Nguyen[1(✉)], Cheng Yin[1],
Emiliano Garcia-Palacios[1], and Sang Q. Nguyen[2]

[1] School of Electronics, Electrical Engineering and Computer Science,
Queen's University Belfast, Belfast, UK
{mstewart,lnguyen04,cyin}@qub.ac.uk, e.garcia@ee.qub.ac.uk
[2] Faculty of Information Technology, Duy Tan University, Da Nang, Vietnam
sangnqdv05@gmail.com

Abstract. In this paper, the secrecy performance of cooperative networks with jamming signals of eavesdroppers are studied under the impacts of unreliable backhaul networks. By proposing a two-phase transmitter/relay selection scheme, the desired signal-to-noise ratio (SNR) of relays is maximized by selected the best transmitter, meanwhile, the jamming signal-to-interference-plus-noise ratio (SINR) of the eavesdroppers is minimized by selected the best relay. The secrecy outage probability is derived in closed-form expressions by following some useful lemmas and theorems. Furthermore, the analysis of asymptotic secrecy outage probability is also performed to explicitly reveal the impacts of unreliable backhaul links on the secrecy performance. By the impact of imperfect backhaul links, the diversity gain is limited as shown in our results.

Keywords: Cooperative networks · Physical layer security
Secrecy performance · Unreliable backhaul

1 Introduction

Physical layer security (PLS) concept has become an emerging feature in wireless communication systems [14] with the serious effects of malicious eavesdroppers on the confidential transmission. In fact, the transmitted information might be vulnerable with the jamming interference by the existing of eavesdroppers [15] in

Supported by the Newton Prize 2017 and by a Research Environment Links grant, ID 339568416, under the Newton Programme Vietnam partnership. The grant is funded by the UK Department of Business, Energy and Industrial Strategy (BEIS) and delivered by the British Council. For further information, please visit www.newtonfund.ac.uk/.

T. Q. Duong and N.-S. Vo (Eds.): INISCOM 2018, LNICST 257, pp. 253–263, 2019.
https://doi.org/10.1007/978-3-030-05873-9_21

the wireless communication systems. For preventing eavesdropping attacks, the message between the source and destination is encrypted/decrypted with sharing the specific secretly key. To this end, PLS performs the information-theoretical methods, which involve the cooperative relaying to secure received messages at the receiver by exploiting the impact of jamming signals, to utilise the benefits of physical characteristics by different upper layer security [3]. On the other hand, cooperative networks (CNs) have grown in network infrastructure for satisfying the excessing increase of data transmission [8]. However, by deploying higher density of small-cell numbers, e.g., relay nodes, one of the critical problems of CNs is unreliable backhaul links [11,15,20]. Besides the threats of jamming interference from eavesdropping, the issues of propagation, e.g., multipath fading, transmission delay and synchronization can impact the reliable backhaul network as well as the network performance in CNs [13,16]. As a result, the unreliable backhaul transmissions significantly depreciate the performance of network [12,20].

Many previous PLS works ensure the success of the private signals from source to the legitimate destination under the assumption of ideal reliability backhaul link. In these studies, the cooperative relay technique is investigated to minimize the jamming interference of the eavesdroppers [4,9] under both decode-and-forward (DF) and amplify-and-forward (AF) relaying schemes [1,3,17]. However, in practice, the presence of unreliable backhaul cannot avoid since this issue strongly impacts the real network performance. Some researches have tackled the unreliable backhaul for enhancing the network performance [5,11,12,20]. For instances, the cooperative frameworks are proposed to scale the performance under the impact of unreliable backhaul links in [5,20] or the extension of spectrum sharing in backhaul communication with the limit interference of primary users [11,12].

From the observations of PLS and unreliable backhaul in CNs, it is obvious that the investigation of PLS under unreliable backhaul links has been completely necessary. In this paper, we study the network performance of a CN with the impact of jamming interference in term of secrecy outage probability. By exploiting a CN scenario, the unreliable backhaul links between a macro-cell and relay nodes (e.g., small-cells) is considered. This scheme is to eliminate the inter-symbol interference (ISI) [6,7]. On the other hand, the transmission information will be reflected by considering frequency selective fading channel [15,17]. Very recently, the performance of physical layer security has been investigated with single carrier system under the impact of imperfect backhaul condition [10]. To the best of our knowledge, all of the previous works on the physical layer security performance under the impact of unreliable backhaul only considered the Rayleigh fading channels. Therefore, in this paper, we take a step further to extend this research into a more general fading channel, namely, Nakagami-m. For the transmission scheme, a two-phase transmitter/relay selection scheme is provided in density multi-relay networks. The best relay selection is obtained in the first phase for maximizing the desired SNR at the relays and minimizing the signal-to-interference-plus-noise ratio (SINR) at the eavesdroppers during the second phase simultaneously. We show that the backhaul reliability is an important factor in PLS system design, which strongly affects the achievable secrecy performance.

Notation: $\mathcal{CN}(\mu, \sigma_n^2)$ denotes the complex Gaussian distribution with mean μ and variance σ_n^2; $\boldsymbol{I}_m$ is an $m \times m$ identity matrix; $\mathbb{C}^{m \times n}$ is vector space of $m \times n$ complex matrices; $X \sim \chi^2(N_X, \alpha_X)$ denotes chi-square distribution with degree of freedom (DoF) N_X and power normalizing constant α_X; $X \sim \mathrm{Ga}(\mu_X, \eta_X)$ denotes Gamma distribution with shape μ_X and rate η_X. $F_\lambda(\gamma)$ and $f_\lambda(\gamma)$ denote the cumulative distribution function (CDF) and probability density function (PDF) of the random variable (RV) λ, respectively; $\mathbb{E}_\lambda \{f(\gamma)\}$ denotes the expectation of $f(\gamma)$ with respect to the RV λ. In addition, $\binom{\tau_1}{\tau_2} = \dfrac{\tau_1!}{\tau_2!(\tau_1 - \tau_2)}$ denotes the binomial coefficient.

2 Network and Channel Models

We consider a CN in where a macro-cell base station (Macro BS) connected to the core network as shown in Fig. 1. Meanwhile, K small-cell transmitters T_k, $k \in \{1, 2, \cdots, K\}$ connect to the Macro-BS via unreliable backhaul links and communicate with a destination (D) via M relays R_m, $m \in \{1, 2, \cdots, M\}$ applying DF scheme. Otherwise, a single jammer J and N eavesdroppers E_n, $n \in \{1, 2, \cdots, N\}$ exist in the network. Due to poor channel conditions, the direct link from T_k to destination and between T_k and E_n $\forall k \in K, \forall n \in N$ are neglected. On the other hand, the interference at the eavesdroppers is caused by J while the eavesdroppers cooperate to overhear the transmissions between R_m and D. In this work, we also assume that both of transmitters and receivers use half-duplex mode and exploit a single antenna.

For a cooperative system, we make the following assumptions for the channel model. Nakagami-m fading is considered in the channel model of the system. Therein, the channel between T_k and R_m, $\forall k, m$ is denoted by $h^{k,m} \sim \mathrm{Ga}(\mu_k^m, \tilde{\eta}_k^m)$. Meanwhile, $g^{m,n} \sim \mathrm{Ga}(\mu_m^n, \tilde{\eta}_m^n)$ denotes the channel between R_m and E_n, $\forall m, n$. The path loss component corresponding to $h^{k,m}$ and $g^{m,n}$ are denoted as $\alpha_T^{k,m}$ and $\alpha_E^{m,n}$, respectively. The channel between R_m and D, $\forall m$ and the channel between J and E_n, $\forall n$ are defined by $f^m \sim \mathrm{Ga}(\mu_d^m, \tilde{\eta}_d^m)$ and $q^n \sim \mathrm{Ga}(\mu_j^n, \tilde{\eta}_j^n)$. The path loss component corresponding to f^m and q^n are represented by α_D^m and α_J^n, respectively. We also assume that $\mathbb{E}[x] = \mathbb{E}[v] = 0$.

For all active links, the channel state information (CSI) is common assumed to be perfectly known at the relays, J, and D in PLS literature [3,15,18]. Also, the information can be measured by J from the eavesdroppers in the network [2]. The transmit symbol block $\boldsymbol{x}$, which is transmitted from the Macro-BS, must pass through the dedicated wireless backhaul links. The success/failure transmission represents the reception status at the K transmitters. Hence, a Bernoulli process $\mathbb{I}_k$ can be applied to the reliability of the wireless backhaul links, i.e., the message is successfully received at the receivers with a successful probability of λ_k. The failure probability is accordingly given by $1 - \lambda_k$ [5,12].

The received signal at R_m from T_k is expressed as

$$y_R^{k,m} = \sqrt{\mathcal{P}_t \alpha_T^{k,m}} h^{k,m} \mathbb{I}_k x + n_R^{k,m}, \tag{1}$$

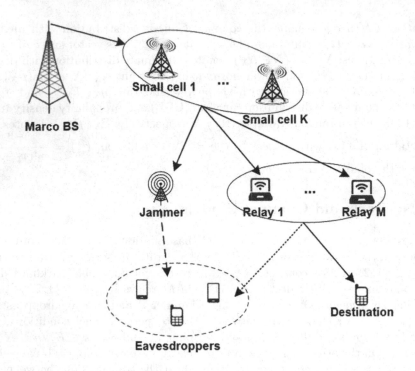

Fig. 1. System model.

where $\mathcal{P}_t$ is the transmit power and $n_R^{k,m} \sim \mathcal{CN}\left(0, \sigma_n^2\right)$ is an additive noise vector at R_m. $\mathbb{I}_k$ recalls the backhaul reliability which is modeled as a Bernoulli process. This model is common for the representation of the backhaul reliability with canonically success/failure transmission [5,12,15,20]. In the first time slot, the instantaneous SNR between T_k and R_m from (1) can be expressed as

$$\gamma_R^{k,m} = \frac{\mathcal{P}_t \alpha_T^{k,m} \|h^{k,m}\|^2}{\sigma_n^2} \mathbb{I}_k = \widetilde{\alpha}_T^{k,m} \|h^{k,m}\|^2 \mathbb{I}_k = \lambda^{k,m} \mathbb{I}_k, \tag{2}$$

where $\widetilde{\alpha}_T^{k,m} \triangleq \dfrac{\mathcal{P}_t \alpha_T^{k,m}}{\sigma_n^2}$, $\lambda^{k,m} \sim \mathrm{Ga}(\mu_k^m, \eta_k^m)$ with $\eta_k^m = \dfrac{\widetilde{\alpha}_T^{k,m} \, \mathbb{E}\left\{\|h^{k,m}\|^2\right\}}{\mu_k^m}$.

The received signals at E_n and D from R_m are given by

$$y_E^{m,n} = \sqrt{\mathcal{P}_r \alpha_E^{m,n}} \, g^{m,n} x + \sqrt{\mathcal{P}_j \alpha_j^n} \, q^n v + n_E^{m,n},$$
$$y_D^m = \sqrt{\mathcal{P}_r \alpha_D^m} \, f^m x + n_D^m, \tag{3}$$

where $\mathcal{P}_r$ and $\mathcal{P}_j$ are the transmit powers at the relays and J, respectively, $n_E^{m,n} \sim \mathcal{CN}(0, \sigma_n^2)$ and $n_D^m \sim \mathcal{CN}(0, \sigma_n^2)$ define the noise at E_n and D.

Therefore, the instantaneous SINR between R_m and E_n can be formulated as

$$\gamma_E^{m,n} = \frac{\mathcal{P}_r \alpha_E^{m,n} ||g^{m,n}||^2}{\sigma_n^2 + \mathcal{P}_j \alpha_j^n ||q^n||^2} = \frac{\widetilde{\alpha}_E^{m,n} ||g^{m,n}||^2}{1 + \widetilde{\alpha}_j^n ||q^n||^2} = \frac{\lambda^{m,n}}{1 + \lambda_j^n}, \tag{4}$$

where $\widetilde{\alpha}_E^{m,n} \triangleq \dfrac{\mathcal{P}_r \alpha_E^{m,n}}{\sigma_n^2}, \widetilde{\alpha}_j^n \triangleq \dfrac{\mathcal{P}_j \alpha_j^n}{\sigma_n^2}, \lambda^{m,n} \sim \mathrm{Ga}(\mu_m^n, \eta_m^n), \lambda_j^n \sim \mathrm{Ga}(\mu_j^n, \eta_j^n)$ with
$\eta_m^n = \dfrac{\widetilde{\alpha}_E^{m,n} \, \mathbb{E}\left\{||g^{m,n}||^2\right\}}{\mu_m^n}, \eta_j^n = \dfrac{\widetilde{\alpha}_j^n \, \mathbb{E}\left\{||q^n||^2\right\}}{\mu_j^n}.$

In the remaining time slot, the instantaneous SNR between R_m and D can be expressed as

$$\gamma_D^m = \frac{\mathcal{P}_r \alpha_D^m ||f^m||^2}{\sigma_n^2} = \widetilde{\alpha}_D^m ||f^m||^2 = \lambda_D^m, \tag{5}$$

where $\widetilde{\alpha}_D^m \triangleq \dfrac{\mathcal{P}_r \alpha_D^m}{\sigma_n^2}, \lambda_D^m \sim \mathrm{Ga}(\mu_d^m, \eta_d^m)$ with $\eta_d^m = \dfrac{\widetilde{\alpha}_D^m \, \mathbb{E}\left\{||f^m||^2\right\}}{\mu_d^m}.$

Moreover, the channels are distributed according to chi-square distribution since they are assumed to undergo Nakagami-m fading. Thus, the CDF and PDF of RV $\chi \sim \mathrm{Ga}(\mu_\chi, \eta_\chi)$, where $\chi \in \{h^{k,m}, g^{m,n}, f^m, q^n\}$ are given, respectively,

$$f_\chi(x) = \frac{1}{(\mu_\chi - 1)!(\eta_\chi)^{\mu_\chi}} x^{\mu_\chi - 1} e^{(-x/\eta_\chi)},$$

$$F_\chi(x) = 1 - e^{(-x/\eta_\chi)} \sum_{i=0}^{\mu_\chi - 1} \frac{1}{i!} (x/\eta_\chi)^i, \tag{6}$$

respectively. We assume that the unreliable backhaul links are independent from the indices of the K transmitters, i.e., $\lambda_k = \lambda, \forall k$ and the positive fading severity parameter μ_χ and η_χ are identically varied among the K transmitters, M relays and N eavesdroppers, i.e., $\mu_T = \mu_k^m, \mu_E = \mu_m^n, \mu_J = \mu_j^n, \mu_D = \mu_d^m$, and $\eta_T = \eta_k^m, \eta_E = \eta_m^n, \eta_J = \eta_j^n, \eta_D = \eta_d^m, \forall k, m, n.$

3   Secrecy Performance Analysis

3.1   Two-Phase Transmitter/Relay Selection Scheme

By providing a two-phase selection scheme, our target is to achieve high PLS level. This scheme is to maximizes the achievable performance and minimizes the undesired performance at the eavesdroppers simultaneously. Following this intuition, in the first phase, each relay chooses the best transmitter among the K small-cells to maximize their achievable SNR. The selected scheme in the first phase can be mathematically expressed as

$$\text{Phase 1:} \quad k^* = \arg \max_{k=1,\ldots,K} \gamma_R^{k,m}, \tag{7}$$

where $\gamma_R^{k,m}$ recalls the instantaneous SNR at R_m via T_k. From (7), the statistical property of the instantaneous SNR via the best transmitter T_{k^*} is given in the following theorem.

Theorem 1. *Given K independent and identical unreliable backhaul connections, the CDF of the received SNR at R_m via the best transmitter is given as*

$$F_{\gamma_R^{k^*,m}}(x) = 1 + \sum_{k=1}^{K} \sum_{\omega_1,...,\omega_{\mu_R}} \binom{K}{k} \left(\frac{k!}{\omega_1!...\omega_{\mu_R}!} \right)$$

$$\frac{(-1)^k \lambda^k}{\prod_{t=0}^{\mu_R-1} (t!(\eta_T)^t)^{\omega_{t+1}}} x^{\sum_{t=0}^{\mu_R-1} t\omega_{t+1}} e^{-kx/\eta_T}. \tag{8}$$

Proof. The proof will show in the journal version by the limit of conference version.

The relay selection will be implemented in the second phase, which is to minimize the SINR between the particular relay and N eavesdroppers. This scheme can be formulated as

$$\text{Phase 2:} \quad m^* = \arg \min_{m=1,...,M} \gamma_E^{m,n^*}, \tag{9}$$

where $\gamma_E^{m,n^*} = \max(\gamma_E^{m,1}, ..., \gamma_E^{m,N})$ is the maximum instantaneous SINR between R_m and N eavesdroppers. The CDF of the RV γ_E^{m,n^*} is given in the following lemma and theorems in [10].

3.2 Secrecy Outage Probability

In this subsection, we focus on the secrecy outage probability, where the eavesdroppers's CSI is assumed unavailable in the considered network. In this case, the transmitters encode and send the confidential message with the constant secrecy rate of θ. If the instantaneous secrecy capacity, denoted by $\mathcal{C}_S$ in bits/s/Hz, is greater than θ, the secrecy gain is guaranteed. To study the asymptotic behavior, the asymptotic secrecy outage probability will be analysed in this work.

Following [19], the secrecy capacity $\mathcal{C}_S$ can be expressed as

$$\mathcal{C}_S = \frac{1}{2} \left[\log_2(1 + \widetilde{\gamma}_{DF}^{m^*}) - \log_2(1 + \gamma_E^{m^*,n^*}) \right]^+, \tag{10}$$

where $\log_2(1 + \widetilde{\gamma}_{DF}^{m^*})$ is the instantaneous capacity at D respect to the m^*-th relay and $\log_2(1 + \gamma_E^{m^*,n^*})$ is the instantaneous capacity of the m^*-th relay and n^*-th eavesdropper link.

Thus, the secrecy outage probability, i.e. the probability when the secrecy capacity (10) falls below the given rate threshold, can be expressed as [17]

$$\mathcal{P}_{out}(\theta) = Pr(\mathcal{C}_S < \theta)$$

$$= \int_0^\infty F_{\widetilde{\gamma}_{DF}^{m^*}} \left(2^{2\theta}(1+x) - 1 \right) f_{\gamma_E^{m^*,n^*}}(x) \, dx. \tag{11}$$

$$\mathcal{P}_{out}(\theta) = 1 + \mathcal{Q}\widetilde{\sum_{D}\sum_{E}}\sum_{\alpha=0}^{\beta}\binom{\beta}{\alpha}(\Upsilon-1)^{\beta-\alpha}(\Upsilon)^{\alpha}\eta_{E}^{\tilde{\varphi}_{3}}e^{-\Phi(\Upsilon-1)}(\mathcal{O}_{1}-\mathcal{O}_{2}+\mathcal{O}_{3}), \quad (12)$$

where $\Upsilon \overset{\triangle}{=} 2^{2\theta}$, $\epsilon \overset{\triangle}{=} \dfrac{\eta_{E}}{\eta_{J}}$ and

$$\mathcal{O}_{1} = \mathcal{B}_{1}\Gamma(\tilde{\varphi}_{2}+\alpha+1)\epsilon^{\tilde{\varphi}_{2}+\alpha+1-\tilde{\varphi}_{3}}\Psi\left(\tilde{\varphi}_{2}+\alpha+1, \tilde{\varphi}_{2}+\alpha+2-\tilde{\varphi}_{3}, \epsilon(\Phi\Upsilon+\tilde{\varphi}_{1})\right),$$
$$\mathcal{O}_{2} = \mathcal{B}_{2}\Gamma(\tilde{\varphi}_{2}+\alpha)\epsilon^{\tilde{\varphi}_{2}+\alpha-\tilde{\varphi}_{3}}\Psi\left(\tilde{\varphi}_{2}+\alpha, \tilde{\varphi}_{2}+\alpha+1-\tilde{\varphi}_{3}, \epsilon(\Phi\Upsilon+\tilde{\varphi}_{1})\right),$$
$$\mathcal{O}_{3} = \mathcal{B}_{3}\Gamma(\tilde{\varphi}_{2}+\alpha+2)\epsilon^{\tilde{\varphi}_{2}+\alpha+2-\tilde{\varphi}_{3}}\Psi\left(\tilde{\varphi}_{2}+\alpha+2, \tilde{\varphi}_{2}+\alpha+3-\tilde{\varphi}_{3}, \epsilon(\Phi\Upsilon+\tilde{\varphi}_{1})\right).$$

$$\mathcal{P}_{out}^{\infty}(\theta) \overset{\eta_{D}\to\infty}{=} 1 + \mathcal{Q}\widetilde{\sum_{D^{\infty}}\sum_{E}}\sum_{\alpha=0}^{\tilde{\beta}}\binom{\tilde{\beta}}{\alpha}(\Upsilon-1)^{\tilde{\beta}-\alpha}(\Upsilon)^{\alpha}\eta_{E}^{\tilde{\varphi}_{3}}e^{-\tilde{\Phi}(\Upsilon-1)}(\widehat{\mathcal{O}_{1}}-\widehat{\mathcal{O}_{2}}+\widehat{\mathcal{O}_{3}}),$$

$$(13)$$

where $\quad \tilde{\beta} \quad = \quad \sum_{t=0}^{\mu_{R}-1} t\omega_{t+1}, \quad \tilde{\Phi} \quad = \quad \dfrac{k}{\eta_{T}}, \quad \widetilde{\sum_{D^{\infty}}} \quad =$

$\sum_{k=1}^{K}\sum_{\omega_{1},...,\omega_{\mu_{R}}}\binom{K}{k}\left(\dfrac{k!}{\omega_{1}!...\omega_{\mu_{R}}!}\right)\dfrac{(-1)^{k-1}\lambda^{k}}{\prod_{t=0}^{\mu_{R}-1}(t!(\eta_{T})^{t})^{\omega_{t+1}}}$, and

$$\widehat{\mathcal{O}_{1}} = \mathcal{B}_{1}\Gamma(\tilde{\varphi}_{2}+\alpha+1)\epsilon^{\tilde{\varphi}_{2}+\alpha+1-\tilde{\varphi}_{3}}\Psi\left(\tilde{\varphi}_{2}+\alpha+1, \tilde{\varphi}_{2}+\alpha+2-\tilde{\varphi}_{3}, \epsilon(\tilde{\Phi}\Upsilon+\tilde{\varphi}_{1})\right),$$
$$\widehat{\mathcal{O}_{2}} = \mathcal{B}_{2}\Gamma(\tilde{\varphi}_{2}+\alpha)\epsilon^{\tilde{\varphi}_{2}+\alpha-\tilde{\varphi}_{3}}\Psi\left(\tilde{\varphi}_{2}+\alpha, \tilde{\varphi}_{2}+\alpha+1-\tilde{\varphi}_{3}, \epsilon(\tilde{\Phi}\Upsilon+\tilde{\varphi}_{1})\right),$$
$$\widehat{\mathcal{O}_{3}} = \mathcal{B}_{3}\Gamma(\tilde{\varphi}_{2}+\alpha+2)\epsilon^{\tilde{\varphi}_{2}+\alpha+2-\tilde{\varphi}_{3}}\Psi\left(\tilde{\varphi}_{2}+\alpha+2, \tilde{\varphi}_{2}+\alpha+3-\tilde{\varphi}_{3}, \epsilon(\tilde{\Phi}\Upsilon+\tilde{\varphi}_{1})\right).$$

The closed-from expression for the secrecy outage probability in (11) is given in the following theorem.

Theorem 2. *For the CN with unreliable backhaul links, the secrecy outage probability with two-phase transmitter/relay selection scheme is given as in (12) at the top of next page.*

Proof. The proof will be provided in the journal version by the limit of conference version.

To study the impacts of unreliable backhaul connections, the asymptotic expression for the secrecy outage probability is given in the following theorem.

Theorem 3. *Given the fixed set $\{\eta_{T}, \eta_{E}, \eta_{J}\}$, the asymptotic expression for secrecy outage probability is given as (13) at the top of next page.*

Proof. The proof will be provided in the journal version by the limit of conference version.

From (13), due to the unreliable backhaul links, the secrecy diversity gain is not achievable. Furthermore, the limitation on secrecy outage probability is determined as a constant since the asymptotic secrecy outage is independent of DoF of channels between relay and D.

4   Numerical Results

In this section, we validate our analysis and investigate the secrecy outage probability of the considered system model by providing the numerical results. In our simulation, we deploy the binary phase-shift keying (BPSK) modulation with transmission block size $S = 64$ symbols. For convenience, we define Ex and An as the results of link-level simulations and the analytical results, respectively. In the following, we investigate the network performance with various parameters to examine the effects of the degrees of cooperative transmission, DoFs, and backhaul reliability.

4.1   Secrecy Outage Probability

Fig. 2 illustrates the secrecy outage probability for various M and N. The network parameters are set as $K = 3, \lambda = 0.995, \{\mu_R, \mu_E, \mu_J, \mu_D\} = \{2, 2, 2, 3\}$, and $\{\eta_T, \eta_E, \eta_J\} = \{10, 10, 10\}$ dB. It can be observed that the number of relays/eavesdroppers strongly affects the secrecy outage probability. For example, when $N = 1$, the secrecy outage probability becomes lower when more relays help with the cooperative transmission. Differently, when $M = 1$, the secrecy outage probability becomes higher when the number of eavesdroppers increases. This is due to the fact that when the number of relays increases, the secrecy rate becomes higher as a result of the reduction in the wiretap channel capacity. Similarly, the secrecy rate decreases proportionally to the increase in the number of eavesdroppers. We further see that our analysis precisely matches the simulations and our analysis approaches the asymptotic results, presented in Theorem 3, in the high SNR regime.

In Fig. 3, we investigate the effects of DoFs and $\{\eta_T, \eta_E, \eta_J\}$ on the secrecy outage probability. In the settings, we set $K = 3, M = 1, N = 2, \lambda = 0.98$, and $\{\mu_R, \mu_E, \mu_J\} = \{2, 2, 4\}$. As μ_D increases, we observe that the lower secrecy outage probability is achieved. We also observe that as η_T and η_J increase, the secrecy outage probability becomes lower while the increase in η_E results in high achievable secrecy outage. It is clearly to see that the increase in η_T results in a high received power at the receiver while the increase in η_J reduces the SINR of the eavesdroppers.

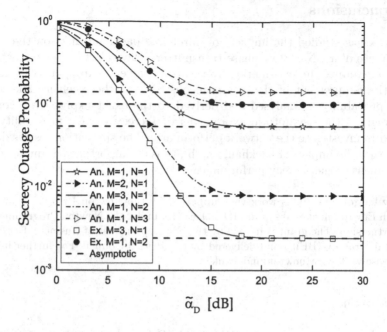

Fig. 2. Secrecy outage probability for various M, N of the proposed network.

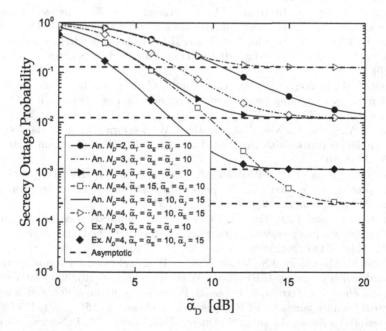

Fig. 3. Secrecy outage probability for various DoFs and $\{\eta_T, \eta_E, \eta_J\}$ of the proposed network.

5 Conclusions

This work has studied the impacts of unreliable backhaul links on the secrecy performance of a CN. A two-phase transmitter/relay selection scheme was proposed to minimize the information overheard of the eavesdroppers with guaranteeing the requirement of the instantaneous SNR at the receiver. The secrecy outage probability is analysed and derived to the exact expressions. For high SNRs regime, the asymptotic expressions of secrecy outage probability were attained to investigate the network performance. The simulation results demonstrated that the impact of backhaul reliability is an important parameter for the improvement of the secrecy performance.

Acknowledgement. This work was supported by the Newton Prize 2017 and by a Research Environment Links grant, ID 339568416, under the Newton Programme Vietnam partnership. The grant is funded by the UK Department of Business, Energy and Industrial Strategy (BEIS) and delivered by the British Council. For further information, please visit www.newtonfund.ac.uk/

References

1. Bao, V.N.Q., Linh-Trung, N., Debbah, M.: Relay selection schemes for dual-hop networks under security constraints with multiple eavesdroppers. IEEE Trans. Wirel. Commun. **12**(12), 6076–6085 (2013)
2. Bloch, M., Barros, J., Rodrigues, M.R., McLaughlin, S.W.: Wireless information-theoretic security. IEEE Trans. Inf. Forensics Secur. **54**(6), 2515–2534 (2008)
3. Dong, L., Han, Z., Petropulu, A.P., Poor, H.V.: Improving wireless physical layer security via cooperating relays. IEEE Trans. Signal Process. **58**(3), 1875–1888 (2010)
4. Hoang, T.M., Duong, T.Q., Vo, N.S., Kundu, C.: Physical layer security in cooperative energy harvesting networks with a friendly jammer. IEEE Wirel. Commun. Lett. **6**(2), 174–177 (2017)
5. Khan, T.A., Orlik, P., Kim, K.J., Heath, R.W.: Performance analysis of cooperative wireless networks with unreliable backhaul links. IEEE Commun. Lett. **19**(8), 1386–1389 (2015)
6. Kim, K.J., Duong, T.Q., Elkashlan, M., Yeoh, P.L., Poor, H.V., Lee, M.H.: Spectrum sharing single-carrier in the presence of multiple licensed receivers. IEEE Trans. Wirel. Commun. **12**(10), 5223–5235 (2013)
7. Kim, K.J., Duong, T.Q., Tran, X.N.: Performance analysis of cognitive spectrum-sharing single-carrier systems with relay selection. IEEE Trans. Signal Process. **60**(12), 6435–6449 (2012)
8. Krikidis, I., Thompson, J.S., Mclaughlin, S.: Relay selection for secure cooperative networks with jamming. IEEE Trans. Wirel. Commun. **8**(10), 5003–5011 (2009)
9. Liu, W., Zhou, X., Durrani, S., Popovski, P.: Secure communication with a wireless-powered friendly jammer. IEEE Trans. Wirel. Commun. **15**(1), 401–415 (2016)
10. Nguyen, H.T., Zhang, J., Yang, N., Duong, T.Q., Hwang, W.J.: Secure cooperative single carrier systems under unreliable backhaul and dense networks impact. IEEE Access **5**, 18310–18324 (2017)

11. Nguyen, H.T., Duong, T.Q., Hwang, W.J.: Multiuser relay networks over unreliable backhaul links under spectrum sharing environment. IEEE Commun. Lett. **21**, 1–4 (2017). Accepted and appeared on IEEE Xplorer
12. Nguyen, H.T., Ha, D.B., Nguyen, S.Q., Hwang, W.J.: Cognitive heterogeneous networks with unreliable backhaul connections. J. Mobile Netw. Appl. (MONET) (2017)
13. Pantisano, F., Bennis, M., Saad, W., Debbah, M., Latva-Aho, M.: On the impact of heterogeneous backhauls on coordinated multipoint transmission in femtocell networks. In: Proceedings of IEEE International Conference on Communications, Ottawa, Canada, pp. 5064–5069, June 2012
14. Rodriguez, L.J., Tran, N.H., Duong, T.Q., Le-Ngoc, T., Elkashlan, M., Shetty, S.: Physical layer security in wireless cooperative relay networks: state of the art and beyond. IEEE Commun. Mag. **53**(12), 32–39 (2015)
15. Shafie, A.E., Duong, T.Q., Al-Dhahir, N.: QoS-aware enhanced-security for TDMA transmissions from buffered source nodes. IEEE Trans. Wirel. Commun. **16**(2), 1051–1065 (2017)
16. Simeone, O., Somekh, O., Erkip, E., Poor, H.V., Shitz, S.S.: Robust communication via decentralized processing with unreliable backhaul links. IEEE Trans. Inf. Theory **57**(7), 4187–4201 (2011)
17. Wang, L., Kim, K.J., Duong, T.Q., Elkashlan, M., Poor, H.V.: Security enhancement of cooperative single carrier systems. IEEE Trans. Inf. Forensics Secur. **10**(1), 90–103 (2015)
18. Yang, J., Kim, I.M., Kim, D.I.: Optimal cooperative jamming for multiuser broadcast channel with multiple eavesdroppers. IEEE Trans. Wirel. Commun. **12**(6), 2840–2852 (2013)
19. Yang, N., Suraweera, H.A., Collings, I.B., Yuen, C.: Physical layer security of TAS/MRC with antenna correlation. IEEE Trans. Inf. Forensics Secur. **8**(1), 254–259 (2013)
20. Yin, C., Nguyen, H.T., Kundu, C., Kaleem, Z., Garcia-Palacios, E., Duong, T.Q.: Secure energy harvesting relay networks with unreliable backhaul connections. IEEE Access **6**, 12074–12084 (2018)

Outage Probability of Cognitive Heterogeneous Networks with Multiple Primary Users and Unreliable Backhaul Connections

Cheng Yin[1]([✉]), Jingxian Xie[1], Emiliano Garcia-Palacios[1],
and Hien M. Nguyen[2]

[1] Queen's University Belfast, Belfast, UK
{cyin01,jxie08}@qub.ac.uk, e.garcia@ee.qub.ac.uk
[2] Duy Tan University, Da Nang, Vietnam
nguyenminhhien2501@gmail.com

Abstract. A cognitive heterogeneous network with unreliable backhaul connections is studied in this paper. In this system, a macro-base station connected to cloud transmits information to multiple small cells via backhaul links. In addition, multiple small cells acting as secondary transmitters send information to a receiver by sharing the same spectrum with multiple primary users. Bernoulli process is adopted to model the backhaul reliability. Selection combining protocol is used at the receiver side to maximize the received signal-to-noise ratio. We investigate the impacts of the number of small-cells, the number of primary users as well as the backhaul reliability on the system performance, i.e., outage probability in Rayleigh fading channels. Closed-form expressions are derived and asymptotic analysis is also provided.

Keywords: Cognitive radio network · Wireless unreliable backhaul Heterogeneous network · Multiple primary users

1 Introduction

In order to satisfy the increasing data traffic demand, future networks are expected to be more dense and heterogeneous [1]. To cope with increasing demand at the access, millimeter wave band can be exploited [2]. In addition, another approach is heterogeneous networks (HetNets), low power small cells including microcells, picocells, femtocells etc. are deployed within the high power macrocells coverage area to achieve substantial gains in coverage and capacity [3–5]. In HetNets, the conventional wired backhaul provides solid connections between macrocells and small cells. However, when a large number of small cells is needed to cover dense scenarios, the cost for the deployment and maintenance is high. To overcome the disadvantage, wireless backhaul has emerged as a suitable solution. However, wireless backhaul is not as reliable as wired backhaul

T. Q. Duong and N.-S. Vo (Eds.): INISCOM 2018, LNICST 257, pp. 264–275, 2019.
https://doi.org/10.1007/978-3-030-05873-9_22

due to wireless channel impairments such as non-line-of-sight (nLOS) propagation and channel fading [6].

Several previous works have studied the impact of unreliable backhaul on system performance [1,6–16]. In [6,10,11,14–16], the impact of unreliable backhaul on cooperative relay systems was investigated. In [11], the outage probability of finite-sized selective relaying systems with unreliable backhaul was studied, and the transmitter-relay pair providing the highest end-to-end signal-to-noise ratio (SNR) was selected for transmission. In the aforementioned studies about unreliable backhaul connections, backhaul reliability has been shown as a key factor for the system performance.

The increasing wireless demands for frequencies have caused the spectrum to be exhausted [17]. In HetNets, frequency sharing is essential to increase the spectral efficiency and system capacity, thus to achieve better system performance. Cognitive radio (CR) technology is considered as a promising solution to solve the spectrum scarcity [18]. In the cognitive radio network (CRN), a secondary user (SU) is allowed to use the spectrum that is prior allocated to a primary user (PU) if the interference caused by SUs to the PUs is within an acceptable tolerance level [19]. In [20], the outage probability of the cognitive radio network has been evaluated and the impact of a single PU on the SU's systems has been studied. In [21], other aspects such as the impact of the PU on the energy harvesting CRN was also studied. However, in the CRN, SUs cooperating with a single PU is not sufficient to exploit the cooperation benefits. Recently, some works have investigated a CRN with multiple PUs, which is more practical and realistic [22,23]. However, all of the research related to CR technology [20,22–26] ignored the impact of unreliable backhaul.

The very recent works [8,9,12] introduced CR technology to HetNets and examined the impact of unreliable backhaul on CRNs. In [8], a single transmitter acting as a small cell was considered in the system. However, in the real scenarios, there can be large number of transmitters rather than a single transmitter. There is likely to be several small cells in HetNets to cooperate and achieve better system performance. Therefore, in this research, we assume a cognitive HetNet system with multiple small cells that can be accounted for more scenarios. To the best of our knowledge, there is no previous research that study backhaul reliability in a CRN with multiple PUs. Therefore, we propose a cognitive heterogeneous network with multiple small cells acting as secondary transmitters and multiple PUs that limit the transmit power of secondary transmitters in the system. Our main contributions are summarized as follows:

- For the first time, we propose a cognitive heterogeneous network with multiple small cell transmitters and primary users. In addition, the reliability of the backhaul is modeled as Bernoulli process [6].
- Selection combining is used to choose the best small cell that has the maximum SNR at the destination. The impacts of backhaul reliability, the number of small cells and primary users on the system performance are examined.

– Closed-form and asymptotic expressions for outage probability are derived. Moreover, numerical results are conducted to verify the system performance using Monte Carlo simulations.

The remainder of the paper is organized as follows. System and channel models are described in Sect. 2. Derivation of the SNR distributions in the proposed system is obtained in Sect. 3. The closed-form expressions for outage probability are carried out in Sect. 4, while numerical results are presented in Sect. 5. Finally, the paper is concluded in Sect. 6.

Notation: $P[\cdot]$ is the probability of occurrence of an event. For a random variable X, $F_X(\cdot)$ denotes its cumulative distribution function (CDF) and $f_X(\cdot)$ denotes the corresponding probability density function (PDF). In addition, $\max(\cdot)$ and $\min(\cdot)$ denote the maximum and minimum of their arguments, respectively.

2   System and Channel Models

We consider a cognitive heterogeneous network consisting of a macro-base station (BS) connected to cloud, K small cells as the secondary transmitters $\{SC_1...SC_k,...SC_K\}$, a secondary receiver $(SU - D)$ as the destination and N primary users $\{PU_1...PU_n,...PU_N\}$, as shown in Fig. 1. The BS is connected to K small cells by unreliable wireless backhaul links. The K small cells send information to the destination while using the same spectrum of N primary users. All nodes are supposed to be equipped with a single antenna. Assuming all the

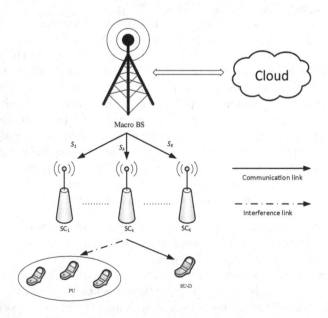

Fig. 1. A cognitive heterogeneous network with multiple primary users and multiple small cells acting as secondary users

channels are Rayleigh fading and are independent and identically distributed, in which the channel power gains are exponential distributed with parameter λ_X for $X = \{\lambda_{kp}, \lambda_{ks}\}$. The link $SC - PU$ follows exponential distribution with parameter λ_{kp}, and the link $SC - SU$ follows exponential distributed with parameter λ_{ks}. In the CRN, the secondary network consists K secondary transmitters and a $SU - D$, they can operate in the same spectrum licensed to PUs as long as they do not cause any harmful interference to PUs. The maximum tolerable interference power at the PU are I_p. Assuming the transmit powers at the secondary transmitters are limited to P_T [20]. In this way, the transmit power at the secondary transmitters can be written as

$$P_k = \min \left(P_T, \frac{I_p}{\max\limits_{i=1,\ldots,N} |h_{kp_i}|^2} \right), \tag{1}$$

where h_{kp_i}, $i = \{1, \ldots n, \ldots N\}$ donates the channel coefficients of the interference link from SC to PUs. Without considering the backhaul reliability, the instantaneous received SNR of the link SC to $SU - D$ is given as

$$\gamma_{ks} = \min \left(\gamma_P |h_{ks}|^2, \frac{\gamma_I}{\max\limits_{i=1,\ldots,N} |h_{kp_i}|^2} |h_{ks}|^2 \right), \tag{2}$$

where h_{ks} donates the channel coefficients of the interference link from SC to $SU - D$. The average SNR of the primary network is given as $\gamma_I = \frac{I_p}{\sigma_n^2}$, and the average SNR of the secondary networks is given as $\gamma_P = \frac{P_t}{\sigma_n^2}$, where σ_n^2 is the noise variance.

Considering the backhaul reliability, the signal received at the destination $SU - D$ is given as

$$y_{ks} = \sqrt{P_k} h_{ks} \mathbb{I}_k x + n_{ks}, \tag{3}$$

where P_k is given in (1), n_{ks} is the complex additive white Gaussian noise (AWGN) with zero mean and variance σ, i.e., $z \sim CN(0, \sigma)$.

In the first hop, the signal is transmitted from BS to the small cells via unreliable backhaul links. The unreliable backhaul links can perform either success or failure transmission. So the reliability backhaul is modeled as Bernoulli process $\mathbb{I}_k$ with success probability s_k where $P(\mathbb{I}_{k^*} = 1) = s_k$ and $P(\mathbb{I}_{k^*} = 0) = 1 - s_k$. This indicates that the probability of the message successfully delivered over its dedicated backhaul is s_k, however, the failure probability is $1 - s_k$. Assume that x is the desired transmitted signal from BS to $SU - D$.

In the second hop, selection combining protocol is used at the destination $SU - D$ in order to select the best small cell that has the maximum SNR to transmit the signal. The small cell SC_{k^*} is selected as

$$k^* = \max\limits_{k=1,\ldots,K} \arg \left(\gamma_{ks} \mathbb{I}_k \right), \tag{4}$$

In this way, considering the backhaul reliability, the end-to-end SNR at the receiver $SU - D$ can be rewritten as

$$\gamma_s = \min\left(\gamma_P|h_{k^*s}|^2, \frac{\gamma_I}{\max\limits_{i=1,\ldots,N}|h_{k^*p_i}|^2}|h_{k^*s}|^2\right)\mathbb{I}_{k^*}. \tag{5}$$

3   SNR Distributions in Cognitive Heterogeneous Systems

In this section, the distributions of the SNRs are derived, and the system performances are studied based on the derivations in the next section.

From the end-to-end SNR in (5), assume $Y = \max\limits_{i=1,\ldots,N}|h_{kp_i}|^2$, the CDF and PDF of Y can be given as

$$F_Y(y) = [1 - \exp(-\lambda y)]^N, \tag{6}$$

$$f_Y(y) = \lambda N \sum_{i=0}^{N-1}(-1)^i\binom{N-1}{i}\exp\left[-\lambda(i+1)y\right]. \tag{7}$$

Without considering the impact of backhaul reliability, the CDF of the end-to-end SNR given in (2) can be written as,

$$F_{\gamma_{ks}}(x) = 1 + \sum_{n=1}^{N}(-1)^n\binom{N}{n}\exp(-\frac{\gamma_P\gamma_I n}{\gamma_P}) - \exp(-\frac{\lambda_{ks}}{\gamma_P}) - \sum_{n=1}^{N}(-1)^n\binom{N}{n}$$
$$\exp\left(-\frac{\gamma_P\gamma_I n + \lambda_{ks}x}{\gamma_P}\right) + N\sum_{i=0}^{N-1}\frac{(-1)^i}{i+1}\binom{N}{i}\exp\left[-\frac{\gamma_{kp}\gamma_I(i+1)}{\gamma_P}\right] - N\sum_{i=0}^{N-1}(-1)^i$$
$$\binom{N-1}{i}\frac{\lambda_{kp}}{\frac{\lambda_{ks}x}{\gamma_I}+\lambda_{kp}(i+1)}\exp\left[\frac{\gamma_I}{\gamma_P}\left(-\frac{\lambda_{ks}x}{\gamma_I}-\lambda_{kp}(i+1)\right)\right] \tag{8}$$

Proof

$$F_{\gamma_{ks}}(x) = P\left[\min\left(\gamma_P|h_{ks}|^2, \frac{\gamma_I}{Y}|h_{ks}|^2\right) \le x\right]$$
$$= \underbrace{P\left[|h_{ks}|^2 \le \frac{x}{\gamma_P}; \frac{\gamma_I}{Y} \ge \gamma_P\right]}_{J_1} + \underbrace{P\left[\frac{|h_{ks}|^2}{Y} \le \frac{x}{\gamma_I}; \frac{\gamma_I}{Y} \le \gamma_P\right]}_{J_2} \tag{9}$$

For the term J_1, because $|h_{ks}|^2$ and $|h_{kp_i}|^2$ are independent and $Y = \max\limits_{i=1,\ldots,N}|h_{kp_i}|^2$, J_1 can be expanded as

$$J_1 = P\left[|h_{ks}|^2 \le \frac{x}{\gamma_P}; Y \le \frac{\gamma_I}{\gamma_P}\right]$$
$$= F_{\gamma_{ks}}(\frac{x}{\gamma_P})\left[F_{\gamma_{kp}}(\frac{\gamma_I}{\gamma_P})\right]^N. \tag{10}$$

For the term J_2, the concept of probability theory is used and with the help of (7), J_2 is expressed as

$$J_2 = P\left[|h_{ks}|^2 \leq \frac{xY}{\gamma_I}; Y \geq \frac{\gamma_I}{\gamma_P}\right]$$

$$= \int_0^{\frac{xY}{\gamma_I}} f_{|h_{ks}|^2}(y) \int_{\frac{\gamma_I}{\gamma_P}}^{\infty} f_{|h_{kp_i}|^2}(z)dydz. \tag{11}$$

The above Eq. (8) is the CDF of SNR without considering the unreliable backhual, we now take into account the backhaul reliability and derive the CDF of the end-to-end SNR given in (5) as follows.

The PDF of $\gamma_{ks}\mathbb{I}_k$ is modeled by the mixed distribution,

$$f_{\gamma_{ks}\mathbb{I}_k}(x) = (1-s)\delta(x) + s\frac{\partial F_{\gamma_{ks}}(x)}{\partial x}, \tag{12}$$

where $\delta(x)$ is the Dirac delta function. According to (12), the CDF of the $\gamma_{ks}\mathbb{I}_k$ is given as

$$F_{\gamma_{ks}\mathbb{I}_k}(x) = \int_0^x f_{\gamma_{ks}\mathbb{I}_k}(x)dx. \tag{13}$$

With the help of [27, Eq. (3.353.2)], the CDF is expressed as

$$F_{\gamma_{ks}\mathbb{I}_k}(x) = 1 - s\exp\left(-\frac{\lambda_{ks}x}{\gamma_P}\right) - s\sum_{n=1}^{N}(-1)^n\binom{N}{n}\exp\left(-\frac{\lambda_{kp}\gamma_I n + \lambda_{ks}x}{\gamma_P}\right)$$

$$+ s\sum_{n=1}^{N}(-1)^n\binom{N}{n}\exp\left(-\frac{\lambda_{kp}\gamma_I n}{\gamma_P}\right) + sN\sum_{i=0}^{N-1}\frac{(-1)^i}{i+1}\binom{N-1}{i}\exp\left[-\frac{\gamma_I\lambda_{kp}(i+1)}{\gamma_P}\right]$$

$$- sN\sum_{i=0}^{N-1}(-1)^i\binom{N-1}{i}\frac{\lambda_{kp}}{\frac{x\lambda_{ks}}{\gamma_I}+\lambda_{kp}(i+1)}\exp\left[-\frac{\gamma_I\lambda_{kp}(i+1)+\lambda_{ks}x}{\gamma_P}\right]. \tag{14}$$

According to (4), k is selected when $\gamma_{ks}\mathbb{I}_k$ achieves the maximum value, since for all random variables $\gamma_{ks}\mathbb{I}_k$ are independent and identically distributed. The CDF of SNR γ_s can be written as

$$F_{\gamma_s}(x) = F^K_{\gamma_{ks} \mathbb{I}_k}(x)$$

$$= 1 - \sum_{k=1}^{K} (-1)^k \binom{K}{k} s^k \sum_{j=0}^{k} \binom{k}{j} \sum_{m=0}^{k-j} \binom{k-j}{m} \exp\left[-\frac{\lambda_{ks} x (k-j-m)}{\gamma_\mathsf{P}}\right]$$

$$\sum_{p=0}^{m} (-1)^p \binom{m}{p} \exp\left(-\frac{\lambda_{ks} x p}{\gamma_\mathsf{P}}\right) \sum_{a_1,\dots a_N} \binom{m}{a_1 \dots a_N} \prod_{t=1}^{N} \left[\binom{N}{t}\right]^{a_t} (-1)^{t a_t}$$

$$\exp\left(-\frac{\lambda_{kp} \gamma_\mathsf{I} t a_t}{\gamma_\mathsf{P}}\right) \sum_{q=0}^{k} (N)^q \binom{k}{q} \exp\left(-\frac{\lambda_{ks} x q}{\gamma_\mathsf{P}}\right) \sum_{b_0,\dots b_{N-1}} \binom{q}{b_0 \dots b_{N-1}} \prod_{r=0}^{N-1}$$

$$\left[\binom{N-1}{r}\right]^{b_r} (-1)^{r b_r} \exp\left[-\frac{\lambda_{kp} \gamma_\mathsf{I}(r+1) b_r}{\gamma_\mathsf{P}}\right] \left[\frac{\lambda_{kp}}{\frac{\lambda_{ks} x}{\gamma_\mathsf{I}} + \lambda_{kp}(r+1)}\right]^{b_r} (-N)^{k-q}$$

$$\sum_{c_0,\dots c_{N-1}} \binom{k-q}{c_0 \dots c_{N-1}} \prod_{d=0}^{N-1} \left[\binom{N-1}{d}\right]^{c_d} \frac{(-1)^{c_d} d}{d+1} \exp\left[-\frac{\lambda_{kp} \gamma_\mathsf{I}(d+1) c_d}{\gamma_\mathsf{P}}\right].$$

$$(15)$$

4   Performance Analysis of the Proposed System

This section studies the performance of outage probability utilizing the SNR distributions obtained in the previous section. Closed-form expressions are derived and asymptotic calculation is also provided to evaluate the system performance.

4.1   Outage Probability Analysis

The outage probability is defined as the probability that the SNR falls below a certain threshold γ_{th},

$$P_{out}(\gamma_{\mathsf{th}}) = P(\gamma_\mathsf{s} \leq \gamma_{\mathsf{th}}) = F_{\gamma_\mathsf{s}}(\gamma_{\mathsf{th}}). \tag{16}$$

The outage probability closed-form expressions of the proposed system,

$$P_{\gamma_s}(\gamma_{\mathsf{th}}) = 1 - \sum_{k=1}^{K} (-1)^k \binom{K}{k} s^k \sum_{j=0}^{k} \binom{k}{j} \sum_{m=0}^{k-j} \binom{k-j}{m} \exp\left[-\frac{\lambda_{ks} \gamma_{\mathsf{th}}(k-j-m)}{\gamma_\mathsf{P}}\right]$$

$$\sum_{p=0}^{m} (-1)^p \binom{m}{p} \exp\left(-\frac{\lambda_{ks} \gamma_{\mathsf{th}} p}{\gamma_\mathsf{P}}\right) \sum_{a_1,\dots a_N} \binom{m}{a_1 \dots a_N} \prod_{t=1}^{N} \left[\binom{N}{t}\right]^{a_t} (-1)^{t a_t} \exp\left(-\frac{\lambda_{kp} \gamma_\mathsf{I} t a_t}{\gamma_\mathsf{P}}\right)$$

$$\sum_{q=0}^{k} (N)^q \binom{k}{q} \exp\left(-\frac{\lambda_{ks} \gamma_{\mathsf{th}} q}{\gamma_\mathsf{P}}\right) \sum_{b_0,\dots b_{N-1}} \binom{q}{b_0 \dots b_{N-1}} \prod_{r=0}^{N-1} \left[\binom{N-1}{r}\right]^{b_r} (-1)^{r b_r}$$

$$\exp\left[-\frac{\lambda_{kp} \gamma_\mathsf{I}(r+1) b_r}{\gamma_\mathsf{P}}\right] \left[\frac{\lambda_{kp}}{\frac{\lambda_{ks} \gamma_{\mathsf{th}}}{\gamma_\mathsf{I}} + \lambda_{kp}(r+1)}\right]^{b_r} (-N)^{k-q} \sum_{c_0,\dots c_{N-1}} \binom{k-q}{c_0 \dots c_{N-1}}$$

$$\prod_{d=0}^{N-1} \left[\binom{N-1}{d}\right]^{c_d} \frac{(-1)^{c_d} d}{d+1} \exp\left[-\frac{\lambda_{kp} \gamma_\mathsf{I}(d+1) c_d}{\gamma_\mathsf{P}}\right].$$

$$(17)$$

Asymptotic Analysis. In the high SNR regime, when $\gamma_\mathsf{P} \to \infty$ in the proposed cognitive heterogeneous network, the asymptotic is given by

$$P_{out}^{Asy}(\gamma_\mathsf{th}) = (1-s)^K. \tag{18}$$

5   Numerical Results and Discussions

In this section, numerical results of the outage probability are studied to evaluate the impacts of backhaul reliability, the number of primary transmitters and secondary transmitters on the system performance. The 'Sim' curves are the simulation results, 'Ana' curves indicate analytical results and 'Asy' curves donate the asymptotic results. In the figures, we can observe that both the simulation curves and analytical curves match very well. In this section, the threshold of outage probability is fixed at $\gamma_\mathsf{th} = 3$ dB and the location of the nodes are $SC = (0.5, 0)$, $SU - D = (0, 0)$, $PU = (0.5, 0.5)$ in Cartesian coordinate system respectively. Hence, the distance between two nodes can be found as $d_{AB} = \sqrt{(x_A - x_B)^2 + (y_A - y_B)^2}$, where A and B have the co-ordinates (x_A, y_A) and (x_B, y_B) and $A, B = \{SC, PU, SU - D\}$. It is assumed that average SNR of each link is dependent on the path loss as $1/\lambda_X = 1/d_X^{pl}$, where pl is the path loss exponent and $pl = 4$ is assumed. Moreover, we also assume that the average SNR $\gamma_\mathsf{P} = \gamma_\mathsf{I}$.

5.1   Outage Probability Analysis

The figures in this section show the impacts of backhaul reliability s, the number of primary users PUs and secondary transmitters SCs on the system performance. In Fig. 2, s is fixed at 0.99 and the number of PUs is 3. Assuming the number of SCs is $K = 1$, $K = 2$, $K = 3$ to evaluate the impact of the number of SCs on system performance. In the figures, when the number of SCs increase, the outage probability decreases and the system can achieve a better performance due to the correlation of multiple signals at the receiver. Also, all the curves converage to the asymptotic limitation.

In Fig. 3, the outage probability behaviour at different backhaul reliability is investigated. $N = 3$ and $K = 3$ is assumed in this scenario. We assume that $s = 0.99$, $s = 0.90$ and $s = 0.80$ to evaluate the impact of backhaul reliability on the system performance. In Fig. 3, when s increases, the system performs better as the outage probability decreases. This is because when the probability of the information successfully delivered over the backhaul links gets higher, the system can achieve a better performance.

In Fig. 4, the outage probability with different number of PUs is investigated. we asume that $s = 0.99$ and $K = 3$. We can observe that in low-SNR regime, when N increases, the system performance gets worse. This is because when the number of PUs increases, SCs must satisfy the power constraints of all the PUs. The power constraints would get tighter when the number of PUs increases. The transmit power of SCs would reduce due to the increasing power constraints.

However, in high SNR regime, increasing the number of PUs does not have any effect on the system performance, as is shown in (18). According to Figs. 2, 3 and 4, in high SNR regime, only the backhaul reliability and the number of SCs can affect the system performance.

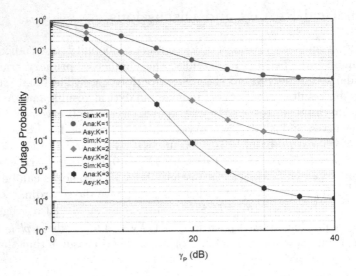

Fig. 2. Outage probability with different number of secondary transmitters at a fixed backhaul reliability ($s = 0.99$) and a fixed number of primary users ($N = 3$)

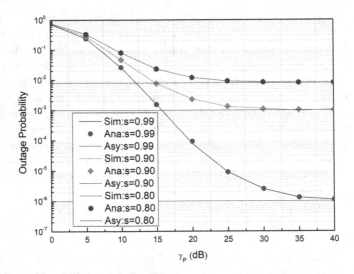

Fig. 3. Outage probability with different backhaul reliability at a fixed number of secondary transmitters ($K = 3$) and a fixed number of primary users ($N = 3$)

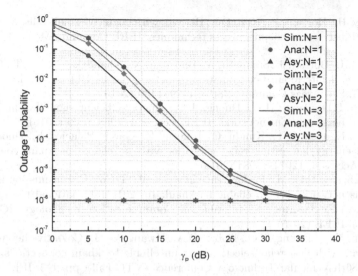

Fig. 4. Outage probability with different number of primary users at a fixed number of secondary transmitters ($K = 3$) and a fixed backhaul reliability ($s = 0.99$)

6 Conclusions

In this paper, we propose a cognitive heterogeneous network with multiple small cell transmitters and primary users to investigate the impacts of backhaul reliability, the number of small cells and primary users on the system performance. The backhaul reliability is modeled as Bernoulli process. Selection combining is used to choose the best small cell, having the maximum SNR at the destination. Closed-form expressions for outage probability are derived and asymptotic analysis is provided. It has been shown that, when the number of cooperative nodes and backhaul reliability increase, the system performs better. Moreover, the increase of the primary users' number can decrease the system performance within low-SNR regime.

Acknowledgement. This work was supported by the Newton Prize 2017 and by a Research Environment Links grant, ID 339568416, under the Newton Programme Vietnam partnership. The grant is funded by the UK Department of Business, Energy and Industrial Strategy (BEIS) and delivered by the British Council. For further information, please visit www.newtonfund.ac.uk/.

References

1. Kim, K.J., Khan, T.A., Orlik, P.V.: Performance analysis of cooperative systems with unreliable backhauls and selection combining. IEEE Trans. Veh. Technol. **66**(3), 2448–2461 (2017)
2. Scott-Hayward, S., Garcia-Palacios, E.: Channel time allocation PSO for gigabit multimedia wireless networks. IEEE Trans. Multimed. **16**(3), 828–836 (2014)

3. ElSawy, H., Hossain, E., Kim, D.I.: Hetnets with cognitive small cells: user offloading and distributed channel access techniques. IEEE Commun. Mag. **51**(6), 28–36 (2013)
4. Madan, R., Borran, J., Sampath, A., Bhushan, N., Khandekar, A., Ji, T.: Cell association and interference coordination in heterogeneous LTE-A cellular networks. IEEE J. Sel. Areas Commun. **28**(9), 1479–1489 (2010)
5. Nguyen, L.D., Tuan, H.D., Duong, T.Q.: Energy-efficient signalling in QoS constrained heterogeneous networks. IEEE Access **4**, 7958–7966 (2016)
6. Yin, C., Nguyen, H.T., Kundu, C., Kaleem, Z., Garcia-Palacios, E., Duong, T.Q.: Secure energy harvesting relay networks with unreliable backhaul connections. IEEE Access **6**, 12074–12084 (2018)
7. Ali, M.S., Synthia, M.: Performance analysis of JT-CoMP transmission in heterogeneous network over unreliable backhaul. In: 2015 International Conference on Electrical Engineering and Information Communication Technology (ICEEICT), pp. 1–5. IEEE (2015)
8. Nguyen, H.T., Duong, T.Q., Dobre, O.A., Hwang, W.-J.: Cognitive heterogeneous networks with best relay selection over unreliable backhaul connections. In: 2017 IEEE 86th Vehicular Technology Conference (VTC-Fall), pp. 1–5. IEEE (2017)
9. Nguyen, H.T., Ha, D.-B., Nguyen, S.Q., Hwang, W.-J.: Cognitive heterogeneous networks with unreliable backhaul connections. Mob. Netw. Appl., 1–14 (2017)
10. Liu, H., Kim, K.J., Tsiftsis, T.A., Kwak, K.S., Poor, H.V.: Secrecy performance of finite-sized cooperative full-duplex relay systems with unreliable backhauls. IEEE Trans. Signal Process. **65**(23), 6185–6200 (2017)
11. Liu, H., Kwak, K.S.: Outage probability of finite-sized selective relaying systems with unreliable backhauls. In: 2017 International Conference on Information and Communication Technology Convergence (ICTC), pp. 1232–1237. IEEE (2017)
12. Nguyen, H.T., Duong, T.Q., Hwang, W.-J.: Multiuser relay networks over unreliable backhaul links under spectrum sharing environment. IEEE Commun. Lett. **21**(10), 2314–2317 (2017)
13. Khan, T.A., Orlik, P., Kim, K.J., Heath, R.W.: Performance analysis of cooperative wireless networks with unreliable backhaul links. IEEE Commun. Lett. **19**(8), 1386–1389 (2015)
14. Kim, K.J., Yeoh, P.L., Orlik, P.V., Poor, H.V.: Secrecy performance of finite-sized cooperative single carrier systems with unreliable backhaul connections. IEEE Trans. Signal Process. **64**(17), 4403–4416 (2016)
15. Nguyen, H.T., Zhang, J., Yang, N., Duong, T.Q., Hwang, W.-J.: Secure cooperative single carrier systems under unreliable backhaul and dense networks impact. IEEE Access **5**, 18310–18324 (2017)
16. Kim, K.J., Orlik, P.V., Khan, T.A.: Performance analysis of finite-sized cooperative systems with unreliable backhauls. IEEE Trans. Wirel. Commun. **15**(7), 5001–5015 (2016)
17. Kim, K.J., Wang, L., Duong, T.Q., Elkashlan, M., Poor, H.V.: Cognitive single-carrier systems: joint impact of multiple licensed transceivers. IEEE Trans. Wirel. Commun. **13**(12), 6741–6755 (2014)
18. Deng, Y., Kim, K.J., Duong, T.Q., Elkashlan, M., Karagiannidis, G.K., Nallanathan, A.: Full-duplex spectrum sharing in cooperative single carrier systems. IEEE Trans. Cogn. Commun. Netw. **2**(1), 68–82 (2016)
19. Yin, C., Doan, T.X., Nguyen, N.-P., Mai, T., Nguyen, L.D.: Outage probability of full-duplex cognitive relay networks with partial relay selection. In: International Conference on Recent Advances in Signal Processing, Telecommunications & Computing (SigTelCom), pp. 115–118. IEEE (2017)

20. Duong, T., Bao, V.N.Q., Zepernick, H.-J.: Exact outage probability of cognitive AF relaying with underlay spectrum sharing. Electron. Lett. **47**(17), 1001–1002 (2011)
21. Zhang, J., Nguyen, N.-P., Zhang, J., Garcia-Palacios, E., Le, N.P.: Impact of primary networks on the performance of energy harvesting cognitive radio networks. IET Commun. **10**(18), 2559–2566 (2016)
22. Huang, H., Li, Z., Si, J.: Multi-source multi-relay underlay cognitive radio networks with multiple primary users. In: 2015 IEEE International Conference on Communications (ICC), pp. 7558–7563. IEEE (2015)
23. Feng, X., Gao, X., Zong, R.: Cooperative jamming for enhancing security of cognitive radio networks with multiple primary users. China Commun. **14**(7), 1–15 (2017)
24. Duong, T.Q., Yeoh, P.L., Bao, V.N.Q., Elkashlan, M., Yang, N.: Cognitive relay networks with multiple primary transceivers under spectrum-sharing. IEEE Signal Process. Lett. **19**(11), 741–744 (2012)
25. Tran, H., Duong, T.Q., Zepernick, H.-J.: Performance analysis of cognitive relay networks under power constraint of multiple primary users. In: 2011 IEEE Global Telecommunications Conference (GLOBECOM 2011), pp. 1–6. IEEE (2011)
26. Duong, T., Bao, V.N.Q., Tran, H., Alexandropoulos, G.C., Zepernick, H.-J.: Effect of primary network on performance of spectrum sharing AF relaying. Electron. Lett. **48**(1), 25–27 (2012)
27. Jeffrey, A., Zwillinger, D.: Table of Integrals, Series, and Products. Academic press, Cambridge (2007)

Impact of Direct Communications on the Performance of Cooperative Spectrum-Sharing with Two-Way Relays and Maximal Ratio Combining

Tu Lam Thanh[1], Tiep M. Hoang[2](✉), Vo Nguyen Quoc Bao[3],
and Hien M. Nguyen[4]

[1] Universite Paris-Saclay, Paris, France
`lamthanh.tu@l2s.centralesupelec.fr`
[2] Queen's University Belfast, Belfast, UK
`mhoang02@qub.ac.uk`
[3] Posts and Telecommunications Institute of Technology,
Ho Chi Minh City, Vietnam
`baovnq@ptithcm.edu.vn`
[4] Duy Tan University, Duy Tan, Vietnam
`nguyenminhhien2501@gmail.com`

Abstract. In this paper, we investigate a three-phase two-way (TW) amplify-and-forward (AF) relaying for cognitive radio networks. By utilizing the direct communications, the end user can employ maximal ratio combining to achieve the full diversity. We derive the closed-form and asymptotic expressions for user and system outage probabilities which allow us to highlight the advantage of cooperative cognitive communications. The numerical results, obtained through compact forms of these outage probabilities, yield that the cognitive TW AF relaying scheme can significantly enhance the reliability of unlicensed networks in which the transmit power at each secondary user is strictly governed.

1 Introduction

With a tremendous growth of wireless multimedia services and the number of customers, it has been an enormous pressure on available frequency bands and spectrum allocation policies. However, most of frequency bands are under-utilization according to the report of Federal Communication Commission (FCC). To get around this troublesome, cognitive radio (CR) technique has been proposed to allow the unlicensed user can utilize the licensed spectrum band [1]. The principal idea of CR networks is that the secondary users (SUs) are able to use the spectrum bands of primary users (PUs) provided that the quality of service (QoS) of licensed networks is not compromised. Several CR schemes have been introduced in the literature to implement the CR network. In particular, for interweave paradigm, the unlicensed users are not allowed to

© ICST Institute for Computer Sciences, Social Informatics and Telecommunications Engineering 2019
Published by Springer Nature Switzerland AG 2019. All Rights Reserved
T. Q. Duong and N.-S. Vo (Eds.): INISCOM 2018, LNICST 257, pp. 276–285, 2019.
https://doi.org/10.1007/978-3-030-05873-9_23

occupy the spectrum bands if PU activities are detected. As such, the transmission of CR network strictly relies on the primary system. On the other hand, the underlay spectrum-sharing paradigm allows SUs to transmit its information simultaneously with PUs as long as the maximal interference does not exceed the predefined threshold. For this approach, it ensures the stable transmission for SUs at an expense of limited coverage area and low QoS. One efficient way to alleviate the disadvantages of CR underlay scheme is to combine CR with relay networks [2,3], where the latter is known as an efficient approach for combating the effect of fading channels and expanding the communication range through the assistance of the third party named relay. In particular, relay node helps source node to transmit its signal by adopting one of relaying techniques, i.e., amplify and forward (AF) [4] and decode and forward (DF) [5].

Although the one-way cognitive relay network can overcome both the impact of fading channels and the drawbacks of underlay scheme, the spectral efficiency of this system is still constrained by multiple time-slots owing to half-duplex relaying protocol [6]. More recently, two-way relaying (TWR) technique [7] has drawn a lot of attention due to fully compensating this loss by permitting two users concurrently transmit its signal to each other with the help of half-duplex relays [8]. Despite getting higher spectral efficiency than the traditional cognitive one-way relay networks, only several works investigated the performance of two-way counterpart [9–14]. The performance of two-way relaying with single and multiple relays has been reported in [9] and [10], respectively. In [10], the exact outage probability of opportunistic two-way relaying with spectrum-sharing has been presented. Moreover, it has been proved that system performance largely depends on the number of relay nodes and the location of relay nodes together with primary user. While the topic of energy harvesting in two-way networks has been investigated in [13,14]. In [11], the tight lower bound of user outage performance in multiple primary users environment has been obtained in two cases, i.e., two users and two group of users. The optimal relay selection for two-way cognitive relay networks has been discussed in [12]. In addition, relay selection combined with power allocation for two-way relaying in the presence of imperfect channel state information (CSI) were studied in [15,16]. Note that previous works have only considered DF relaying and neglected the impact of direct communication.

Different from the above works, in this paper, we investigate the two-way AF relaying for underlay spectrum-sharing with the existence of the direct link between two users. Generally, there are two distinct two-way relaying schemes depending on the number of required time-slots to complete the communication [17]: (i) time division broadcast (TDBC) or three-phase two-way relay (3P-TWR) and (ii) multiple access broadcast (MABC) or two-phase TWR (2P-TWR) [18]. For CR networks, the performance of SU is limited due to the fact that its transmit power is governed by the maximally allowable interference power constraint at PU. As such, in this paper, we exploit the direct link in 3P-TWR where the communication reliability is enhanced via diversity combining between the direct and relaying links. Our considered scheme can enhance both spectral efficiency for cognitive relay networks while keeping the desired QoS of secondary networks satisfactorily. Our main contribution in this paper is summarized as follows: We consider the cognitive two-way relay networks in the presence of direct communication under the peak interference power constraint impinged on the licensed user. We investigate the spatial diversity gain for cognitive two-way relay networks by employing MRC technique between relaying and direct links. We characterize the statistics for the end-to-end SNR of cognitive two-way AF relay networks with MRC by deriving the exact cumulative distribution function (CDF). Utilizing this result, the exact closed-form expressions for both user and system outage probability.

2   System Model

We consider a CR network in which two secondary users (S_1 and S_2) exchange information with each other with the help of a non-regenerative relay R as shown in Fig. 1. The secondary network co-exists with the primary network that represents by one PU receiver. All nodes are operated in half-duplex mode and equipped with one antenna. In addition, all channels are assumed to be Rayleigh flat fading, time-invariant and reciprocal while exchanging data. Let us denote h_m and f_n, with ($m \in \{0, 1, 2\}$, $n \in \{1, 2, r\}$), as fading coefficients of data links and interference links. Particularly, h_0, h_1 and h_2 are data links between $S_1 \leftrightarrow S_2$, $S_1 \leftrightarrow R$ and $S_2 \leftrightarrow R$, respectively. Similarly, f_1, f_2 and f_r are interference links between $S_1 \leftrightarrow PU$, $S_2 \leftrightarrow PU$ and $R \leftrightarrow PU$, successively. As a consequence, the channel gains, i.e., $|h_m|^2$ and $|f_n|^2$, are exponential random variables (RVs) with parameter λ_m and ω_n.

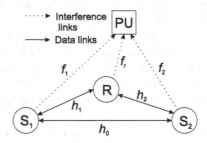

Fig. 1. Two-way relaying in cognitive cooperative communications.

On the other hand, owing to adopting underlay approach, the transmit power of secondary users could not exceed the maximal tolerable interference level $\mathcal{I}_p$. Mathematically, we have

$$P_n = \mathcal{I}_p / |f_n|^2. \tag{1}$$

The communication between S_1 and S_2 is taken over three phase. In the first phase, user S_1 transmits its modulated signal x_1 to user S_2 and relay R. Followed by, S_2 send its signal x_2 to user S_1 and relay R in the second phase. The received signals at R and S_j in i-phase $(i, j \in \{1, 2\}, i \neq j)$ can be given by

$$y_{R,i} = \sqrt{P_i} h_i x_i + n_{R,i}$$
$$y_{S_j,i} = \sqrt{P_i} h_0 x_i + n_{S_j,i}, \tag{2}$$

where n is a circular symmetric complex Gaussian random variable with zero mean and variance $\mathcal{N}_0$. In the third phase, R broadcasts the scaling version of two previous received signals as

$$x_{R,3} = G (y_{R,1} + y_{R,2}), \tag{3}$$

where $G = \sqrt{\frac{P_r}{P_1|h_1|^2 + P_2|h_2|^2 + 2\mathcal{N}_0}}$ is the amplifying gain. Signal is received by user S_j in the third phase after canceling self-interference term is given as follows:

$$y_{S_j,3} = G\sqrt{P_i} h_i h_j x_i + G h_j (n_{R,1} + n_{R,2}) + n_{S_j,3}. \tag{4}$$

3 Performance Analysis

3.1 Maximal Ratio Combining

For MRC technique, two end-users will combine two links, namely, direct and indirect link, linearly. The end to end signal to noise ratios (SNRs) at user S_j denoted as γ_{ij} is obtained as

$$\gamma_{ij} = P_i |h_0|^2 + \frac{P_r P_i |h_i|^2 |h_j|^2}{2P_r|h_j|^2 + P_i|h_i|^2 + P_j|h_j|^2 + 2}. \tag{5}$$

The upper bound of Eq. (5) is given by

$$\gamma_{ij} \leq P_i |h_0|^2 + \min \left(\frac{P_r P_i |h_i|^2}{2} P_r + P_j, P_r |h_j|^2 \right). \tag{6}$$

User Outage Probability (UOP). In this subsection, we study the outage probability (OP) of each user with MRC is used at the secondary users. OP at user S_j occurs when the information flow from node $i \rightarrow j$ is below the target rate $\mathcal{R}$. Mathematically, we have

$$\text{UOP}^j_{\text{MRC}} = \Pr \left[\frac{1}{3} \log_2 (1 + \gamma_{ij}) < \mathcal{R} \right] = \Pr \left[\gamma_{ij} < \gamma_{\text{th}} \right], \tag{7}$$

where $\gamma_{\text{th}} = 2^{3\mathcal{R}} - 1$. Due to sharing the same variable, i.e., $|f_i|^2$ the direct and indirect link are not independent. As a result, (7) is rewritten as

$$\text{UOP}^j_{\text{MRC}} = \int_0^\infty \int_0^\gamma F_{\gamma_R||f_i|^2} (\gamma - y) f_{\gamma_0||f_i|^2} (y) f_{|f_i|^2} (x) dy dx. \tag{8}$$

As can be observed in (8), we need to find out the cumulative distribution function (CDF) of two links before evaluating the user outage probability. The CDF of indirect link under condition $|f_i|^2$ is given by $F_{\gamma_R|x}(\gamma) =$

$$1 - \frac{\overline{\gamma}\lambda_j}{\gamma\omega_r + \overline{\gamma}\lambda_j} \exp\left(-2\frac{\gamma x}{\lambda_i \overline{\gamma}}\right) + x\frac{\overline{\gamma}(\lambda_j)^2 \gamma\omega_r}{\lambda_i \omega_j (\gamma\omega_r + \overline{\gamma}\lambda_j)^2} \exp\left[-x\left(\frac{2\gamma}{\lambda_i \overline{\gamma}} - \frac{\gamma\lambda_j\omega_r}{\lambda_i \omega_j(\gamma\omega_r + \overline{\gamma}\lambda_j)}\right)\right] \times$$

$E_1\left(x\frac{\gamma\lambda_j\omega_r}{\lambda_i\omega_j(\gamma\omega_r+\overline{\gamma}\lambda_j)}\right)$. where $\overline{\gamma} = \frac{I_p}{N_0}$ denotes as an average SNRs of system and $E_1(x)$ is exponential integral function, defined in [19, Eq. 8.211].

The probability density function (PDF) of $|f_i|^2$ and $\gamma_{0||f_i|^2}$ are given as

$f_{|f_i|^2}(x) = \omega_i^{-1}\exp\left(-x\omega_i^{-1}\right)$ and $f_{\gamma_0|x}(y) = x(\overline{\gamma}\lambda_0)^{-1}\exp\left(-yx(\overline{\gamma}\lambda_0)^{-1}\right)$.
Finally, the outage probability at S_j is given by

$$UOP_{MRC}^j = J_1(\omega_i, \overline{\gamma}\lambda_0, \gamma_{th}) - \frac{\lambda_j}{\lambda_0\omega_i}J_2\left(\gamma_{th}\omega_r + \overline{\gamma}\lambda_j, \omega_r, \frac{2\gamma_{th}}{\overline{\gamma}\lambda_i} + \frac{1}{\omega_i}, \frac{2}{\overline{\gamma}\lambda_i} - \frac{1}{\overline{\gamma}\lambda_0}, \gamma_{th}\right)$$

$$+ \frac{(\lambda_j)^2}{\lambda_0\lambda_i\omega_i\omega_j\omega_r}J_3\left(\frac{2\gamma}{\overline{\gamma}\lambda_i} + \frac{1}{\omega_i}, \frac{2}{\overline{\gamma}\lambda_i} - \frac{1}{\overline{\gamma}\lambda_0}, \frac{\lambda_j}{\lambda_i\omega_j}, \gamma + \frac{\overline{\gamma}\lambda_j}{\omega_r}, \gamma\right). \quad (9)$$

where $J_1(a,b,c) = \frac{ac}{b+ac}$, $J_2(a,b,c,d,g) = A_1\log\left(1 - \frac{gb}{a}\right) + B_1\log\left(1 - \frac{gd}{c}\right) + \frac{gB_2d^2}{c(c-dg)}$, $A_1 = -\frac{1}{bd^2\left(\frac{a}{b}-\frac{c}{d}\right)^2}$, $B_2 = -\frac{1}{bd^2\left(\frac{c}{d}-\frac{a}{b}\right)}$ and $B_1 = \frac{1}{bd^2\left(\frac{c}{d}-\frac{a}{b}\right)^2}$.

Besides, $J_3(a,b,c,d,g) = -\frac{8\sqrt{2}\pi a_N a_I}{b^3}\sum_{n=1}^{N+1}\sum_{i=1}^{I+1}\sum_{o=1}^{3}\sqrt{b_n}\left[C_o J_4\left(\gamma, \frac{E+F}{2}, o\right) + \right.$

$$\left. D_o J_4\left(\gamma, \frac{E-F}{2}, o\right)\right], \quad J_4(a,b,n) = \begin{cases} \log\left(1 - \frac{a}{b}\right) & ;n=1 \\ \frac{(-b)^{1-n} - (a-b)^{1-n}}{(n-1)} & ;n \neq 1 \end{cases}, \quad E =$$

$\frac{a+bd-c(1-4b_nb_i)}{b}$, $F = \left(E^2 - 4\left(\frac{ad}{b} - \frac{cg}{b}(1-4b_nb_i)\right)\right)^{1/2}$,

$C_o = \frac{1}{(3-o)!}\frac{d^{(3-o)}}{dy}\left[\frac{(g-y)(y-d)}{\left(y-\frac{E+F}{2}\right)^3}\right]\Bigg|_{y=\frac{E-F}{2}}$, $D_o = \frac{1}{(3-o)!}\frac{d^{(3-o)}}{dy}\left[\frac{(g-y)(y-d)}{\left(y-\frac{E-F}{2}\right)^3}\right]\Bigg|_{y=\frac{E+F}{2}}$.

Here a_N, a_I, b_n and b_i are calculated similar in [20]. The remain UOP at S_i gets easily by applying the similar steps.

System Outage Probability (SOP). The system outage probability (SOP) appears when one of two user's data is under the threshold, γ_{th}. Mathematically, we have

$$SOP_{MRC} = \underbrace{Pr\left[T_1 \leq \gamma_{th}, P_i \leq P_j\right]}_{\Omega_1} + \underbrace{Pr\left[T_2 \leq \gamma_{th}, P_j \leq P_i\right]}_{\Omega_2},$$

$$T_1 = P_i|h_0|^2 + \frac{P_r\min\left(P_i|h_i|^2, P_j|h_j|^2\right)}{2P_r + P_j}, \quad (10)$$

$$T_2 = P_j|h_0|^2 + \frac{P_r\min\left(P_i|h_i|^2, P_j|h_j|^2\right)}{2P_r + P_i}. \quad (11)$$

By using the same approach as UOP, the CDF of indirect links of Ω_i is given by

$$F_{\gamma_R^{\text{sys}}|x}(\gamma)$$

$$= 1 - e^{-\frac{x}{\omega_j}} - \frac{\overline{\gamma}\lambda_j}{\overline{\gamma}\lambda_j\omega_j + \gamma\omega_r\omega_j} e^{-(x\frac{2\gamma}{\overline{\gamma}\lambda_i})} \left\{ \frac{\overline{\gamma}\lambda_j\omega_j}{\overline{\gamma}\lambda_j + 2\gamma\omega_j} \left[1 - \exp\left(-x\frac{\overline{\gamma}\lambda_j + 2\gamma\omega_j}{\overline{\gamma}\lambda_j\omega_j}\right)\right]\right.$$

$$-x\frac{\gamma\omega_r\lambda_j}{\lambda_i}\left(\overline{\gamma}\lambda_j + \gamma\omega_r\right)\exp\left(x\frac{\gamma\omega_r(\overline{\gamma}\lambda_j + 2\gamma\omega_j)}{\overline{\gamma}\omega_j\lambda_i(\overline{\gamma}\lambda_j + \gamma\omega_r)}\right)$$

$$\times \left\{ E_1\left(x\frac{\gamma\omega_r(\overline{\gamma}\lambda_j + 2\gamma\omega_j)}{\overline{\gamma}\omega_j\lambda_i(\overline{\gamma}\lambda_j + \gamma\omega_r)}\right) - E_1\left[x\left(\frac{\overline{\gamma}\lambda_j + 2\gamma\omega_j}{\overline{\gamma}\lambda_j\omega_j}\right)\left(\frac{\gamma\omega_r\lambda_j + \lambda_i(\overline{\gamma}\lambda_j + \gamma\omega_r)}{\lambda_i(\overline{\gamma}\lambda_j + \gamma\omega_r)}\right)\right]\right\}\right\}. \quad (12)$$

After that we get Ω_i which is offer in Eq. (13).

$$\Omega_1 = J_1\left(\omega_i, \overline{\gamma}\lambda_0, \gamma_{\text{th}}\right) - \left(1 + \frac{\omega_j}{\omega_i}\right)J_1\left(\left(\frac{1}{\omega_i} + \frac{1}{\omega_j}\right)^{-1}, \overline{\gamma}\lambda_0, \gamma_{\text{th}}\right)$$

$$+ \frac{\overline{\gamma}(\lambda_j)^2}{2\lambda_0\omega_i\omega_j\omega_r}\left[J_5\left(\frac{\overline{\gamma}\lambda_j\omega_j + \gamma\omega_r\omega_j}{\omega_r\omega_j}, \frac{\overline{\gamma}\lambda_j + 2\gamma\omega_j}{2\omega_j}, \frac{2\gamma}{\overline{\gamma}\lambda_i} + \frac{1}{\omega_i}, \frac{1}{\overline{\gamma}\lambda_0} - \frac{2}{\overline{\gamma}\lambda_i}, \gamma_{\text{th}}\right)\right.$$

$$- J_5\left(\frac{\overline{\gamma}\lambda_j\omega_j + \gamma\omega_r\omega_j}{\omega_r\omega_j}, \frac{\overline{\gamma}\lambda_j + 2\gamma\omega_j}{2\omega_j}, \frac{2\gamma}{\overline{\gamma}\lambda_i} + \frac{1}{\omega_i} + \frac{2\gamma}{\overline{\gamma}\lambda_j} + \frac{1}{\omega_j}, \frac{1}{\overline{\gamma}\lambda_0} - \frac{2}{\overline{\gamma}\lambda_i} - \frac{2}{\overline{\gamma}\lambda_j}, \gamma_{\text{th}}\right)\right]$$

$$+ \frac{(\lambda_j)^2}{\lambda_0\lambda_i\omega_i\omega_j\omega_r}\left[J_6\left(\gamma + \frac{\overline{\gamma}\lambda_j}{\omega_r}, \frac{1}{\overline{\gamma}\lambda_0} - \frac{2}{\overline{\gamma}\lambda_i}, \frac{1}{\omega_i} + \frac{2\gamma}{\overline{\gamma}\lambda_i}, \frac{2}{\overline{\gamma}\lambda_i}, \gamma + \frac{\overline{\gamma}\lambda_j}{2\omega_j}, \frac{2}{\overline{\gamma}\lambda_i}, \gamma, \gamma + \frac{\overline{\gamma}\lambda_j}{2\omega_j}, \gamma\right)\right.$$

$$- J_6\left(\gamma + \frac{\overline{\gamma}\lambda_j}{\omega_r}, \frac{1}{\overline{\gamma}\lambda_0} - \frac{2}{\overline{\gamma}\lambda_i}, \frac{1}{\omega_i} + \frac{2\gamma}{\overline{\gamma}\lambda_i}, \frac{2}{\overline{\gamma}\lambda_i}, \gamma + \frac{\overline{\gamma}\lambda_j}{2\omega_j}, \frac{2}{\overline{\gamma}\lambda_i} + \frac{2}{\overline{\gamma}\lambda_i}, \gamma + \frac{\overline{\gamma}\lambda_i\lambda_j}{\omega_r(\lambda_i + \lambda_j)}, \gamma + \frac{\overline{\gamma}\lambda_j}{2\omega_j}, \gamma\right)\right].$$

$$(13)$$

Here, $J_5(a, b, c, d, g) = \frac{1}{d^2}\left[G_1 J_4(g, a, 1) + H_1 J_4(g, b, 1) + K_1 J_4\left(g, -\frac{c}{d}, 1\right) + K_2 J_4\left(g, -\frac{c}{d}, 2\right)\right]$, where $G_1 = \frac{1}{(a-b)\left(a+\frac{c}{d}\right)^2}$, $H_1 = \frac{1}{(b-a)\left(b+\frac{c}{d}\right)^2}$, $K_2 = \frac{1}{\left(\frac{c}{d}+a\right)}$ $\frac{1}{\left(\frac{c}{d}+b\right)}$ and $K_1 = \frac{a+b+\frac{2c}{d}}{\left(a+\frac{c}{d}\right)^2\left(b+\frac{c}{d}\right)^2}$. While $J_6(a, b, c, d, e, f, g, h, i) = \frac{1}{b^3}\sum_{u=1}^{2}\sum_{k=1}^{4}$ $U_{n_k,u}J_4(\gamma_{\text{th}}, n_k, u) + \frac{2}{b^3}\sum_{u=1}^{3}\sum_{k=3}^{4}\left[\sum_{o=3}^{4}V_{n_k,u}J_7(n_o, n_k, \gamma_{\text{th}}, u) - \sum_{o=5}^{6}V_{n_k,u}J_7(n_o, n_k, \gamma_{\text{th}}, u)\right]$ where $U_{n_k,u} = \frac{1}{(2-u)!}\frac{d^{(2-u)}}{dy}\left[\frac{(y-n_k)^2(i-y)(a-y)(S-3Q)}{(y-n_1)^2(y-n_2)^2(y-n_3)^2(y-n_4)^2}\right]\Big|_{y=n_k}$, $V_{n_k,u} = \frac{1}{(3-u)!}\frac{d^{(3-u)}}{dy}\left[\frac{(y-n_k)^3(i-y)(a-y)}{(y-n_3)^3(y-n_4)^3}\right]\Big|_{y=n_k}$, $Q = S + (a-y)\left(y+\frac{c}{b}\right) - \frac{d}{b}y^2 + \frac{d(i+e)}{b}y - \frac{edi}{b}$, $S = \frac{f}{b}y^2 - \frac{f}{b}(g+h)y + \frac{fgh}{b}$. Finally, n_1, n_2 are roots of $S-Q$, n_3, n_4 are roots of Q, n_5, n_6 are roots of S and $J_7(a, b, c, n) = \int_0^c \frac{\log(a-y)}{(y-b)^n}dy = W(a) - W(a-c)$.

Here W is calculated with the support of [19, 2.727.1]. On the other hand, due to the symmetric between Ω_1 and Ω_2, we solely need to obtain Ω_1 then taking similar steps to find out Ω_2. Finally, the SOP$_{\text{MRC}}$ is calculated as SOP$_{\text{MRC}} = \Omega_1 + \Omega_2$.

Asymptotic System Outage Probability (ASOP). In this subsection, we derive the asymptotic system outage probability (ASOP) for discovering the system diversity. As this case, we assume that $\overline{\gamma} \to \infty$ and using the fact that

$$\text{ASOP}_{\text{MRC}} \overset{\overline{\gamma} \to \infty}{=} \text{AUOP}^1_{\text{MRC}} + \text{AUOP}^2_{\text{MRC}} - \text{AUOP}^1_{\text{MRC}}\text{AUOP}^2_{\text{MRC}}$$

$$\text{ASOP}_{\text{MRC}} = \text{AUOP}^1_{\text{MRC}} + \text{AUOP}^2_{\text{MRC}}. \tag{14}$$

Here $\text{AUOP}^i_{\text{MRC}}$, $i \in \{1,2\}$, is the asymptotic of i-th user outage probability. By using binomial expansion [19, Eq. 1.110] and vanishing the second term, the asymptotic OP of S_i is given by

$$\text{AUOP}^i_{\text{MRC}} \overset{\overline{\gamma} \to \infty}{=} \mathcal{E}\left\{ \mathcal{G} - \frac{1}{\overline{\gamma}}\left[\mathcal{A} + \mathcal{B} + \mathcal{C} - \frac{\gamma_{\text{th}}\mathcal{H}}{a_1 a_2^2} \right] - \frac{1}{\overline{\gamma}^2}\left[\frac{\gamma_{\text{th}}(\mathcal{A}+\mathcal{C})}{a_2} + \frac{2\gamma_{\text{th}}(\mathcal{B}-\mathcal{A})}{a_1} \right] \right\} \tag{15}$$

where $\mathcal{E} = \sum_{n=1}^{N+1}\sum_{i=1}^{I+1} \frac{8\sqrt{2b_n}\pi a_{\mathbf{N}}a_{\mathbf{I}}(\lambda_i\lambda_j\lambda_0)^2}{\omega_i\omega_j\omega_r(\lambda_i-2\lambda_0)^3}$, $\mathcal{C} = \frac{a_2 D\gamma_{\text{th}}}{a_1 a_2}$, $\mathcal{B} = \frac{3\left(2a_2\omega_r + (\lambda_j - a_1\omega_r)\right)\gamma_{\text{th}}}{2a_1\omega_r\left[a_2(a_1-4a_2)\right]^2}$,

$a_1 = \frac{-\lambda_0\lambda_j(4b_n b_i - 1)}{\omega_j(\lambda_i - 2\lambda_0)}$, $\mathcal{A} = \frac{\gamma_{\text{th}}(\lambda_j - a_1\omega_r)}{2\omega_r(a_1 a_2)^2(a_1 - 4a_2)}$, $\mathcal{F} = \frac{-2\left(a_1(a_1-4a_2)D+1\right)}{\sqrt[3]{a_1(a_1-4a_2)}}$,

$\mathcal{D} = \frac{3(\lambda_j - a_1\omega_r) - 6a_2\omega_r + \omega_r(a_1-4a_2)}{\omega_r a_1((a_1-4a_2))^2}$, $\mathcal{G} = \mathcal{F}\left[\tanh^{-1}(a_6) - \tanh^{-1}(a_7)\right]$,

$a_6 = \frac{2\gamma_{\text{th}} - a_1\overline{\gamma}}{\overline{\gamma}\sqrt{a_1(a_1-4a_2)}}$, $\mathcal{H} = \frac{2a_2(\lambda_j - \omega_r a_1)(\overline{\gamma})^2 + a_5(\overline{\gamma}) + a_2 a_3 a_4(2\lambda_j - \omega_r a_1)}{(\overline{\gamma})^2\omega_r\left(a_2 + \frac{\gamma_{\text{th}}}{\overline{\gamma}}\right)}$, $a_2 =$

$\frac{\lambda_i\omega_j}{\omega_i\omega_r(4b_n b_i - 1)}$, $a_3 = \frac{\gamma_{\text{th}}(\lambda_j - a_1\omega_r)}{a_2(2\lambda_j - a_1\omega_r)}$, $a_5 = 2a_2(\lambda_j - \omega_r a_1)(a_3 + a_4) +$ $\omega_r(a_2 a_3 + a_1\gamma_{\text{th}} - a_2\gamma_{\text{th}})$, $a_4 = \frac{\gamma_{\text{th}}}{2a_2} - \frac{\gamma_{\text{th}}}{a_1}$, $a_7 = \sqrt{\frac{a_1}{a_1-4a_2}}$. The AUOP of S_j is obtained by applying the similar approach. Finally, the ASOP of MRC is obtained by substituting Eq. (15) into (14). As can be seen in (15), the diversity gain of the system with MRC technique is 2.

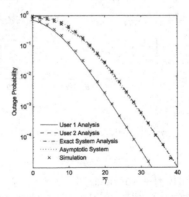

Fig. 2. Outage probability of MRC vs $\overline{\gamma}$ with $\mathcal{R} = 1$ and $\eta = 3$.

4   Numerical Results

Let us consider our simulation model in two-dimensional plane in which user S_1 and S_2 locate at $(0, 0)$ and $(1, 0)$, respectively. Whereas, the position of relay and primary user are (x_R, y_R) and (x_{PU}, y_{PU}), successively. Furthermore, only the location of relay and PU is changeable when two users situation is fixed throughout this section. The channel gain λ_m and ω_n are calculated by a simplified path loss model, i.e., $\lambda_1 = d_{S_1R}^{\eta}$, with d_{ij} is distance from node i to j, η is path loss exponent.

Figure 2 plots UOP and SOP of MRC combining with the location of relay and PU are $(0.4, 0.2)$ and $(0.8, 0.8)$, respectively. As can be seen in Fig. 2, our analyses absolutely match with simulation results. Moreover, the SOP curve is equal to the curve of UOP at user 2 especially in high SNRs region. It shows that the overall outage probability of the network completely depends on the weaker user's rate. In addition, the ASOP also has the same value with exact curve in high SNRs regime. Figure 3 plots the OP versus x_R where $y_R = 0$, and PU $= (0.8, 0.8)$. As we can be seen that when x_R is quite small or it is close to the S_1, the UOP of S_1 is outperform than S_2 and vice versa. In addition, the SOP only leans on the weaker rate while the relative position between relay node and primary user is sufficient large, whereas it has a little gap with the weaker rate. Figure 4 illustrates the impact of PU position on the performance of consider system. Particularly, the location of PU is changing from $(0, 1)$ to $(1, 1)$, it means only the x-axis is vary. In this figure, we see that when the PU is proximity to S_1 or x_{PU} is tiny, the SOP is limited by UOP1 curve and vice versa. It can be explained that the transmit power of a specific user is approach to zero when the PU is quite closely. Thereby, it easily go into outage events.

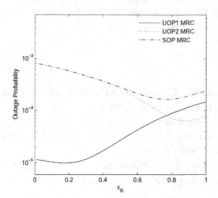

Fig. 3. Outage probability versus the position of relay with $\bar{\gamma} = 30$ dB, $\mathcal{R} = 1$ and $\eta = 3$.

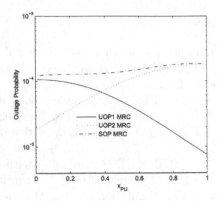

Fig. 4. Outage probability versus the position of primary user with $\bar{\gamma} = 30$, PR = (0.4, 0.2), $\mathcal{R} = 1$ dB and $\eta = 3$.

5   Conclusions

In this paper, outage performance has been studied in TW cognitive spectrum sharing in the presence of the direct link. In particular, the closed-form and asymptotic expression for both user and system OP have been addressed in basically tractable functions. Furthermore, it is proven that full diversity is got by adopting MRC at two end users. The correctness of our analysis is verified through simulation.

Acknowledgement. This work was supported by the Newton Prize 2017 and by a Research Environment Links grant, ID 339568416, under the Newton Programme Vietnam partnership. The grant is funded by the UK Department of Business, Energy and Industrial Strategy (BEIS) and delivered by the British Council. For further information, please visit https://www.newtonfund.ac.uk/.

References

1. Mitola, J., Maguire, G.Q.: Cognitive radio: making software radios more personal. IEEE Pers. Commun. **6**(4), 13–18 (1999)
2. Deng, Y., Wang, L., Elkashlan, M., Kim, K.J., Duong, T.Q.: Generalized selection combining for cognitive relay networks over Nakagami-m fading. IEEE Trans. Sig. Process. **63**(8), 1993–2006 (2015)
3. Liu, Y., Wang, L., Duy, T.T., Elkashlan, M., Duong, T.Q.: Relay selection for security enhancement in cognitive relay networks. IEEE Wirel. Commun. Lett. **4**(1), 46–49 (2015)
4. Duong, T.Q., Suraweera, H.A., Zepernick, H.-J., Yuen, C.: Beamforming in two-way fixed gain amplify-and-forward relay systems with CCI. In: Proceedings of IEEE International Communications Conference (ICC 2012), Ottawa, Canada, June 2012
5. Laneman, J.N., Tse, D.N.C., Wornell, G.W.: Cooperative diversity in wireless networks: efficient protocols and outage behavior. IEEE Trans. Inf. Theory **50**(12), 3062–3080 (2004)

6. Rankov, B., Wittneben, A.: Spectral efficient protocols for half-duplex fading relay channels. IEEE J. Sel. Areas Commun. 25(2), 379–389 (2007)

7. Yan, M., Chen, Q., Lei, X., Duong, T.Q., Fan, P.: Outage probability of switch and stay combining in two-way amplify-and-forward relay networks. IEEE Wirel. Commun. Lett. 1(4), 296–299 (2012)

8. Atapattu, S., Jing, Y., Jiang, H., Tellambura, C.: Relay selection schemes and performance analysis approximations for two-way networks. IEEE Trans. Commun. 61(3), 987–998 (2013)

9. Li, Q., Ting, S.H., Pandharipande, A., Han, Y.: Cognitive spectrum sharing with two-way relaying systems. IEEE Trans. Veh. Technol. 60(3), 1233–1240 (2011)

10. Duy, T.T., Kong, H.Y.: Exact outage probability of cognitive two-way relaying scheme with opportunistic relay selection under interference constraint. IET Commun. 6(16), 2750–2759 (2012)

11. Kim, K.J., Duong, T.Q., Elkashlan, M., Yeoh, P.L., Nallanathan, A.: Two-way cognitive relay networks with multiple licensed users. In: Proceedings of IEEE Global Communications Conference (GLOBECOM 2013), pp. 1014–1019, Atlanta, GA, December 2013

12. Ubaidulla, P., Aissa, S.: Optimal relay selection and power allocation for cognitive two-way relaying networks. IEEE Wireless Commun. Lett. 1(3), 225–228 (2012)

13. Liu, Y., Wang, L., Elkashlan, M., Duong, T.Q., Nallanathan, A.: Two-way relaying networks with wireless power transfer: policies design and throughput analysis. In: Proceedings of IEEE Global Communications Conference (GLOBECOM 2014), Austin, TX, December 2014

14. Nguyen, D.K., Matthaiou, M., Duong, T.Q., Ochi, H.: RF energy harvesting two-way cognitive DF relaying with transceiver impairment. In: Proceedings of IEEE International Communications Conference (ICC 2015), London, UK, June 2015

15. Taghiyar, M.J., Muhaidat, S., Liang, J., Dianati, M.: Relay selection with imperfect CSI in bidirectional cooperative networks. IEEE Commun. Lett. 16(1), 57–59 (2012)

16. Taghiyar, M.J., Muhaidat, S., Liang, J.: Max-min relay selection in bidirectional cooperative networks with imperfect channel estimation. IET Commun. 6(15), 2497–2502 (2012)

17. Popovski, P., Yomo, H.: Wireless network coding by amplify-and-forward for bi-directional traffic flows. IEEE Commun. Lett. 11(1), 16–18 (2007)

18. Krikidis, I.: Relay selection for two-way relay channels with MABC DF: a diversity perspective. IEEE Trans. Veh. Technol. 59(9), 4620–4628 (2010)

19. Gradshteyn, I.S.: Table of Integrals, Series, and Products. Academic press, London (2007)

20. Alkheir, A.A., Ibnkahla, M.: An accurate approximation of the exponential integral function using a sum of exponentials. IEEE Commun. Lett. 17(7), 1364–1367 (2013)

Energy Efficiency Maximization with Per-Antenna Power Constraints for Multicell Networks Using D.C. Programming

Le Ty Khanh[1(✉)], Ha Hoang Kha[1], and Nguyen Minh Hoang[2]

[1] Ho Chi Minh City University of Technology, VNU-HCM, Ho Chi Minh, Vietnam
khanhlety@gmail.com, hhkha@hcmut.edu.vn
[2] Banking University of HCM City, Ho Chi Minh, Vietnam
nmhoang@gmail.com

Abstract. This paper studies the energy efficiency (EE) optimization problem in multicell wireless networks in which each base station (BS) equipped with multiple antennas serves multiple users at the same time and in the same frequency. The problem of interest is to design the precoders to maximize the network EE subject to practical power constraints at physical layers. The resultant optimization design problem is nonconvex fractional programming and, thus, finding its optimal solution is mathematically challenging. In this paper, we use a combination of difference of convex (d.c.) programming and the Dinkelbach algorithm to iteratively solve the optimization problem. Then, by numerical simulations, we verify the convergence characteristics of the iterative algorithm and examine the EE performance of the system as compared to an spectral efficiency (SE) approach.

Keywords: Energy efficiency · Multicell · Precoder design
Per-antenna power constraints · D.C. programming

1 Introduction

In the recent decades, there has been an exponential growth in the number of wireless connected terminals and high data rate transmission applications. As shown in report [1], it is predicted that there will be about 11.6 billion mobile connected devices and about 30.6 exabytes of data per month in 2020. To serve a huge number of wireless devices and increased high data rate applications, the deployment of small cells with higher frequency reuse efficiency is an effective means [2]. The small cells with low transmit power can serve the mobile terminals (MTs) in the small areas and, thus, they can provide high quality of service transmission links. However, with the large number of base stations (BSs) operating in the same frequency, interference will be a crucial factor which significantly

© ICST Institute for Computer Sciences, Social Informatics and Telecommunications Engineering 2019
Published by Springer Nature Switzerland AG 2019. All Rights Reserved
T. Q. Duong and N.-S. Vo (Eds.): INISCOM 2018, LNICST 257, pp. 286–295, 2019.
https://doi.org/10.1007/978-3-030-05873-9_24

affects the network performance. There have been extensive studies on how to handle interference and to improve the spectral efficiency (SE) in small cells. The authors in [3] proposed the optimal design for linear transceivers to maximize the SE in multiple-input multiple-output (MIMO) broadcasting channels. Reference [4] studied the power control and beamforming schemes to minimize interference and maximize the SE in femto-cells.

It is known that the wireless system designs aiming at maximizing the SE may lead to use all transmit power budget and, thus, it may be not power consumption efficiency. Recently, there has been an increasing interest in issues related to energy consumption, e.g., global greenhouse gas emissions from information and communication technology (ICT). According to reference [5], the total energy consumption of ICI is about 3 percent of the total worldwide electric energy consumed. Thus, the energy efficiency (EE) performance metric has been received great attention in both academic and industrial community [6–8]. The EE design systems can not only reduce the power consumption, operating costs but also prolong life of energy-constrained mobile terminals [9]. The EE is measured by the transmission information bits per unit of energy [5,10,11]. The EE designs have been investigated for various wireless communication scenarios in the past decades [12–14]. The authors in [15] studied the tradeoffs between the EE and SE in orthogonal frequency division multiplexing (OFDM). Reference [14] considered the EE for multiuser interference channels. The method in [14] is to use the interference alignment (IA) technique to mitigate interference and, then, adopt the Dinkelbach approach to find the optimized precoders. In multicell wireless communications, the authors in [13] investigated the EE for multiple-input single-output (MISO) systems. The authors in [16] designed the precoders for multicell with full-duplex communications by proposing the path-following algorithm with convex quadratic program in iterations. Similarly, the authors in [17] considered the max-min individual user EE in a multicell MIMO system by using sequential convex approximation.

In this paper, we will extend the single user per cell model in [17] to a more general scenario in which the multiple antenna base stations (BSs) simultaneously serve multiple antenna multi-users at the same time in the same frequency band. Different from [17,18] which considered the power constraints at the BSs, the present paper will investigate the per-antenna power constraints. It is known that the per-antenna power constraints are more realistic than the BS power constraints since in practice the individual RF module can be connected to each transmit antenna [19,20]. More specifically, our major aim is to investigate the network EE (NEE) performance of the multicell multiuser MIMO wireless networks. Motivated by the previous results [17,18], we focus on jointly designing the precoder matrices at the BSs to maximize the total NEE. In [18], the similar multicell multuser scenario was investigated. However, the design problem in the current paper will additionally impose the per-antenna power constraints which are more realistic. Similar to works in [17], the optimization design problem is of nonconvex fractional programming in matrix variables, which makes the optimization problem intractable. Note that in multiuser multicell networks the mutual intracell and intercell interference among users results the coupling objective function, i.e., the EE performance improvement of a particular user can result in a decrease in the EE of other users and the NEE. By exploiting the concave property of the logdet function, the achievable user rates can be rewritten as a difference of convex (d.c.) functions and their lower bound can be obtained by linearizing the convex parts. Then, the resulting objective function becomes a convex-concave fractional programming and, thus, the Dinkelbach method can be applicable. Consequently, we can find the optimal precoders maximizing the NEE subject to practical power constraints by using an iterative algorithm which combines of both d.c. programming and Dinkelbach methods. We perform numerical simulations to examine the convergence characteristics of the iterative algorithm and to investigate the NEE performance. The numerical results show that the iterative algorithm is converged in less than 50 iterations. In addition, the EE optimization provides an NEE performance improvement as compared to the SE approach when the transmit power budget is high.

The remainder of this paper is organized as follows. Section 2 introduces the system model and formulates the design problem as an optimization problem. Section 3 presents the iterative algorithm based on the d.c. procedure and Dinkelbach approach. Section 4 provides numerical simulation results to evaluate the effectiveness of the iterative algorithms in terms of convergence behaviors and the achievable NEE. Finally, concluding remarks are given in Sect. 5.

Notations: We use bold upper letters X and lower case letters x to denote matrices and vectors, respectively. I is an identity matrix with appropriate dimensions. e_i stands for a vector with all zeros except element i equal to 1. $\langle \cdot \rangle$, $|\cdot|$, and $\mathbb{E}(.)$ are trace, determinant, and expectation operations. A random vector x of complex circular Gaussian distribution with mean $\bar{x}$ and covariance matrix R_x is denoted by $x \sim \mathcal{CN}(\bar{x}, R_x)$. $X \succeq 0$ stands for a positive semi-definite matrix.

2   Downlink Multicell Channel Model and Design Problem Formulation

We consider a down-link channel model in K-cell cooperative wireless networks in which BS k equipped with N_k transmit antennas serves L_k users in its cell simultaneously. The set of BSs is denoted by $\mathcal{K} = \{1, 2, ..., K\}$ and the set of antennas at BS k is denoted by $\mathcal{N}_k = \{1, 2, ..., N_k\}$. We define $\mathcal{L}_k = \{k_1, k_2, ...k_{L_k}\}$ as the set of users in cell k and set of all users in the network as $\mathcal{L} = \{k_i | k \in \{1, 2, ..., K\}, i \in \{1, 2, ..., L_K\}\}$. User i in cell k is equipped with M_{k_i} receive antennas. BS k transmits d_{k_i} data streams to user k_i in its cell. To transmit signal $s_{k_i} \in \mathbb{C}^{d_{k_i} \times 1}$ to MT k_i, BS k applies a linear precoder $F_{k_i} \in \mathbb{C}^{N_k \times d_{k_i}}$ to process the signals before transmitting it. The signal transmitted from BS k is

$$x_k = \sum_{i=1}^{L_k} F_{k_i} s_{k_i}. \tag{1}$$

Without loss of generality, it is assumed that $\mathbb{E}\left[s_{k_i} s_{k_i}^H\right] = I$. The channel matrix from BS l to MT k_i is $H_{k_i, l} \in \mathbb{C}^{M_{k_i} \times N_k}$ which stands for the small scale fading while $\eta_{k_i,l}$ is the path loss from BS l to MT k_i. Similar to works in [17,18], we assume that global channel state information (CSI) is perfectly known at the BSs. Then, the received signal at MT k_i is given by

$$r_{k_i} = \underbrace{\sqrt{\eta_{k_i,k}}H_{k_i,k}F_{k_i}s_{k_i}}_{\text{desired signal}} + \underbrace{\sum_{m=1,m\neq i}^{L_k} \sqrt{\eta_{k_i,k}}H_{k_i,k}F_{k_m}s_{k_m}}_{\text{intracell interference}}$$

$$+ \underbrace{\sum_{j=1,j\neq k}^{K} \sum_{l=1}^{L_j} \sqrt{\eta_{k_i,j}}H_{k_i,j}F_{j_l}s_{j_l}}_{\text{intercell interference}} +n_{k_i}, \forall k_i \in \mathcal{L}, \tag{2}$$

where n_{k_i} is noise at receiver k_i and is assumed to be a complex Gaussian distribution $\mathcal{CN}(0,\sigma_{k_i}^2 I)$. It can be observed that the received signal at MT k_i is corrupted by not only interference signals of the users in the same cell but also interference from the other BSs in different cells. By treating interference as noise, the achievable bit rate of user k_i can be given by

$$\mathcal{R}_{k_i}(F) \triangleq \log\left|I + \eta_{k_i,k}H_{k_i,k}F_{k_i}F_{k_i}^H H_{k_i,k}^H R_{k_i}^{-1}(F)\right|, \tag{3}$$

where we have defined $F = \{F_{k_i}|k_i \in \mathcal{L}\}$ and $R_{k_i}(F) = \sum_{l_j\in\mathcal{L},l_j\neq k_i} \eta_{k_i,l}H_{k_i,l}F_{l_j}F_{l_j}^H H_{k_i,l}^H + \sigma_{k_i}^2 I$ is a sum of covariance matrices of interference and noise at receiver k_i.

From Eq. (1), the transmit power at BS k is given by

$$P_{t,k}(F) = \sum_{i=1}^{L_k}\langle F_{k_i}F_{k_i}^H\rangle \tag{4}$$

and the transmit power at antenna j of BS k is

$$P_{a,k_j}(F) = \sum_{i=1}^{L_k}\langle e_j^{kH}F_{k_i}F_{k_i}^H e_j^k\rangle \tag{5}$$

Then, the total power consumption at BS k is

$$P_k(F) = \rho_k \sum_{i=1}^{L_k}\langle F_{k_i}F_{k_i}^H\rangle + P_{c_k} \tag{6}$$

where P_{c_k} is static power consumption of hardware circuits and ρ_k is an inefficiency factor of circuit amplifiers. Then, the NEE is defined as the ratio of the achievable sum rate to the total power consumption given by [21]

$$\eta_{NEE}(F) = \frac{\sum_{k=1}^{K}\sum_{i=1}^{L_k}\mathcal{R}_{k_i}(F)}{\sum_{k=1}^{K}P_k(F)}. \tag{7}$$

Similar to work in [18], we aim to maximize the EE of system by designing the optimal precoders. However, in this paper, we maximize the NEE with taking the per-antenna power constraints into consideration. The design problem of interest can be mathematically stated as:

$$\max_{\boldsymbol{F}} \quad \eta_{NEE}(\boldsymbol{F}) = \frac{\sum_{k=1}^{K} \sum_{i=1}^{L_k} \mathcal{R}_{k_i}(\boldsymbol{F})}{\sum_{k=1}^{K} P_k(\boldsymbol{F})} \tag{8a}$$

$$\text{s.t.} \quad \sum_{i=1}^{L_k} \langle \boldsymbol{F}_{k_i} \boldsymbol{F}_{k_i}^H \rangle \leq P_{k,max}, \ k \in \mathcal{K}, \tag{8b}$$

$$\sum_{i=1}^{L_k} \langle \boldsymbol{e}_j^{k\,H} \boldsymbol{F}_{k_i} \boldsymbol{F}_{k_i}^H \boldsymbol{e}_j^k \rangle \leq P_{k_j,max}, \ k \in \mathcal{K}, j \in \mathcal{N}_k, \tag{8c}$$

where $P_{k,max}$ is a transmit power budget at BS k and $P_{k_j,max}$ is the maximum allowable transmit power at antenna j of BS k. It can been seen that the optimization problem in (8) is a non-convex fractional programming and, therefore, it is intractable to directly solve. In next section, we will apply the d.c. programming technique and Dinkelbach method to solve problem (8).

3   An Iterative Algorithm Using D.C. Programming and Dinkelbach Methods

It can be proved that the constraints in (8) are convex quadratic constraints with respect to matrix variables $\boldsymbol{F}$ while the objective function is the ratio of a nonlinear nonconcave function over a convex quadratic function. Thus, the mathematical challenge to solve (8) is due to the non-concavity of the achievable rate. To overcome this difficulty, we adopt the d.c. expression of the achievable rate as follows [17, 22]

$$\mathcal{R}_{k_i}(\boldsymbol{Q}) = f_{k_i}(\boldsymbol{Q}) - g_{k_i}(\boldsymbol{Q}) \tag{9}$$

with

$$f_{k_i}(\boldsymbol{Q}) = \log \left| \sum_{l_j \in \mathcal{L}} \eta_{k_i,l} \boldsymbol{H}_{k_i,l} \boldsymbol{Q}_{l_j} \boldsymbol{H}_{k_i,l}^H + \sigma_{k_i}^2 \boldsymbol{I} \right|, \tag{10}$$

and

$$g_{k_i}(\boldsymbol{Q}) = \log \left| \sum_{l_j \in \mathcal{L}, l_j \neq k_i} \eta_{k_i,l} \boldsymbol{H}_{k_i,l} \boldsymbol{Q}_{l_j} \boldsymbol{H}_{k_i,l}^H + \sigma_{k_i}^2 \boldsymbol{I} \right|, \tag{11}$$

where we have defined $\boldsymbol{Q}_{k_i} = \boldsymbol{F}_{k_i} \boldsymbol{F}_{k_i}^H$ and $\boldsymbol{Q} = \{\boldsymbol{Q}_{k_i} | k_i \in \mathcal{L}\}$. It is clear that $f_{k_i}(\boldsymbol{Q})$ and $g_{k_i}(\boldsymbol{Q})$ are concave functions and, thus, Eq. (9) is a d.c. function. By exploiting the concavity of $g_{k_i}(Q)$, we can find the lower bound for the user rate at given Q as follows

$$\mathcal{R}_{k_i} \geq f_{k_i}(Q) - g_{k_i}(\tilde{Q}) - \left\langle R_{k_i}^{-1}(\tilde{Q}) \left(Z_{k_i}(\tilde{Q}) - Z_{k_i}(Q) \right) \right\rangle \tag{12}$$

where we have defined $Z_{k_i}(Q) = \sum\limits_{l_j \in \mathcal{L}, l_j \neq k_i} \eta_{k_i,l} H_{k_i,l} Q_{l_j} H_{k_i,l}^H$. Then, the optimization problem can be rewritten as

$$\max_{Q} \quad \frac{\sum\limits_{k=1}^{K} \sum\limits_{i=1}^{L_k} \left(f_{k_i}(Q) - g_{k_i}(\tilde{Q}) - \langle R_{k_i}^{-1}(\tilde{Q}) \left(Z_{k_i}(\tilde{Q}) - Z_{k_i}(Q) \right) \rangle \right)}{\sum\limits_{k=1}^{K} \left(\rho_k \sum\limits_{i=1}^{L_k} \langle Q_{k_i} \rangle + P_{c_k} \right)} \tag{13a}$$

$$\text{s.t.} \quad \sum_{i=1}^{L_k} \langle Q_{k_i} \rangle \leq P_{k,max}, \; k \in \mathcal{K} \tag{13b}$$

$$\sum_{i=1}^{L_k} \langle e_j^{kH} Q_{k_i} e_j^k \rangle \leq P_{k_j,max}, \; k \in \mathcal{K}, j \in \mathcal{N}_k. \tag{13c}$$

The optimization problem (13) maximizes the concave-convex fractional function over a set of convex constraints. Thus, the Dinkelbach algorithm can be solved to find the optimal solution [17,18]. By defining the parameter function

$$\mathcal{G}(\lambda, Q) = \sum_{k=1}^{K} \sum_{i=1}^{L_k} \left(f_{k_i}(Q) - g_{k_i}(\tilde{Q}) - \left\langle R_{k_i}^{-1}(\tilde{Q}) \left(Z_{k_i}(\tilde{Q}) - Z_{k_i}(Q) \right) \right\rangle \right)$$
$$- \lambda \sum_{k=1}^{K} \left(\rho_k \sum_{i=1}^{L_k} \langle Q_{k_i} \rangle + P_{c_k} \right). \tag{14}$$

Given λ, the Dinkelbach method iteratively solves the following problem

$$\max_{Q} \quad \mathcal{G}(\lambda, Q) \tag{15a}$$

$$\text{s.t.} \quad \sum_{i=1}^{L_k} \langle Q_{k_i} \rangle \leq P_{k,max}, \; k \in \mathcal{K} \tag{15b}$$

$$\sum_{i=1}^{L_k} \langle e_j^{kH} Q_{k_i} e_j^k \rangle \leq P_{k_j,max}, \; k \in \mathcal{K}, j \in \mathcal{N}_k \tag{15c}$$

which is a convex optimization and can be efficiently solved by CVX [23]. It is shown in [24] that the optimal solution Q_{opt} to problem (15) is also the optimal solution to (13) if $\mathcal{G}(\lambda_{opt}, Q_{opt}) = 0$ at given λ_{opt}. Thus, the iterative algorithm to solve (8) can be described in Algorithm 1.

Remark: Note that problem (13) is a concave-convex fractional programming, and, thus, the Dinkelbach iterative algorithm is converged [24]. In addition, the

---

Algorithm 1. Iterative algorithm for NEE maximization

1: Inputs: system parameters and channel coefficients.

2: Initialization: $n = 0$, generate feasible precoders $\tilde{\boldsymbol{Q}}^{(0)}$ and $\boldsymbol{Q}^{(0)}$.

3: **repeat**

4:   **repeat**

5:     $\lambda^{(n)} = \eta_{NEE}(\boldsymbol{Q}^{(n)})$ from Eq. (8a) by replacing $\boldsymbol{Q}_{k_i} = \boldsymbol{F}_{k_i}\boldsymbol{F}_{k_i}^H$.

6:     With values $\lambda^{(n)}$ and $\tilde{\boldsymbol{Q}}_{i_k}^{(n)}$, solve problem (15) to obtain $\boldsymbol{Q}_{opt}$.

7:     Update $\boldsymbol{Q}^{(n)} \leftarrow \boldsymbol{Q}_{opt}$.

8:   **until** $|\mathcal{G}(\lambda^{(n)}, \boldsymbol{Q}^{(n)})| \leq \epsilon$.

9:   Update $n \leftarrow n + 1$ and $\tilde{\boldsymbol{Q}}^{(n)} \leftarrow \boldsymbol{Q}^{(n)}$.

10: **until** NEE convergence

11: Outputs: $\boldsymbol{Q}_{opt}$ and $\eta_{NEE}(\boldsymbol{Q}_{opt})$.

---

objective function in (13) is a lower bound of a d.c. function in (8), and, thus, its d.c. iterative algorithm is guaranteed to converge to the locally optimal solution to problem (8) [22]. That is, the objective value is not decreasing over iterations. In addition, the objective is upper-bounded since the set of power constraints. Therefore, convergence of Algorithm 1 is guaranteed, and its convergence behaviors will be further verified in the simulations.

4   Simulation Results

In this section, we investigate the convergence behavior and achievable EE performance of the iterative algorithm for NEE maximization in the multi-cell multi-user MIMO networks. We consider a cluster of $K = 4$ small cells placing at positions $\{(R, R), (-R, R), (-R, -R), (R, -R)\}$ where $R = 50$ m is cell radius. Assume that distance from each MT to its associated BS is not less than 10 m. Each BS is equipped with $N_k = 4$ antennas. There are 2 MTs in each cell and each MT is equipped with $M_{k_i} = 4$ antennas. The BS transmits $d_{k_i} = 4$ data streams to each user in its coverage. To model the small scale fading channels, the elements of the channel matrices $\boldsymbol{H}_{k_i,l}$ are generated from a complex Gaussian distribution with zero mean and unit variance. The large scaling fading factor $\eta_{k_i,l}$ is computed from the pathloss model $\eta_{k_i,l}(\text{dB}) = -140.7 - 36.7 \log_{10}(d_{k_i,l}(\text{km}))$ [8,25]. We assume that $\rho_k = 1$, $P_{c_k} = 35$ dBm, $P_{k,max} = P_t$ and $P_{k_j,max} = 1.1 \cdot P_t/N_k$ for all $k \in \mathcal{K}$.

Example 1: In this example, we examine the convergence characteristic of iterative Algorithm 1. We consider for various BS transmit budget $P_{k,max} = \{10, 30, 40\}$ dBm. The NEE over iterations is plotted in Fig. 1 for different transmit power budget levels. As can be seen from Fig. 1 that that the NEE value is not decreasing over iterations and converged in about tens of iterations. In addition, when the transmit power budget increases from 10 dBm to 30 dBm, the NEE is also increased. However, when the transmit power budget increases from 30 dBm to 40 dBm the NEE is converged to the same value. That is because

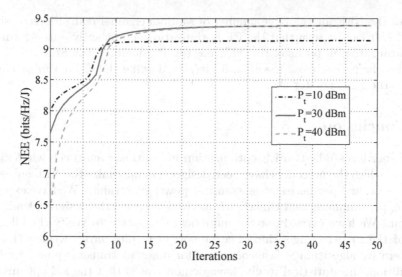

Fig. 1. Convergence behaviors of iterative Algorithm 1.

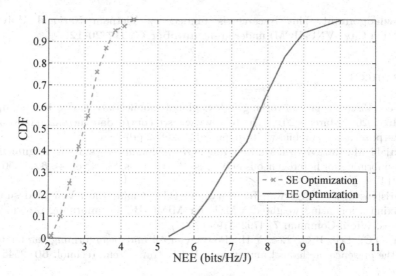

Fig. 2. NEE performance obtained by the EE and SE optimization.

when the power consumption over a certain level will not result in the EE and, thus, at the high levels of power budget, the systems may not use all power budgets.

Example 2: This example evaluates the NEE performance of Algorithm 1 for the EE optimization as compared to that of the SE optimization method. We consider for the case of transmit power budget $P_t = 40$ dBm. To make a statistical EE performance comparison, we plot the cumulative distribution function (cdf) of the NEE for the large number of random channel realizations as shown in Fig. 2. It can be seen that the EE optimization significantly outperforms the SE optimization in terms of the NEE metric. Particularly, the SE optimization provides the NEE almost less than 5 bits/Hz/J while the EE optimization can achieve the NEE greater than 5 bits/Hz/J.

5   Conclusion

This paper has studied an EE optimization in multiuser multi-cell wireless networks in which the precoders have been designed to maximize the NEE subject to per-antenna and per-base station transmit power constraints. We have exploited the d.c. programming and Dinkelbach methods to solve the design optimization problem. We have carried out the numerical simulation to verify the effectiveness of the iterative algorithm. The numerical results have revealed that the d.c. iterative algorithm is fast converged for different transmit power budgets. In addition, the statistical results have demonstrated that the EE optimization outperforms the SP approach in terms of the NEE.

Acknowledgement. This research is funded by Vietnam National University HoChiMinh City (VNU-HCM) under grant **number C2017-20-12**.

References

1. Cisco: Cisco visual networking index: global mobile data traffic forcast update 2015–2020, March 2016. https://www.cisco.com/c/dam/m/en_in/innovation/enterprise/assets/mobile-white-paper-c11-520862.pdf
2. Sanjabi, M., Razaviyayn, M., Luo, Z.Q.: Optimal joint base station assignment and beamforming for heterogeneous networks. IEEE Trans. Sig. Process. **62**, 1950–1961 (2014)
3. Christensen, S.S., Agarwal, R., Carvalho, E.D., Cioffi, J.M.: Weighted sum-rate maximization using weighted MMSE for MIMO-BC beamforming design. IEEE Trans. Wirel. Commun. **7**, 4792–4799 (2008)
4. Oh, D.C., Lee, H.C., Lee, Y.H.: Power control and beamforming for femtocells in the presence of channel uncertainty. IEEE Trans. Veh. Technol. **60**, 2545–2554 (2011)
5. Li, G.Y., et al.: Energy-efficient wireless communications: tutorial, survey, and open issues. IEEE Wirel. Commun. **18**, 28–35 (2011)
6. Jiang, C., Cimini, L.J.: Energy-efficient transmission for MIMO interference channels. IEEE Trans. Wirel. Commun. **12**, 2988–2999 (2013)
7. Huang, Y., Xu, J., Qiu, L.: Energy efficient coordinated beamforming for multicell MISO systems. In: IEEE Global Communication Conference GLOBECOM, pp. 2526–2531 (2013)
8. He, S., Huang, Y., Jin, S., Yu, F., Yang, L.: Max-min energy efficient beamforming for multicell multiuser joint transmission systems. IEEE Commun. Lett. **17**, 1956–1959 (2013)
9. Chen, Y., Zhang, S., Xu, S., Li, G.Y.: Fundamental trade-offs on green wireless networks. IEEE Commun. Mag. **49**, 30–37 (2011)

10. Kwon, H., Birdsall, T.: Channel capacity in bits per joule. IEEE J. Oceanic Eng. **11**, 97–99 (1986)
11. Verdu, S.: On channel capacity per unit cost. IEEE Trans. Inf. Theor. **36**, 1019–1030 (1990)
12. Nguyen, V.D., Nguyen, H.V., Nguyen, C.T., Shin, O.S.: Spectral efficiency of full-duplex multi-user system: beamforming design, user grouping, and time allocation. IEEE Access **5**, 5785–5797 (2017)
13. Nguyen, K.-G., Tran, L.-N., Tervo, O., Vu, Q.-D., Juntti, M.: Achieving energy efficiency fairness in multicell MISO downlink. IEEE Commun. Lett. **19**, 1426–1429 (2015)
14. Vu, T.T., Kha, H.H., Tuan, H.D.: Transceiver design for optimizing the energy efficiency in multiuser MIMO channels. IEEE Commun. Lett. **20**, 1507–1510 (2016)
15. Amin, O., Bedeer, E., Ahmed, M.H., Dobre, O.A.: Energy efficiency spectral efficiency tradeoff: a multiobjective optimization approach. IEEE Trans. Veh. Technol. **65**, 1975–1981 (2016)
16. Sheng, Z., Tuan, H.D., Tam, H.H.M., Nguyen, H.H., Fang, Y.: Energy efficient precoding in multicell networks with full-duplex base stations. EURASIP J. Wirel. Commun. Netw. **2017**(48), 13 (2017)
17. Li, Y., Fan, P., Beaulieu, N.C.: Cooperative downlink max-min energy-efficient precoding for multicell MIMO networks. IEEE Trans. Veh. Technol. **65**, 9425–9430 (2016)
18. Chi, P.H., Kha, H.H.: Optimized energy efficiency for multicell MIMO coopearitive systems with user rate constraints. In: 2017 International Symposium Electrical and Electronics Engineering, pp. 212–216, November 2017
19. Vu, M.: MIMO capacity with per-antenna power constraint. In: 2011 IEEE Global Telecommunications Conference (GLOBECOM), pp. 1–5, December 2011
20. Tran, L.N., Juntti, M., Bengtsson, M., Ottersten, B.: Successive zero-forcing DPC with per-antenna power constraint: optimal and suboptimal designs. In: 2012 IEEE International Conference Communications ICC, pp. 3746–3751, June 2012
21. He, S., Huang, Y., Yang, L., Ottersten, B.: Coordinated multicell multiuser precoding for maximizing weighted sum energy efficiency. IEEE Trans. Sig. Process. **62**, 741–751 (2014)
22. Kha, H.H., Tuan, H.D., Nguyen, H.H.: Fast global optimal power allocation in wireless networks by local D.C. programming. IEEE Trans. Wirel. Commun. **11**, 510–515 (2012)
23. Grant, M., Boyd, S.: CVX: Matlab software for disciplined convex programming, version 2.1. March 2014. http://cvxr.com/cvx
24. Crouzeix, J.-P., Ferland, J.A.: Algorithms for generalized fractional programming. Math. Program. **52**(1), 191–207 (1991)
25. Hegde, G., Ramos-Cantor, O.D., Cheng, Y., Pesavento, M.: Optimal resourceblock allocation and muting in heterogeneous networks. In: 2016 IEEE International Conference Acoustics, Speech and Signal Processing (ICASSP), pp. 3581–3585, March 2016

Development of a Positioning Solution Using FUKS Based on RTS Smoother Combined with FUKF for Vehicle Management Systems

Binh Thanh Ngo[1(✉)], Michele Zucchelli[2], and Francesco Biral[2]

[1] University of Transport and Communications,
No. 3, Cau Giay, Dong Da, Hanoi, Vietnam
ngobinh74@gmail.com
[2] University of Trento, Via Sommarive 9, 38123 Povo, Trento, Italy
francesco.biral@unitn.it

Abstract. This article introduces a new way to improve accuracy of trajectory in transportation management systems, in which it describes a design for integrated INS/GPS device mounted on vehicle and algorithms for trajectories at the station. The significant features of this system are the ways to process data at station by using a flexible unscented Kalman filter algorithm, and a backward retrieval calculation algorithm based on Rauch-Tung-Striebel smoother, called flexible unscented Kalman smoother. This system has the capability of receiving the information in order to locate, monitor hybrid buses more exactly and manage some other motion parameters to improve the quality of monitoring and management transportation system, and also to evaluate driving style of drivers in services and support for smart cities.

Keywords: Integrated INS/GPS system · FUKF · FUKS
Monitoring and management system

1 Introduction

The goals of the system monitoring transportation, especially in service and support for smart cities, are the development of effective management and administration of these means and checking the status of the devices during normal operation as well as when a problem occurs. Quality of the monitoring systems reflects not only in the accuracy of the position and velocity, but also in the monitoring motion parameters of moving object, especially the parameters about fluctuations and some other typical parameters such as accelerometers, vibration angles, direction angles and some parameters on the operation state of

Supported by the Ministry of Education and Training (MoET) under grant number B2016-GHA-02.

T. Q. Duong and N.-S. Vo (Eds.): INISCOM 2018, LNICST 257, pp. 296–305, 2019.
https://doi.org/10.1007/978-3-030-05873-9_25

the vehicle. These systems always use the integrated GPS (Global Positioning System) [1, 8, 10] and INS (Initial Navigation System), especially INS based on MEMS (Micro-Electro-Mechanical Systems).

In GNSS (Global Navigation Satellite Systems), to locate a moving object, they use information from GPS. In case of losing GPS signal, the position of objects will be lost. To solve this problem, also to improve the accuracy of monitoring system, an INS will be integrated with GPS in the object device. In the previous time, they have to accept the accumulated errors of INS data and then use calculator tools to predict the next point of the object based on by using KF (Kalman filter) combining with fuzzy logic [3] or neuron network [4] to limit influence of noise to obtain accurate results. There are some kinds of developed KF methods, including: EKF (Extended Kalman Filter), and UKF (Unscented Kalman Filter), in which UKF has better results than EKF [2]. In fact, drawing lines of orbits on the digital map are not the actual trajectories instead of the zigzag lines of predicting points. During the evaluation, we need the continuous orbits, so the interest in data processing such as smoothing trajectory is always accompanied by positioning problems.

There are some successful research results using UKF and UKS (Unscented Kalman Smoother), but their models that have been investigated and applied in these papers are unchanged in the UKF [2, 12], and only in theoretical models in the UKS [11]. These studies often ignore the asynchronous data in terms of time, for example not computed with data that does not have the same sampling frequency. Therefore, they are forced to use a fixed-type UKF [8, 12]. In recent times, there were some studies of changing in UKF structure in the biological field through the use of parametric estimation based on prior calculations [7, 9]. These UKF methods had been done with pre-made parameter recognition analysis, not real data from input signals with different sampling frequencies. In fact, the input data of the system have different sampling periods, for example INS has very fast sample rate but GPS and other sensors have slower sample rates. Thus, the automatic change of UKF and UKS structures according to the input data is very necessary, ensuring that we can calculate with full sufficient input data, not ignoring important data at the time that calculation cannot be done in case of using a fixed-field model. This article introduces a successfully developed solution of smoothing orbit based on the RTS (Rauch-Tung-Striebel) smoother combined with UKF, in which we develop a UKF algorithm and a UKS algorithm that can change the structure automatically depending on sample rates of input signals, called FUKF (Flexible UKF) and FUKS (Flexible UKS). The results of practical FUKF and FUKS tests have been done for some buses in an experiment transportation system in Trento, Italy.

2 The Vehicle Devices and Structure of System

To improve the quality of MEMS INS, different with using quaternion solution [5], we solved INS additional error problem by DCM (Direction Cosine Matrix) self-correction based on [13] with a development of using two flexible PI. Firstly, we calculated yaw angle from internal magnetometer built-in inside INS based on vector recalculation instead of heading angle from GPS signal [14], or from an external compass [6]. After that, we calculated correction and renormalization values, and pushed them to feedback controller to recalculate DCM. Corrected Euler angles were calculated from updated DCM. The variation of offset caused by supply voltage and temperature is usually rather slow, so that DCM can continually remove the offset and maintain lock. Drift around all three axes could be completely eliminated with DCM even the bus didn't move continuously. Thus, our DCM self-correction method had solved the accumulated errors in INS systems and could help MEMS INS 9-DOF (Degree Of Freedom) systems operate independently.

Next step, we designed and created a vehicle device using that developed DCM algorithm, and also built FUKF and FUKS algorithms at the station for a fleet of hybrid bus in Trento city, Italy during a campaign to monitor vehicle performance in which model of bus described in [1]. With the permission to access to standard J1939 FMS protocol data of Van Hool A330 Hybrid buses and MAN Lion's City Hybrid A37 buses, we have some more parameters of those hybrid buses to process filtering and smoothing data. The integrated vehicle devices of our system have been designed and created from consumer equipment such as MEMS INS 9-DOF Razor Stick, Arduino platform and can be connected to Radar AC20 TRW, datalogger CTAG. The devices are assembled on the bus as shown in Fig. 1, and their positions on the bus are shown in Fig. 2. Their data processing in the management and monitoring system is shown in Fig. 3.

Fig. 1. Diagram and real vehicle devices

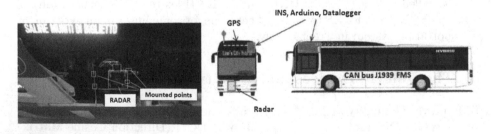

Fig. 2. The positions of devices on the bus

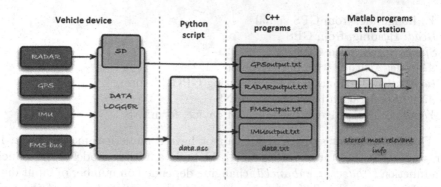

Fig. 3. Data processing in management and monitoring system

3   FUKF and FUKS Implementation

In this article, we introduce a mixed Nonlinear - Linear model, in which the model state of system is expressed as a nonlinear function, and the measurement model is expressed in terms of numerical matrices. The new proposals take advantage of the kinetic description of the object on Matlab with powerful tools to support mathematics of it.

Equation of state:

$$\dot{x}(t) = F\,x\,(t) + Bu(t) + v_x(t) \tag{1}$$
$$\dot{y}(t) = C\,x\,(t) + Du(t) + v_y(t)$$

For monitoring and management system, we have B = 0 and D = 0.
Discrete states equation of model:

$$x_k = f\left(x_{k-1}, u_{k-1}\right) + w_{k-1} \tag{2}$$

$$x_k = \left\{ \begin{array}{c} Distance_k \\ X_k \\ Y_k \\ Head_k \\ Speed_k \\ Altitude_k \\ Inclination_k \\ ALon_k \\ YawRate_k \end{array} \right\}$$

Where:

- *Distance*: Measured distances, from wheel encoder (FMS Output)
- *X*: Latitude, from GPS signal

- Y: Longitude, from GPS signal
- *Head*: Heading, from GPS
- *Speed*: Speed, from GPS
- *Altitude*: Altitude with our set-point
- *Inclination*: Inclination from GPS using GPGGA form
- *Alon*: Acceleration in z-axis, form INS
- *YawRate*: Derivative of yaw angle of device, form INS

This is non-linear system, so we cannot have a numerical matrix for it. In our program at the station, based on [1], we build a flexible model of the vehicle as a function (*@buses_f_enhanced*) changing depended on number of input data (3 or 9 input signals). This function computes the predicted state of the next step using Euler Integration.

$$x_k = x_{k-1} + dt \cdot \widehat{x\_dot}$$

In addition, the filter needs the matrix Q, undetermined states covariance computed as [11], shown as below:

$$Q = L \cdot Qc \cdot L' \cdot dt$$

$$Qc = \begin{bmatrix} std \wedge 2(inclination\_der) & 0 & 0 \\ 0 & std \wedge 2(ALon\_der) & 0 \\ 0 & 0 & std \wedge 2(YawRate\_der) \end{bmatrix}$$

Measurement matrix in our system:

$$y_k = H \cdot x_k + r_k \tag{3}$$

$$Y_k = \begin{Bmatrix} Distance_{FMS,k} \\ X_{gps,k} \\ Y_{gps,k} \\ Head_{gps,k} \\ Speed_{gps,k} \\ Altitude_{gps,k} \\ Inclination_{gps,k} \\ ALon_{IMU,k} \\ YawRate_{IMU,k} \end{Bmatrix}$$

Where:

- $\dot{d} = V$
- $\dot{x} = V \times \cos(\psi)$
- $\dot{y} = -V \times \sin(\psi)$
- $\dot{\psi} = YawRate$

- $\dot{V} = Alon - g \times \sin(\alpha)$
- $\dot{z} = V \times \sin(\alpha)$
- α is inclination calculated from GPS
- *Alon*: is derivation of Acc in z axis
- *YawRate*: Derivative of Yaw of device.

$$H = \begin{bmatrix} 1 & 0 & 0 & 0 & 0 & 0 & 0 & 0 & 0 \\ 0 & 1 & 0 & 0 & 0 & 0 & 0 & 0 & 0 \\ 0 & 0 & 1 & 0 & 0 & 0 & 0 & 0 & 0 \\ 0 & 0 & 0 & 1 & 0 & 0 & 0 & 0 & 0 \\ 0 & 0 & 0 & 0 & 1 & 0 & 0 & 0 & 0 \\ 0 & 0 & 0 & 0 & 0 & 1 & 0 & 0 & 0 \\ 0 & 0 & 0 & 0 & 0 & 0 & 1 & 0 & 0 \\ 0 & 0 & 0 & 0 & 0 & 0 & 0 & 1 & 0 \\ 0 & 0 & 0 & 0 & 0 & 0 & 0 & 0 & 1 \end{bmatrix} ; r_k = \begin{bmatrix} std(Yk(1))^2 \\ std(Yk(2))^2 \\ std(Yk(3))^2 \\ std(Yk(4))^2 \\ std(Yk(5))^2 \\ std(Yk(6))^2 \\ std(Yk(7))^2 \\ std(Yk(8))^2 \\ std(Yk(9))^2 \end{bmatrix}$$

Where: std is standard deviation.

Depended on sample rates of input signals, we develop a UKF algorithm that its structure can be changed automatically, called FUKF. Its algorithm and operation modes are shown in Figs. 4 and 5, as below:

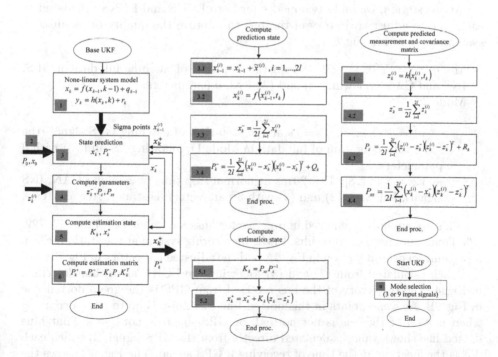

Fig. 4. The FUKF algorithm diagram

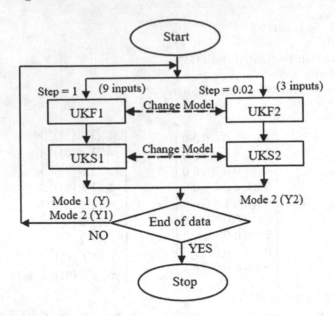

Fig. 5. Selection of operating function diagram

At the station, we build two modes for both FUKF and FUKS calculated at each step to adjust trajectory of the bus to improve the quality of monitoring system, as shown in Fig. 5.

- Mode 1: $Y = Y[0, ..., 8]$; Ts = 1s at the step-time of catching full data of GPS, INS and CAN, following (2) and (3) with 9 parameters.
- Mode 2:

  + $Y1 = Y[0, ..., 8]$; Ts = 1s at the step-time of catching GPS signal (the same with operation of full data in Mode 1), following (2) and (3) with 9 parameters;
  + $Y2 = Y[0, 7, 8]$; Ts = 20 ms at normal step-time of catching CAN, INS signals, following (2) and (3) with 3 parameters: Distance, Acc and Yaw.

This system have been used in practice for busses number 725, 726, 727, 729, 725-Trento and have good results. In the monitoring system at the station, Fig. 6 represents an orbital motion in Fig. 6A and the adjusted trajectory in Fig. 6B and C, which calculated from (2) and (3) following the FUKF algorithm described in Fig. 4. The trajectory of the bus received from GPS is the green dotted line in Fig. 6B. At some points in this dotted line, the noise is quite large occurring when motion of the bus is not normal, or GPS signal is not good. That blue dotted line shows values calculated directly from the GPS signal, in which each dot is the location at the time of receiving a GPS signal. The Fig. 6C shows the adjustment of the values at the time of calculation, from several millimeters to several meters depended on many factors, such as: setting-up time of the system, position of the bus, weather, intensity of GPS signals... This process is recorded continuously at each computational step through the store values by saving text data file creating images. The set of Fig. 6B and C forms a video file recorded in the database showing clearly the process of filtration and correction trajectory of the bus.

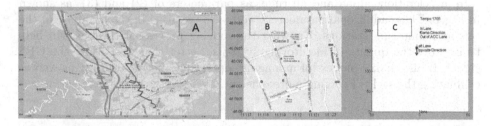

Fig. 6. A full trajectory of the bus and adjustment of trajectory at each step (Color figure online)

These values were skewed away from real trajectory and made the fake trajectory of the bus in Fig. 7. Figure 7A and B show clearly the result of adjustment of vehicle trajectory following the FUKF. That is the black zig-zag line, showing much better accuracy than GPS trajectory as blue dots, especially at some chaos points (Fig. 7A) and curved stretch calculated to shift trajectory of the bus at the step time of catching GPS signal (Fig. 7B).

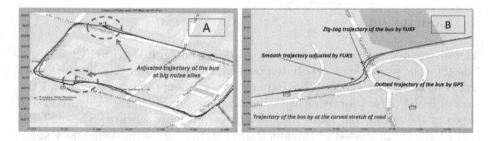

Fig. 7. GPS points and adjusted trajactories at big noise sites of road (Color figure online)

Large orbital deviations occur when the bus has abnormal motion positions such as: at some points of changing speed and direction suddenly, at some points of shot-stop with working engine, including at bus-stops, at the station, at the traffic lights, and at the positions beneath the high building (Fig. 7A). The new solution in the article filtered and gave out the exact trajectory of the bus. Result is evident in these positions. Parts of the trajectory deviation were removed and corrected to get the orbit of the bus close to reality as possible. The pink line is backward calculation FUKS, described in Fig. 5. This is a smooth trajectory, which avoids shifting trajectory adjustment of FUKF (the black line), and is closest to the actual trajectory of the bus (the pink line).

In addition to the usual parameters such as position and velocity, the monitoring system described in this article also shows some typical parameters of the vehicle, which are also parameters of (2) and (3), including distance, yaw, yaw rate, acceleration, alpha and altitude, as parameters of (2) and (3), as shown in Fig. 8. These additional parameters were also recorded in the database of the bus at monitoring stations. Knowing the moving process of the bus will help to improve the quality of monitoring systems, better than just using the GPS signal. It also makes process of managing, operating and monitoring, as well as evaluating the style of driving more efficient.

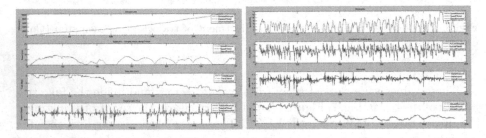

Fig. 8. Parameters of buses including: distance, velocity, yaw, yaw rate, distance, velocity, yaw, yaw rate, speed, acc, alpha and altitude

4   Conclusion

The distributed solution is a new way to design integrated GPS/INS systems. Based on the principle of distributed control, the solution in this article divides the task of solving drifting INS data at device mounted on the bus and the task of FUKF filter processing at the station in order to enhance the processing speed of the device and make the data processed more efficient. These FUKF and FUKS are processed at the monitoring station to put aside memory space for other processing and to speed up the embedded system mounted on vehicles.

The noise for GPS systems leading to large deviations of trajectory occurs in case of the motion of the bus is abnormal, such as: sudden changing speed and direction, strong vibration; or also in case of the positions of the bus are in the areas beneath high buildings, at bus-stops and at the traffic lights. Monitoring results showed that the system filtered, eliminated false parts and gave accurate trajectory of bus. In addition to the parameters of velocity and position, the monitoring system using FUKF and FUKS monitors specific parameters for movement of the bus, such as: moving distance, speed, yaw angle, yaw rate, acceleration, slope angle (alpha), altitude and some other parameters, which can be used to evaluate driving style of drivers. This solution can also be used some more added devices to transmit data to station directly. That will create fully worked-out IoT devices for vehicle management and monitoring system supporting efficiently for smart cities.

References

1. Biral, F., Galvani, M., Zucchelli, M., Giacomelli, G.: Objective performance evaluation on mountain routes of diesel-electric hybrid busses. In: 2013 IEEE International Conference on Mechatronics Conference, ICM 2013, pp. 394–399. IEEE Xplore (2013)
2. Ngo, T.B., Le, H.L., Nguyen, T.H.: Survey of Kalman filters and their application in signal processing. In: 2009 International Conference on Artificial Intelligence and Computational Intelligence, Conference ACCI 2009, Shanghai, China, vol. 3, pp. 335–339. IEEE Xplore (2009)
3. Abdel-Hamid, W., Noureldin, A., El-Sheimy, N.: Adaptive fuzzy prediction of low-cost inertial-based positioning errors. IEEE Trans. Fuzzy Syst. **15**(3), 519–529 (2007)
4. Chai, L., Yuan, J., Fang, Q., Kang, Z., Huang, L.: Neural network aided adaptive kalman filter for multi-sensors integrated navigation. In: Yin, F.-L., Wang, J., Guo, C. (eds.) ISNN 2004. LNCS, vol. 3174, pp. 381–386. Springer, Heidelberg (2004). https://doi.org/10.1007/978-3-540-28648-6_60
5. Mahony, R., Hamel, T., Pflimlin, J.M.: Nonlinear complementary filters on the special orthogonal group. IEEE Trans. Autom. Control **5**(53), 1203–1218 (2008)
6. Phuong, N.H.Q., Kang, H.-J., Suh, Y.-S., Ro, Y.-S.: A DCM based orientation estimation algorithm with an inertial measurement unit and a magnetic compass. J.USC - J. Univers. Comput. Sci. **4**(15), 859–876 (2009)
7. Kandepu, R., Foss, B., Imsland, L.: Applying the unscented Kalman filter for nonlinear state estimation. J. Process. Control **18**(18), 753–768 (2008)
8. Biswas, S., Qiao, L., Dempster, A.: Computationally efficient unscented Kalman filtering techniques for launch vehicle navigation using a space-borne GPS receiver. J. ION GNSS+ (2016). https://dblp.uni-trier.de/rec/bibtex/journals/corr/BiswasQD16
9. Junker, B.H., Baker, S.M., Poskar, C.H.: Unscented Kalman filter with parameter identifiability analysis for the estimation of multiple parameters in kinetic models. EURASIP J. Bioinform. Syst. Biol. **1**(7), 99–110 (2011)
10. Chiang, K.-W., Duong, T.T., Liao, J.-K.: The performance analysis of a real-time integrated INS/GPS vehicle navigation system with abnormal GPS measurement elimination. Sensors **2**(13), 10599–10622 (2013)
11. Hartikainen, J., Solin, A., Sarkka, S.: Optimal Filtering with Kalman Filters and Smoothers a Manual for the Matlab Toolbox EKF-UKF, 2nd edn. GNU, Helsinki (2011)
12. van der Merwe, R.: Sigma-point Kalman filters for probabilistic inference in dynamic state-space models. Ph.D. thesis in Oregon Health & Science University (2004)
13. Bartz, P.: Github Razor-AHRS. https://github.com/Razor-AHRS/razor-9dof-ahrs. Accessed 14 May 2018
14. Dalutal, S.: Starllno Electronics. http://www.starlino.com/dcm_tutorial.html. Accessed 14 May 2018

Modeling Spatially-Correlated Cellular Networks by Using Inhomogeneous Poisson Point Processes

Marco Di Renzo[✉], Shanshan Wang, and Xiaojun XI

Paris-Saclay University & CNRS, Signals and Systems Laboratory (UMR 8506),
CentraleSupelec, Paris, France
{marco.direnzo,shanshan.wang,xiaojun.xi}@l2s.centralesupelec.fr

Abstract. In this paper, we introduce the Inhomogeneous Double Thinning (IDT) approach, which allows us to analyze the performance of downlink cellular networks in which the Base Stations (BSs) constitute a stationary Point Process (PP) that exhibits some degree of spatial repulsion (i.e., inhibition). The accuracy of the proposed IDT approach is substantiated by using empirical data for the spatial distribution of the BSs.

Keywords: Cellular networks · Stochastic geometry
Inhomogeneous point processes · Spatial inhibition

1   Introduction

Poisson Point Processes (PPPs) have been extensively used for modeling the locations of cellular Base Stations (BSs) [1]. Examples include Heterogeneous Cellular Networks (HCNs) [2], millimeter-wave cellular HCNs [3], Multiple-Input-Multiple-Output (MIMO) HCNs [4], and the energy efficiency of cellular networks [5]. General mathematical frameworks that account for the impact of antenna radiation patterns, spatial blockages, and the network load are available as well [6].

Using the PPPs has the inherent advantage of mathematical tractability. However, the locations of the Base Stations (BSs) are usually spatially correlated. Therefore, some authors have recently studied the performance of cellular networks by using other point processes, e.g., [7–9], where Determinantal Point Processes (DPP), Ginibre Point Processes (GPP), and Log-Gaussian Cox Processes (LGCP) are considered, respectively. The resulting analytical frameworks, however, are less tractable than the PPP.

Motivated by these considerations, we introduce a new approach for modeling and analyzing the performance of spatially-correlated cellular networks, which is based on Inhomogeneous PPPs (I-PPPs). The choice of using I-PPPs is motivated by their analytical tractability and by the fact that·empirical evidence

T. Q. Duong and N.-S. Vo (Eds.): INISCOM 2018, LNICST 257, pp. 306–313, 2019.
https://doi.org/10.1007/978-3-030-05873-9_26

shows that spatial correlations may be difficult to disentangle from spatial inhomogeneities [10, Sect. 7.3.5.2]. It is worth noting that the spatial models in [7–9] are all based on stationary point processes. On the other hand, I-PPPs are non-stationary point processes, which implies that the performance of a randomly chosen user is no longer independent on its actual location. In our approach, we introduce an equivalent I-PPP whose inhomogeneity is created from the point of view of the typical user of the original motion-invariant point process. This makes our approach applicable for approximating several point processes. The proposed approach is referred to as Inhomogeneous Double Thinning (IDT) approach.

The rest of the paper is organized as follows. In Sect. 2, we present the system model. In Sect. 3, the IDT approach is introduced. In Sect. 4, the mathematical framework of the coverage probability is provided. In Sect. 5, the accuracy of the IDT approach is validated by using empirical data sets. Finally, Sect. 6 concludes the paper.

Notation: The notation used in the paper can be found in Table 1.

Table 1. Notation

Symbol/function	Definition
$\mathbb{E}\{\cdot\}$, $\mathbb{1}(\cdot)$	Expectation operator, indicator function
$\mathrm{E}^{!x}\{\cdot\}$	Expectation (reduced Palm measure)
$f_X(\cdot)$	Probability density function (PDF) of X
$\mathcal{M}_{\mathrm{I},X}(\cdot)$	Laplace functional conditioned on X
$_2F_1(\cdot,\cdot,\cdot,\cdot)$	Gauss hypergeometric function
$\max\{x,y\}$, $\min\{x,y\}$	Maximum and minimum of x and y
(a_F, b_F, c_F), (a_K, b_K, c_K)	Parameters of the approximating I-PPPs

2   System Model

The BSs are modeled as points of a stationary point process, Ψ_{BS}, of density λ_{BS}. The locations of the BSs are denoted by $x \in \Psi_{\mathrm{BS}} \subseteq \mathbb{R}^2$. The Mobile Terminals (MTs) are distributed uniformly at random in $\mathbb{R}^2$. A fully loaded network is considered. The serving BS is denoted by BS_0, and its location is $x_0 \in \Psi_{\mathrm{BS}}$. The interfering BSs constitute another point process, denoted by $\Psi_{\mathrm{BS}}^{(\mathrm{I})}$. Each BS transmits with power $\mathrm{P_{tx}}$. The power of the Gaussian noise is σ_{N}^2.

For each BS-to-MT link, path-loss and fast-fading are considered. The path-loss is $l(x) = \kappa\|x\|^\gamma$, where κ and $\gamma > 2$ are the path-loss constant and the path-loss slope. The power gain, g_x, due to small-scale fading is an exponential random variable with unit mean.

The cell association criterion is based on the highest average received power. The location, x_0, of the serving BS, BS_0, is obtained as follows:

$$x_0 = \arg\max_{x \in \Psi_{BS}} \{1/l(x)\} = \arg\max_{x \in \Psi_{BS}} \{1/L_x\} \tag{1}$$

where $L_x = l(x)$ is a shorthand. As for the intended link, we have $L_0 = l(x_0) = \min_{x \in \Psi_{BS}} \{L_x\}$.

The coverage probability P_{cov} is defined as follows:

$$P_{cov} = \Pr\left\{\frac{P_{tx}g_0/L_0}{\sigma_N^2 + \sum_{x \in \Psi_{BS}^{(I)}} P_{tx}g_x/L_x} > T\right\} \tag{2}$$

where $\Psi_{BS}^{(I)} = \Psi_{BS} \backslash x_0$, and T is the decoding threshold.

Lemma 1. *The coverage probability in (2) can be expressed as follows:*

$$P_{cov} = \int_0^{+\infty} \exp\left(-\xi T\sigma_N^2/P_{tx}\right) \mathcal{M}_{I,L_0}(\xi; T) f_{L_0}(\xi) d\xi \tag{3}$$

where $f_{L_0}(\cdot)$ is the probability density function of L_0 and $\mathcal{M}_{I,L_0}(\cdot; \cdot)$ is the Laplace functional of the point process, $\Psi_{BS}^{(I)} = \Psi_{BS} \backslash x_0$, of interfering BSs:

$$\mathcal{M}_{I,L_0}(\xi = L_0; T) = \mathbb{E}_{\Psi_{BS}}^{!x_0}\left\{\prod_{x \in \Psi_{BS}^{(I)}} (1 + T(\xi/l(x)))^{-1}\right\} \tag{4}$$

Proof. It directly follows from [1].   □

From (4) and (5), we conclude that to compute the coverage probability we need $f_{L_0}(\cdot)$, which depends on the Contact Distance Distribution (CDD) of the point process, and $\mathcal{M}_{I,L_0}(\cdot; \cdot)$, which depends on the reduced Palm distribution of the point process [12]. However, the CDD and reduced Palm distribution of a stationary point process may not be known or may not be mathematically tractable.

3   Inhomogeneous Double Thinning

The proposed approach for computing P_{cov} consists of introducing an equivalent system model based on I-PPPs. More precisely, the BSs are modeled as the superposition of the points of two *independent* isotropic I-PPPs, denoted by $\Phi_{BS}^{(F)}$ and $\Phi_{BS}^{(K)}$, with intensity measures $\Lambda_{\Phi_{BS}^{(F)}}(\cdot)$ and $\Lambda_{\Phi_{BS}^{(K)}}(\cdot)$, respectively. Since I-PPPs are not motion-invariant, we are interested in computing the coverage probability from the point of view of a *probe* MT located at the origin. The serving BS and interfering BSs are assumed to belong to $\Phi_{BS}^{(F)}$ and $\Phi_{BS}^{(K)}$, respectively, and are defined as follows:

$$x_0^{(F)} = \arg\max_{x \in \Phi_{BS}^{(F)}} \{1/l(x)\}$$

$$\Phi_{BS}^{(I)} = \Phi_{BS}^{(I)}\left(x_0^{(F)}\right) = \left\{x \in \Phi_{BS}^{(K)} : l(x) > L_0^{(F)} = l\left(x_0^{(F)}\right)\right\} \tag{5}$$

It is worth noting that the I-PPPs $\Phi_{BS}^{(F)}$ and $\Phi_{BS}^{(I)}$ are only *conditionally independent*.

The coverage probability of the probe MT located at the origin can be formulated as follows:

$$\widetilde{P}_{cov}^{(o)} = Pr\left\{\frac{P_{tx}g_0 \big/ L_0^{(F)}}{\sigma_N^2 + \sum_{x\in\Phi_{BS}^{(I)}} P_{tx}g_x/l\,(x)} > T\right\} \tag{6}$$

where the superscript (o) denotes that (6) holds for the probe MT at the origin.

Lemma 2. *The coverage probability in (6) can be computed as follows:*

$$\widetilde{P}_{cov}^{(o)} = \int_0^{+\infty} \exp\left(-\xi T\sigma_N^2/P_{tx}\right) \widetilde{\mathcal{M}}_{I,L_0^{(F)}}\,(\xi;T)\, \widetilde{f}_{L_0^{(F)}}\,(\xi)\, d\xi \tag{7}$$

where $\widetilde{f}_{L_0^{(F)}}\,(\cdot)$ is the probability density function of $L_0^{(F)}$ and $\widetilde{\mathcal{M}}_{I,L_0^{(F)}}\,(\cdot;\cdot)$ is the Laplace functional of $\Phi_{BS}^{(I)}$ given in (8) at the top of this page, and $\Lambda_{(\cdot)}^{(1)}\,(\mathcal{B}\,(0,r)) = d\Lambda_{(\cdot)}\,(\mathcal{B}\,(0,r))/dr$ is the first-order derivative of the intensity measure.

Proof. It follows from [6]. $\qquad\square$

$$\widetilde{f}_{L_0^{(F)}}\,(\xi) = \left(\frac{\xi}{\kappa}\right)^{1/\gamma}\frac{1}{\gamma\xi}\Lambda_{\Phi_{BS}^{(F)}}^{(1)}\left(\mathcal{B}\left(0,\left(\frac{\xi}{\kappa}\right)^{1/\gamma}\right)\right)\exp\left(-\Lambda_{\Phi_{BS}^{(F)}}\left(\mathcal{B}\left(0,\left(\frac{\xi}{\kappa}\right)^{1/\gamma}\right)\right)\right)$$

$$\widetilde{\mathcal{M}}_{I,L_0^{(F)}}\,(\xi;T) = \exp\left(-\int_\xi^{+\infty}\left(1+\frac{z}{T\xi}\right)^{-1}\left(\frac{z}{\kappa}\right)^{1/\gamma}\frac{1}{\gamma z}\Lambda_{\Phi_{BS}^{(K)}}^{(1)}\left(\mathcal{B}\left(0,\left(\frac{z}{\kappa}\right)^{1/\gamma}\right)\right)dz\right) \tag{8}$$

In order to apply the IDT approach, the intensity measures $\Lambda_{\Phi_{BS}^{(F)}}\,(\cdot)$ and $\Lambda_{\Phi_{BS}^{(K)}}\,(\cdot)$ are determined with the objective of ensuring $\widetilde{P}_{cov}^{(o)} \approx P_{cov}$. This is elaborated in the next section.

3.1   Intensity Measures of the I-PPPs

The intensity function of an I-PPP determines its intensity measure [11, Sect. 2.2]. Let $\lambda_{BS}^{(F)}\,(\cdot)$ and $\lambda_{BS}^{(K)}\,(\cdot)$ be the intensity functions of $\Phi_{BS}^{(F)}$ and $\Phi_{BS}^{(K)}$. The following holds:

$$\Lambda_{\Phi_{BS}^{(F)}}\,(\mathcal{B}\,(0,r)) = 2\pi\int_0^r \lambda_{BS}^{(F)}\,(\zeta)\,\zeta d\zeta$$
$$\Lambda_{\Phi_{BS}^{(K)}}\,(\mathcal{B}\,(0,r)) = 2\pi\int_0^r \lambda_{BS}^{(K)}\,(\zeta)\,\zeta d\zeta \tag{9}$$

Let (a_F, b_F, c_F) and (a_K, b_K, c_K) be two triplets of non-negative real numbers, where $c_F \geqslant b_F \geqslant 1$ and $b_K \leqslant c_K \leqslant 1$. The following intensities are proposed to model the spatial inhibition among the locations of the BSs:

$$\lambda_{BS}^{(F)}\,(r) = \lambda_{BS}c_F\min\{(a_F/c_F)\,r + b_F/c_F, 1\}$$
$$\lambda_{BS}^{(K)}\,(r) = \lambda_{BS}\min\{a_K r + b_K, c_K\} \tag{10}$$

The intensity measures in (9) can be formulated as $\Lambda_{\Phi_{\mathrm{BS}}^{(\cdot)}}(\mathcal{B}(0,r)) = \Upsilon(r; a_{(\cdot)}, b_{(\cdot)}, c_{(\cdot)})$, where $\Upsilon(r; a, b, c)$ and its first-order derivative $\Upsilon^{(1)}(r; a, b, c)$ are as follows:

$$
\begin{aligned}
\Upsilon(r; a, b, c) &= 2\pi\lambda_{\mathrm{BS}}\left((a/3)\,r^3 + (b/2)\,r^2\right)\mathbb{1}\left(r \leqslant (c-b)/a\right) \\
&+ 2\pi\lambda_{\mathrm{BS}}\left((c/2)\,r^2 - (c-b)^3\big/6a^2\right)\mathbb{1}\left(r > (c-b)/a\right)
\end{aligned}
\tag{11}
$$

$$
\begin{aligned}
\Upsilon^{(1)}(r; a, b, c) &= 2\pi\lambda_{\mathrm{BS}}\left(ar^2 + br\right)\mathbb{1}\left(r \leqslant (c-b)/a\right) \\
&+ 2\pi\lambda_{\mathrm{BS}}cr\,\mathbb{1}\left(r > (c-b)/a\right)
\end{aligned}
$$

We propose to compute the triplets of parameters that determine the intensity measures $\Lambda_{\Phi_{\mathrm{BS}}^{(F)}}(\cdot)$ by solving the following minimization problem:

$$
(a_{\mathrm{F}}, b_{\mathrm{F}}, c_{\mathrm{F}}) = \arg\min_{(a,b,c)\in\Omega_{\mathrm{F}}}\left\{\left|F_{\Psi_{\mathrm{BS}}}(r) - F_{\Phi_{\mathrm{BS}}^{(F)}}(r; a, b, c)\right|^2\right\}
\tag{12}
$$

where $F_{\Phi_{\mathrm{BS}}^{(F)}}(r; a, b, c)$ is the CDD of $\Phi_{\mathrm{BS}}^{(F)}$ and $\Omega_{\mathrm{F}} = \{(a_{\mathrm{F}}, b_{\mathrm{F}}, c_{\mathrm{F}}) : c_{\mathrm{F}} \geqslant b_{\mathrm{F}} \geqslant 1\}$. $\Lambda_{\Phi_{\mathrm{BS}}^{(K)}}(\cdot)$ can be obtained in the similar way, but by approximating the so-called "non-regularized" Ripley's K-function of the point process [13].

4   Coverage Probability by Using the IDT Approach

The following theorem provides a tractable expression of $\widetilde{\mathrm{P}}_{\mathrm{cov}}^{(\mathrm{o})}$ in (6). Two case studies are considered: (i) the network is infinitely large and (ii) the network has a finite size of radius R_{A}.

Theorem 1. $\widetilde{\mathrm{P}}_{\mathrm{cov}}^{(\mathrm{o})}$ in (6) can be formulated as follows:

$$
\begin{aligned}
\widetilde{\mathrm{P}}_{\mathrm{cov}}^{(\mathrm{o})} &= \int_0^{\kappa(d_{\mathrm{F}})^{\gamma}} \exp\left(-\xi\mathrm{T}\sigma_{\mathrm{N}}^2/\mathrm{P}_{\mathrm{tx}}\right)\exp\left(-\mathcal{I}(\xi)\right)\mathcal{U}_{\mathrm{IN}}(\xi)\,d\xi \\
&+ \int_{\kappa(d_{\mathrm{F}})^{\gamma}}^{\Theta} \exp\left(-\xi\mathrm{T}\sigma_{\mathrm{N}}^2/\mathrm{P}_{\mathrm{tx}}\right)\exp\left(-\mathcal{I}(\xi)\right)\mathcal{U}_{\mathrm{OUT}}(\xi)\,d\xi
\end{aligned}
\tag{13}
$$

where $d_{\mathrm{F}} = \frac{c_{\mathrm{F}}-b_{\mathrm{F}}}{a_{\mathrm{F}}}$, $d_{\mathrm{K}} = \frac{c_{\mathrm{K}}-b_{\mathrm{K}}}{a_{\mathrm{K}}}$, $\Theta \to \infty$ and $\mathcal{I}(\xi) = \mathcal{I}_{\infty}(\xi)$ for network with infinite size, $\Theta \to \kappa\mathrm{R}_{\mathrm{A}}^{\gamma}$ and $\mathcal{I}(\xi) = \mathcal{I}_{\mathrm{R}_{\mathrm{A}}}(\xi)$ for network with finite size, and $\mathcal{I}_{\infty}(\cdot)$, $\mathcal{I}_{\mathrm{R}_{\mathrm{A}}}(\cdot)$, $\mathcal{U}_{\mathrm{IN}}(\cdot)$, $\mathcal{U}_{\mathrm{OUT}}(\cdot)$ are defined in (14).

Proof. The proof is omitted due to space limitations.   □

$$\mathcal{U}_{\mathrm{IN}}(\xi) = 2\pi\lambda_{\mathrm{BS}}\left(\frac{a_{\mathrm{F}}}{\gamma\xi}\left(\frac{\xi}{\kappa}\right)^{3/\gamma} + \frac{b_{\mathrm{F}}}{\gamma\xi}\left(\frac{\xi}{\kappa}\right)^{2/\gamma}\right)\exp\left(-2\pi\lambda_{\mathrm{BS}}\left(\frac{a_{\mathrm{F}}}{3}\left(\frac{\xi}{\kappa}\right)^{3/\gamma} + \frac{b_{\mathrm{F}}}{2}\left(\frac{\xi}{\kappa}\right)^{2/\gamma}\right)\right)$$

$$\mathcal{U}_{\mathrm{OUT}}(\xi) = 2\pi\lambda_{\mathrm{BS}}\frac{c_{\mathrm{F}}}{\gamma\xi}\left(\frac{\xi}{\kappa}\right)^{2/\gamma}\exp\left(-2\pi\lambda_{\mathrm{BS}}\left(\frac{c_{\mathrm{F}}}{2}\left(\frac{\xi}{\kappa}\right)^{2/\gamma} - \frac{(c_{\mathrm{F}}-b_{\mathrm{F}})^3}{6a_{\mathrm{F}}^2}\right)\right)$$

$$\mathcal{I}_1(\xi) = \frac{a_{\mathrm{K}}d_{\mathrm{K}}^3}{3}{}_2F_1\left(1,\frac{3}{\gamma},1+\frac{3}{\gamma},-\frac{\kappa}{T\xi}d_{\mathrm{K}}^\gamma\right)\mathbb{1}\left(\xi \leqslant \kappa d_{\mathrm{K}}^\gamma\right),$$

$$\mathcal{I}_2(\xi) = \frac{b_{\mathrm{K}}d_{\mathrm{K}}^2}{2}{}_2F_1\left(1,\frac{2}{\gamma},1+\frac{2}{\gamma},-\frac{\kappa}{T\xi}d_{\mathrm{K}}^\gamma\right)\mathbb{1}\left(\xi \leqslant \kappa d_{\mathrm{K}}^\gamma\right),$$

$$\mathcal{I}_3(\xi) = -\frac{a_{\mathrm{K}}}{3}\left(\frac{\xi}{\kappa}\right)^{3/\gamma}{}_2F_1\left(1,\frac{3}{\gamma},1+\frac{3}{\gamma},-\frac{1}{T}\right)\mathbb{1}\left(\xi \leqslant \kappa d_{\mathrm{K}}^\gamma\right),$$

$$\mathcal{I}_4(\xi) = -\frac{b_{\mathrm{K}}}{2}\left(\frac{\xi}{\kappa}\right)^{2/\gamma}{}_2F_1\left(1,\frac{2}{\gamma},1+\frac{2}{\gamma},-\frac{1}{T}\right)\mathbb{1}\left(\xi \leqslant \kappa d_{\mathrm{K}}^\gamma\right),$$

$$\mathcal{I}_5(\xi) = -\frac{c_{\mathrm{K}}d_{\mathrm{K}}^2}{2}\left(1-{}_2F_1\left(1,-\frac{2}{\gamma},1-\frac{2}{\gamma},-\frac{T\xi}{\kappa}d_{\mathrm{K}}^{-\gamma}\right)\right)\mathbb{1}\left(\xi \leqslant \kappa d_{\mathrm{K}}^\gamma\right),$$ (14)

$$\mathcal{I}_6(\xi) = -\frac{c_{\mathrm{K}}}{2}\left(\frac{\xi}{\kappa}\right)^{2/\gamma}\left(1-{}_2F_1\left(1,-\frac{2}{\gamma},1-\frac{2}{\gamma},-T\right)\right)\mathbb{1}\left(\xi \geqslant \kappa d_{\mathrm{K}}^\gamma\right),$$

$$\mathcal{I}_7(\xi) = \frac{c_{\mathrm{K}}}{2}R_{\mathrm{A}}^2{}_2F_1\left(1,\frac{2}{\gamma},1+\frac{2}{\gamma},-\frac{\kappa}{T\xi}R_{\mathrm{A}}^\gamma\right)\mathbb{1}\left(\xi \leqslant \kappa d_{\mathrm{K}}^\gamma\right),$$

$$\mathcal{I}_8(\xi) = -\frac{c_{\mathrm{K}}d_{\mathrm{K}}^2}{2}{}_2F_1\left(1,\frac{2}{\gamma},1+\frac{2}{\gamma},-\frac{\kappa}{T\xi}d_{\mathrm{K}}^\gamma\right)\mathbb{1}\left(\xi \leqslant \kappa d_{\mathrm{K}}^\gamma\right),$$

$$\mathcal{I}_9(\xi) = \frac{c_{\mathrm{K}}}{2}R_{\mathrm{A}}^2{}_2F_1\left(1,\frac{2}{\gamma},1+\frac{2}{\gamma},-\frac{\kappa}{T\xi}R_{\mathrm{A}}^\gamma\right)\mathbb{1}\left(\xi \geqslant \kappa d_{\mathrm{K}}^\gamma\right),$$

$$\mathcal{I}_{10}(\xi) = -\frac{c_{\mathrm{K}}}{2}\left(\frac{\xi}{\kappa}\right)^{2/\gamma}{}_2F_1\left(1,\frac{2}{\gamma},1+\frac{2}{\gamma},-\frac{1}{T}\right)\mathbb{1}\left(\xi \geqslant \kappa d_{\mathrm{K}}^\gamma\right),$$

$$\mathcal{I}_\infty(\xi) = 2\pi\lambda_{\mathrm{BS}}\sum_{k=1}^{6}\mathcal{I}_k(\xi),\ \mathcal{I}_{\mathrm{R}_{\mathrm{A}}}(\xi) = 2\pi\lambda_{\mathrm{BS}}\left(\sum_{k=1}^{4}\mathcal{I}_k(\xi) + \sum_{k=7}^{10}\mathcal{I}_k(\xi)\right)$$

5   Numerical Validation

In this section, some numerical results that substantiate the applicability of the IDT approach are illustrated. The parameters of the considered network deployments are summarized in Table 2. The triplets of parameters used to apply the IDT approach are obtained by solving (12).

Table 2. Empirical point process (ISD = Inter-Side Distance).

Point process	Parameters
Lattice point process	ISD $= \{100, 300\}$ m
GPP [8] (Urbàn, $\beta = 0.900$)	$\lambda_{\mathrm{BS}} = 31.56\,\mathrm{BS/km^2}$,
	Area $= 3.784^2\pi\,\mathrm{km^2}$, $\gamma = 3.5$
GPP [8] (Rural, $\beta = 0.375$)	$\lambda_{\mathrm{BS}} = 0.03056\,\mathrm{BS/km^2}$,
	Area $= 124.578^2\pi\,\mathrm{km^2}$, $\gamma = 2.5$

The numerical results are reported in Figs. 1 and 2. The markers represent Monte Carlo simulations and the solid lines denote the framework in *Theorem* 1. There are three curves in the figure: (i) "Empirical (R)" is obtained from the data sets listed in Table 2 (generated with R [10]); (ii) "PPP-IDT" is obtained from the IDT approach with the triplets of parameters obtained from (12); and

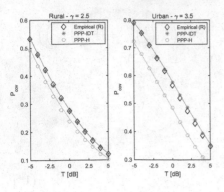 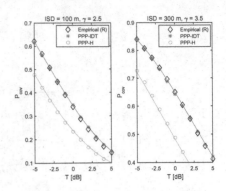

Fig. 1. P_{cov} of GPP-Rural ($\beta = 0.375$) and GPP-Urban ($\beta = 0.9$).

Fig. 2. P_{cov} of square-lattice (ISD $= 100$ m and ISD $= 300$ m).

(iii) "PPP-H" gives the reference curve that corresponds to using homogeneous PPPs. From Figs. 1 and 2, we conclude that the IDT approach is accurate, tractable and capable of reproducing the spatial correlation of several point processes widely used in the literature.

6   Conclusion

In this paper, we have proposed a new mathematically tractable approach to model and analyze cellular networks in which the locations of the BSs are spatially correlated. The proposed approach is based on the theory of I-PPPs. The tractability and accuracy of the proposed methodology have been substantiated by using data sets available in the literature. Based on the obtained results, we conclude that IDT approach have wide applicability to the analysis and design of cellular networks. An extended version of the present paper is available in [13].

References

1. Andrews, J.G., Baccelli, F., Ganti, R.K.: A tractable approach to coverage and rate in cellular networks. IEEE Trans. Commun. **59**(11), 3122–3134 (2011)
2. Di Renzo, M., Guidotti, A., Corazza, G.E.: Average rate of downlink heterogeneous cellular networks over generalized fading channels - a stochastic geometry approach. IEEE Trans. Commun. **61**(7), 3050–3071 (2013)
3. Di Renzo, M.: Stochastic geometry modeling and analysis of multi-tier millimeter wave cellular networks. IEEE Trans. Wirel. Commun. **14**(9), 5038–5057 (2015)
4. Di Renzo, M., Guan, P.: Stochastic geometry modeling and system-level analysis of uplink heterogeneous cellular networks with multi-antenna base stations. IEEE Trans. Commun. **64**(6), 2453–2476 (2016)
5. Di Renzo, M., Zapppone, A., Lam, T.T., Debbah, M.: System-level modeling and optimization of the energy efficiency in cellular networks - a stochastic geometry framework. IEEE Trans. Wirel. Commun. **17**(4), 2539–2556 (2018)

6. Di Renzo, M., Lu, W., Guan, P.: The intensity matching approach: a tractable stochastic geometry approximation to system-level analysis of cellular networks. IEEE Trans. Wirel. Commun. **15**(9), 5963–5983 (2016)
7. Li, Y., Baccelli, F., Dhillon, H., Andrews, J.: Statistical modeling and probabilistic analysis of cellular networks with determinantal point processes. IEEE Trans. Commun. **63**(9), 3405–3422 (2015)
8. Deng, N., Zhou, W., Haenggi, M.: The Ginibre point process as a model for wireless networks with repulsion. IEEE Trans. Wirel. Commun. **14**(1), 107–121 (2015)
9. Kibilda, J., Galkin, B., DaSilva, L.A.: Modelling multi-operator base station deployment patterns in cellular networks. IEEE Trans. Mob. Comput. **15**(12), 3087–3099 (2016)
10. Baddeley, A., Rubak, E., Turner, R.: Spatial Point Patterns: Methodology and Applications with R. Chapman and Hall/CRC, New York (2015)
11. Streit, R.L.: Poisson Point Processes: Imaging, Tracking, and Sensing. Springer, Boston (2010). https://doi.org/10.1007/978-1-4419-6923-1
12. Haenggi, M.: Stochastic Geometry for Wireless Networks. Cambridge University Press, Cambridge (2012)
13. Di Renzo, M., Wang, S., Xi, X.: Inhomogeneous double thinning–modeling and analysis of cellular networks by using inhomogeneous Poisson point processes. IEEE Trans. Wirel. Commun. **17**(8), 5162–5182 (2018)

Author Index